Practical Algorithms for Image Analysis
Second Edition

In classic "cookbook style," this book offers guided access for researchers and practitioners to techniques for the digital manipulation and analysis of images, ranging from the simplest steps to advanced functions. Drawing on their long experience as users and developers of image analysis algorithms and software, the authors present a description and implementation of the most suitable procedures in easy-to-use form. Each self-contained section treats a single operation, describing typical situations requiring that operation and discussing the algorithm and implementation. Sections start with a "before" and "after" pictorial example and a ready-reference listing typical applications, keywords, and related procedures.

This new edition has additional sections on Gabor filtering and on threshholding by connectivity, plus an expanded program listing and suggested projects for classroom use. And now the accompanying CD-ROM contains C programs not only as source code for carrying out the book's procedures but also as executables with a graphical user interface for Windows and Linux.

Lawrence O'Gorman is a Research Scientist at Avaya Labs in Basking Ridge, New Jersey. He was formerly a Distinguished Member of Technical Staff at Bell Laboratories, Murray Hill, New Jersey, where he performed research in pattern recognition and image processing. Dr. O'Gorman is a Fellow of the IEEE and a Fellow of the International Association for Pattern Recognition.

Michael J. Sammon is a Research Scientist at Avaya Labs in Basking Ridge, New Jersey. Mr. Sammon started his career at Bell Laboratories in Murray Hill, New Jersey, where he developed image analysis software to investigate pattern formation in organic thin films.

Michael Seul is President of BioArray Solutions, a high-tech venture he founded to develop miniaturized systems for on-chip chemical analysis and biomedical diagnostics. Previously, during a decade-long tenure at AT&T Bell Laboratories, Dr. Seul established an internationally recognized research program with a focus on the physical chemistry of membranes, interfaces, and colloidal systems as well as pattern formation in thin organic and magnetic films.

Practical Algorithms for Image Analysis

DESCRIPTION, EXAMPLES, PROGRAMS, AND PROJECTS

Second Edition

Lawrence O'Gorman
Avaya Labs
Basking Ridge, New Jersey

Michael J. Sammon
Avaya Labs
Basking Ridge, New Jersey

Michael Seul
BioArray Solutions
Warren, New Jersey

APR 1 4 2011

CAMBRIDGE
UNIVERSITY PRESS

CAMBRIDGE UNIVERSITY PRESS
Cambridge, New York, Melbourne, Madrid, Cape Town, Singapore,
São Paulo, Delhi, Dubai, Tokyo

Cambridge University Press
32 Avenue of the Americas, New York, NY 10013-2473, USA

www.cambridge.org
Information on this title: www.cambridge.org/9780521884112

First published 2008
Reprinted 2009

Printed in the United States of America

A catalog record for this publication is available from the British Library.

Library of Congress Cataloging in Publication Data

O'Gorman, Lawrence.
Practical algorithms for image analysis : description, examples, programs, and projects /
Lawrence O'Gorman, Michael J. Sammon, Michael Seul. – 2nd ed.
    p.   cm.
Seul's name appears first on the earlier ed.
Includes bibliographical references and index.
ISBN 978-0-521-88411-2 (hardback)
1. Image processing – Digital techniques. 2. Algorithms. 3. Image analysis –
Data processing. I. Sammon, Michael J. II. Seul, Michael. III. Title.
TA1637.S38   2007
621.367–dc22   2007042377

ISBN 978-0-521-88411-2 Hardback

Contents

Practical Algorithms for Image Analysis
Second Edition

1 Introduction

1.1 Introduction

In this book, we offer guided access to a collection of algorithms for the digital manipulation of images. Our goal is to facilitate the solution of practical problems, addressing users whose interest is in an informed how-to approach, whether in a technical or in a casual setting. Rather than attempting to be exhaustive, we address salient practical considerations, guiding the selection of a particular approach to commonly encountered image processing and analysis tasks, and we present an implementation of our choice of the most suitable procedures. This selection of "Practical Algorithms" reflects our own experience as long-time users and developers of algorithms and software implementations to process and analyze images in areas as diverse as magnetic domain pattern formation and document analysis.

HOW TO USE THIS BOOK

Organization of Chapters and Sections

This book contains nine chapters and an appendix. Following this introduction, in Section 1.2, an annotated section overview is presented, and, in Section 1.3, a guide to the use of the book the accompanying collection of algorithms are given. Chapters 2 through 7 present the material of the book in self-contained sections of identical format. Chapters 8 and 9, added in this new edition, respectively provide: synposes of all programs, and suggested class or term projects which draw on the material presented in the book. The appendix serves as a review of fundamental concepts to which we refer throughout the text and provides reference material to the technical literature.

Each section contains a header that illustrates the nature of the topic of interest by describing typical applications, identifying key words, and providing cross-references to related topics treated in other sections. Next, the topic of interest is introduced by a description of typical situations requiring a particular processing step or analytical operation; the effect of the operation is illustrated by a pictorial example that comprises a pair of before and after images. Possible strategies of implementation are then discussed, and a particular approach is selected for implementation. Annotated references provide an introduction to further technical literature. Appended to each section is a display of program usage for the code introduced in the section.

Single-Step Procedures

Each section treats a single primary operation (histogram evaluation, low-pass filtering, edge detection, region detection, etc.) and introduces requisite algorithms. Each of the algorithms performs a single transformation on a given input image ("inimg") to produce a modified output image ("outimg") and, in some cases, output data. Sections are self-contained to enable and encourage random access to the most suitable single-step procedure that solves the particular task of interest. Thus, a reader interested in an exposition of simple edge detection techniques would open the book to Section 3.4, while a reader interested in the methodology of Fourier filtering would proceed directly to Section 7.2.

Multistep Procedures

The analysis of images usually requires the successive application of multiple transformations. These may include simple preprocessing steps (noise removal, flat fielding, and feature detection), followed by more complex analytical steps (object shape analysis, line pattern analysis, and point pattern analysis). Multiple individual transformations must be concatenated into multi-step procedures.

To facilitate the flexible design of multi-step procedures, sections are grouped into chapters that bundle common types of analyses (global, local, and frequency domain) and common types of operations according to the images (gray scale and binary) or classes of patterns within images (lines and points) to which they are applied. Each section provides extensive cross references to enable a reader to construct a logical flow of related operations.

Chapters reflect the order in which procedures are ordinarily applied to any given image. For example, histogram analysis will precede binarization: filtering and/or flat fielding will precede object shape analysis, line coding will precede line pattern analysis, and Voronoi analysis will precede statistical analysis of point patterns. Spatial frequency analysis combines several operations in an alternative approach to the analysis of images in the spatial domain, and we choose to present it last. Within chapters, the order of sections reflects increasing levels of task complexity. This organization suggests an overall progression from simple and general to complex and specific tasks.

Code

The book is accompanied by a collection of C programs implementing the algorithms we discuss. All programs operate on existing images; the acquisition of images is not discussed here. Versions for two platforms are supplied, as shown below.

Code Platforms

PLATFORM	C COMPILER
Win32	Microsoft Visual C++, v. 6.0
LINUX kernel v. 2.6.18	GNU gcc, v. 4.1.1

Code Organization

Source code is organized in a directory structure such that modules required for building a particular program reside in a subdirectory corresponding to the book section in which the program is first invoked. For example, **xconv** would be stored in a sub-directory referring to Section 3.1.

Corresponding LINUX Makefiles and Visual C++ Workspaces are provided to generate executables. Details relating to program compilation and installation are described in a README.TXT file contained on the code distribution disk. In some chapters, additional utilities are provided in a separate subdirectory to generate test images or to analyze output data files.

In addition, we provide three libraries:

- **LIBTIFF**, containing functions to handle input and output of image files in an uncompressed Tagged Image File (TIF) format; *LIBTIFF* 3.4, written by Sam Leffler and made available by Silicon Graphics, Inc., was modified to remove support for LZW compression;
- **LIBIMAGE**, containing **getopt()**, a command line parser, as well as a collection of higher-level graphics functions to handle drawing, filling, and character generation; portions of *LIBIMAGE* invoke *gd 1.2*, written by Tom Boutell and made available by Quest Protein Database Center, Cold Spring Harbor Labs;
- **LIBIP**, containing image analysis functions such as **poly_moments()** and **find_area_hist()** to handle common analytical tasks invoked by multiple programs.

Program Execution

Command Line Interface   All programs can be executed from the command line from which a variety of arguments and options can be supplied to adjust and optimize program performance. Some programs prompt for additional input during run time. When executed without arguments, each program displays the command with arguments and options defining program usage. This usage header also is reproduced at the end of each section for programs introduced in that section. In some programs a – I option was added as a command line parameter to invert the input image before processing.

Graphical User Interface   In this new edition, programs can be executed using a graphical user interface (GUI), as illustrated in the "Guided Tour" of Section 1.3. The GUI accommodates command line arguments in a separate field and displays program output, in the form of an image and/or text written to standard output, within separate windows. Output image files also are created within a specified directory while output data can be saved to text files by standard "cutting and pasting" from the corresponding display window. The GUI also provides a listing of all programs, either arranged by book chapters and sections or in alphabetical order. Details regarding the operation of the GUI are provided in a "ReadMe" file on the CD accompanying this book. Executing the GUI requires JAVA 1.5 or later (a Java Runtime Environment (JRE) is not included in the code distribution but may be obtained, for example, from SUN Microsystems). By default, programs, as provided on the accompanying CD, operate from within the GUI; for some programs, operation from the command line will require recompilation in accordance with instructions provided in the corresponding README.TXT files. When programs executed from within the GUI produce standard output, this is automatically displayed within a separate GUI window.

Notices

This source code is distributed under a limited-use license that one may view by invoking an appropriate option on the command line of each program. Unless otherwise indicated in the source code, copyright is jointly held by the authors, as indicated by a copyright notice such as Copyright (C) 1997, 1998, 1999, M. Seul, L. O'Gorman, and M. J. Sammon. In some instances, we use a third-party code in versions available from public sources. In those instances the original authors, as identified in the source code, retain copyright.

Except when otherwise stated in writing, the copyright holders and/or other parties provide each program as is without any warranty of any kind, either expressed or implied, including, but not limited to, the implied warranties of merchantability and fitness for particular purpose. The entire risk as to the quality and performance of the programs is with the user. Should the program prove defective, the user assumes the cost of all necessary servicing, repair, or correction.

In no event shall any copyright holder be liable for damages including any direct, indirect, general, special, incidental, or consequential damages arising from the use or inability to use the programs (including, but not limited to, loss of use or data, data being rendered inaccurate, losses sustained by the user or third parties, or a failure of the programs to operate with any other programs), however caused, and under any theory of liability, whether in contract, strict liability, or tort.

Acknowledgments

We would like to thank those individuals, most notably our editor, Lauren Cowles, who have contributed to this book. This publication includes images from Corel Stock Photo Library, which are protected by the copyright laws of the United States, Canada, and elsewhere and are used under license. Images and other printed materials have been reproduced with permission from the following sources, listed in the order of first appearance and further identified in the referenced sections: [O'Gorman et al. 85] – Figs. 2.1.1, 2.2.2, 2.2.5, 5.2.4, 5.3.1, 5.6.1; [CSPL – Corel Stock Photo Library/Corel Group] – Figs. 2.2.1, 2.2.6, 2.5.1, 3.6.1, 3.7.1, 4.9.2, 5.5.1, 7.1.6, 7.1.7; [Morgan and Seul 95] – Figs. 2.3.1, 3.3.1; [Woodward and Zasadzinski 96] – Fig. 2.4.1; [The Bell Labs Technical Journal] – Fig. 2.4.6; [Costello] – Figs. 3.1.1B, 7.1.1; [Gonzalez and Woods 92] – Figs. 3.1.3, 3.1.4, 3.4.2; [Lucent Technologies – Bell Laboratories Physical Sciences Image Library] – Figs. 3.2.1, 3.2.4, 4.3.1; [Seul et al. 92] – Figs. 3.3.2, 5.3.2, 5.7.2–5; [Canny 86] – Fig. 3.5.2; [Boie et al. 86] – Figs. 3.5.3, 3.5.4; [Seul et al. 91] – Figs. 4.3.2, 4.3.3, 4.5.2, 4.5.3; [Attneave 54] – Fig. 5.4.2; [Seul and Wolfe 92B] – Fig. 5.7.1; [Seul and Wolfe 92A] – Fig. 5.7.6; [Preparata and Shamos 85] – Figs. 6.1.2., 6.1.4; [Ashcroft and Mermin 76] – Fig. 6.1.2; [Fortune 87] – Fig. 6.1.5; [Sire and Seul 95] – Figs. 6.3.1-3, 6.4.1; [Seul and Chen 93] – Fig. 7.1.8; [Goodman 68] – Figs. 7.2.1, A.3; [Bracewell 86] – Figs. A.1, A.2; [Sessions 89] – Fig. A.4; [Kruse et al. 91] – Fig. A.5.

1.2 Annotated Section Overview

Chapter 1. Introduction

SECTION TITLE	SYNOPSIS
1.1 Introduction	Describes organization of chapters and sections and code
1.2 Annotated Section Overview	Functions as table of contents (section titles) with annotation in the form of section headers (Typical Application(s), Key Words, Related Topics)
1.3 A Guided Tour	Illustrates use of code to construct multistep sequences of image processing and analysis operations

Chapter 2. Global Image Analysis

SECTION TITLE	TYPICAL APPLICATION	KEY WORDS	RELATED TOPICS†
2.1 Intensity Histogram: Global Features	image quality test, object location	intensity histogram, global image features	global enhancement (2.2), binarization (3.10)
2.2 Histogram Transformation: Global Enhancement	enhancement of image contrast	histogram transformations, expansion, equalization; mapping function	global features (2.1), local operations (3.1)
2.3 Combining Images	image overlays, background subtraction; XOR binary operation to control composite image	addition, subtraction; AND, OR, XOR	histogram operations (2.1), color image transformations (2.5), flat fielding (3.3), binarization (3.10), morphological and cellular processing (4.1)
2.4 Geometric Image Transformations	scaling; rotation	interpolation, resampling; image magnification and reduction; rotation	geometric image transformations (2.4), subsampling (3.6), multiresolution analysis (3.7), morphology and cellular processing (4.1), sampling (A.4)
2.5 Color Image Transformations		color, bases transformations, intensity, hue, saturation, RGB, IHS, YIQ	gray-scale analysis (Chap. 3)

†The numbers in parentheses indicate the section in which these topics can be found.

Chapter 3. Gray-Scale Image Analysis

SECTION TITLE	TYPICAL APPLICATION	KEY WORDS	RELATED TOPICS
3.1 Local Image Oerations: Convolution	image contrast manipulation by means of filtering, required in many of the applications discussed in subsequent sections	filter mask, kernel; smoothing, sharpening; edge and point detection; transfer function; cyclic convolution	smoothing/noise reduction (3.2), feature enhancement (3.3), edge detection (3.4), filter mask design (7.2), correlation (A.1)
3.2 Noise Reduction	reduction of extraneous image features; reduction of noise introduced by imaging system	noise removal, filtering, low-pass filter, median filter, speckle noise, smoothing, blurring	subsampling (3.6), binary noise removal (4.2), line noise reduction (5.2), frequency domain filtering (7.2)
3.3 Edge Enhancement and Flat Fielding	emphasis of localized features such as contours; elimination of inhomogeneities in scene illumination	high-pass filter, sharpening; unsharp mask, flat fielding	convolution (3.1), edge detection (3.4); shape analysis (4.4), Hough transform (4.10); frequency domain filtering (7.2)
3.4 Edge and Peak Point Detection	detection of lines and points; locate intensity peaks of extended pointlike objects; image segmentation	gradient, Laplacian; region peak detection, converging squares	edge enhancement (3.3), advanced edge detection (3.5), multiresolution analysis (3.7), template matching (3.8); binary region detection (4.3), shape analysis (4.4), Hough transform (4.10)

Section	Purpose	Techniques	Related Sections
3.5 Advanced Edge Detection	image segmentation in the presence of noise	optimal detection, matched filter, Wiener filter	edge enhancement (3.3), edge detection (3.4), multiresolution analysis (3.7), template matching (3.8), shape analysis (4.4), Hough transform (4.10)
3.6 Subsampling	scale reduction of images to be subjected to object or feature detection	subsampling, image size reduction, resolution adjustment	geometric interpolation (2.4); noise reduction (3.2), multiresolution analysis (3.7); frequency domain filtering (7.2)
3.7 Multiresolution Analysis	object detection	scale-space processing, multiresolution analysis, multiresolution pyramids	noise reduction (3.2), edge enhancement (3.3), subsampling (3.6), template matching (3.8)
3.8 Template Matching	detection of objects of known shape in noisy environment	matched filtering, template matching, cross correlation	binary region detection (4.3), shape analysis (4.4), spectral shape analysis (4.6), Hough transform (4.10); critical point detection (5.4)
3.9 Gabor Wavelet Analysis	texture and pattern detection	texture, pattern, wavelets	multiresolution analysis (3.7), template matching (3.8), Fourier Transform (7.1), Frequency Domain Filtering (7.2)
3.10 Binarization	conversion of gray-scale to binary image	thresholding, local, global, contextual	image intensity histogram (2.1), histogram transformations (2.2), edge detection (3.4)

Chapter 4. Binary Image Analysis

SECTION TITLE	TYPICAL APPLICATION	KEY WORDS	RELATED TOPICS
4.1 Morphological and Cellular Processing	modification of region shapes in binary images	mathematical morphology, cellular logic; erosion, dilation; shrink, expand; region growing; structuring element, binary filtering	template matching (3.8), binary noise reduction (4.2), thinning (4.7), linewidth determination (4.8)
4.2 Binary Noise Removal	reduction of noise in binary images	"salt-and-pepper" noise; kFill	noise reduction (3.2)
4.3 Region Detection	segmentation of images by delineation of regions and encoding of contours	region detection, segmentation, contour representation, cumulative angular bend, curvature point, region filling or coloring, connected component labeling; contour detection, polygonal representation, tangential and radial contour representation	peak detection (3.4); object shape analysis (4.4), thinning (4.7)
4.4 Shape Analysis: Geometrical Features and Moments	shape descriptors for objects, particularly those of near-circular shape, e.g., phospholipid vesicles and erythrocytes, and for domains formed in a wide variety of physical–chemical systems	global shape descriptor, curvature energy, moments, moments of inertia, moment invariants, recursive evaluation	edge and peak detection (3.4); spectral shape analysis (4.5), convex hull (4.6); polygonalization (5.3), critical point detection (5.4)

4.5 Advanced Shape Analysis: Fouries Descriptors	analysis of object shape, based either on object area or on object contour	shape analysis, spectral shape analysis, Fourier descriptors	region growing (4.3), shape descriptors (4.4)
4.6 Convex Hull of Polygons	delineation of polygonal region; association of a shape with a group of points forming the vertices of a polygon	extreme points, convex hull, shape	shape analysis (4.4), (4.5)
4.7 Thinning	thinning ("skeletonization") of elongated regions, lines, and contours	skeleton, medial axis transform	polygonalization (5.3), line fitting (5.5)
4.8 Linewidth Determination	thinning of an image with simultaneous retention of linewidth information	augmented thinning, line image reconstruction	thinning (4.7)
4.9 Global Features and Image Profiles	global analysis of collections of multiple objects	statistical features, image moments, image projection profiles, intensity signatures	global image features (2.1), multiresolution analysis (3.7), shape analysis (4.4), two-dimensional Fourier transform (7.1)
4.10 Hough Transform	detection of lines (to a lesser degree, of other shapes such as circles) in noisy images	Hough transform, line fitting	template matching (3.8); shape features (4.4); line fitting (5.5); two-dimensional Fourier transform (7.1)

Chapter 5. Analysis of Lines and Line Patterns

SECTION TITLE	TYPICAL APPLICATION	KEY WORDS	RELATED TOPICS
5.1 Chain Coding	efficient representation of line patterns such as contour maps, engineering diagrams, fingerprints, and magnetic domain patterns	directional coding, chain code	region detection (4.3), thinning (4.7)
5.2 Line Features and Noise Reduction	recording of line pattern features such as branch and end points; removal of spurious line features from thinned patterns	matched line filters, chain code, primitives chain code (PCC), thin line code (TLC)	noise reduction (3.2); binary noise removal (4.2), thinning (4.7); chain code (5.1)
5.3 Polygonalization	smoothing and parameterization of noisy contours or edges	straight-line approximation, curve representation	thinning (4.7); chain coding (5.1), critical point detection (5.4), line fitting (5.5)
5.4 Critical Point Detection	curve shape description, identification of curvature maxima along contours	critical points, dominant points, curvature, curvature plot, curvature maxima and minima, corner detection, difference of slopes (DoS), k-curvature	shape features (4.4); polygonalization (5.3), line fitting (5.5)

5.5 Straight-Line Fitting	detection and parameterization of straight lines, especially in diagrams	line fitting, straight-line fitting, least-squares fit, regression fit, eigenvector line fitting, principal-axis line fitting	Hough transform (4.10); polygonalization (5.3), critical point detection (5.4)
5.6 Cubic Spline Fitting	approximation of curves and contours by smooth polynomial, noise reduction, smoothing	spline fit, B-splines, cardinal splines, approximating splines, interpolating splines, third-order polynomial fit	polygonalization (5.3), critical point detection (5.4), line fitting (5.5)
5.7 Morphology and Topology of Line Patterns	direct-space analysis of line patterns to ascertain local ordering and to derive a quantitative description of morphological determinants; identification of topological (point) defects	parallelism, overlap, and adjacency; segment clusters; cluster geometry and global descriptors; line pattern topology: branch and end points, disclinations	convex hull (4.6), thinning (4.7); chain coding (PCC) (5.1), noise reduction (TLC) (5.2), polygonalization (5.3)

Chapter 6. Analysis of Point Patterns

SECTION TITLE	TYPICAL APPLICATION	KEY WORDS	RELATED TOPICS
6.1 The Voronoi Diagram of Point Patterns	tesselation of planar patterns of point particles, such as those in images of atoms, molecules, or cells adsorbed to surfaces, layers of colloidal spheres in suspension, domain patterns in a wide variety of physical–chemical systems or stars and galaxies	planar graph, tesselation, triangulation, proximity, nearest neighbor (NN)	region detection (4.3), medial-axis transform (thinning) (4.7), point pattern analysis (Chap. 6)
6.2 Spatial Statistics of Point Patterns: Distribution Functions	evaluation of distribution functions for pair distances and coordination numbers to asses degree of randomness	distance statistics, angle statistics; random, clustered, ordered patterns	point detection (3.4),
6.3 Topology and Geometry of Cellular Patterns	statistics of cellular patterns and polygonal networks; analysis of pattern coarsening dynamics	coordination number, topological charge, Lewis law, Aboav–Weaire law charge compensation	k-NN shells (6.4)
6.4 The k-Nearest-Neighbor (k-NN) Problem	partitioning of a given point set into clusters; range finding, particle tracking; evaluation of pattern statistics as a function of increasing index, k	nearest-neighbor (NN) shell, fractal measure, range finding, clustering	proximity problem (6.1), cellular patterns (6.3)

Chapter 7. Frequency Domain Analysis

SECTION TITLE	TYPICAL APPLICATION	KEY WORDS	RELATED TOPICS
7.1 The Two-Dimensional Discrete Fourier Transform	segmentation and recognition by global pattern or texture; evaluation of diffraction patterns; filtering; convolution	discrete Fourier transform (DFT), fast Fourier transform (FFT); inverse Fourier transform; correlation, power spectrum; sampling rate, resolution, maximum frequency, minimum size; aliasing, "jaggies"; windowing	convolution (3.1), subsampling (3.6), multiresolution analysis (3.7); Fourier descriptors (4.5)
7.2 Frequency Domain Filtering	smoothing, edge detection, texture segmentation, pattern segmentation	filtering: low-pass, high-pass, bandpass, band-stop; cutoff frequency; Gaussian filter, Butterworth filter	convolution (3.1), noise reduction (3.2), edge enhancement (3.3), subsampling (3.6), multiresolution analysis (3.7)

Appendix. Synopsis of Important Concepts

SECTION TITLE	INCLUDED TOPICS
A.1 The Fourier Transform: Interconversion between Spatial Domain and Frequency Domain	properties of the Fourier Transform
A.2 Linear Systems: Impulse Response, Convolution, and Transfer Function	convolution, impulse response, transfer function
A.3 Special-Purpose Filters	matched filter, Wiener filter
A.4 The Whittaker-Shannon Sampling Theorem	sampling and reconstruction of band-limited functions
A.5 Commonly Used Data Structures	lists and trees

1.3 A Guided Tour

Programs in each section implementing single transformations are self-contained to enable the concatenation of multiple single-step operations. As illustrated in the Guided Tour below, readers are encouraged to compose their own sequences of operations, stepping from section to section by following cross references and proceeding in the general direction from lower to higher chapters.

Applied to the original image, the set of operations listed below produces the series of images shown in this section. We describe each step in turn with reference to specific sections in the book and reproduce a corresponding command line (with comments).

Original The starting point for our tour is a raw image in the form of a micrograph depicting a set of small, roughly circular dark domains embedded in a bright background (Fig. 1.3.1). These domain patterns form in certain organic films floating on water, similar to soap films; different regions within the films are made visible by fluorescent dye molecules that avoid the interiors of the domains, which therefore remain dark; contrast in these images tends to be low.

Reference Image The first processing step is to remove the significant spatial nonuniformities in the illumination profile that would otherwise introduce bias into the intensity histogram and interfere with subsequent global processing steps such as binarization. Nonuniformities in scene illumination are the rule, not the exception: for example, microscope illuminators are notoriously nonuniform. The requisite correction requires a reference image showing the slow background variations (but not the objects of interest). Unless otherwise available, for example, in the form of a separately stored

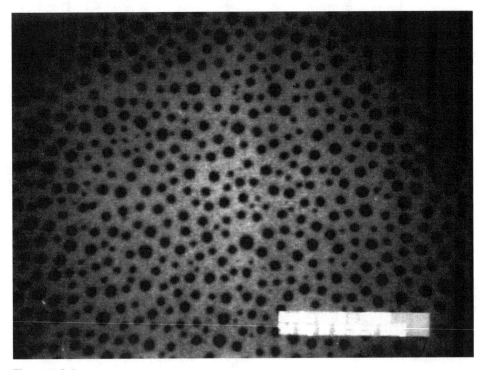

Figure 1.3.1.

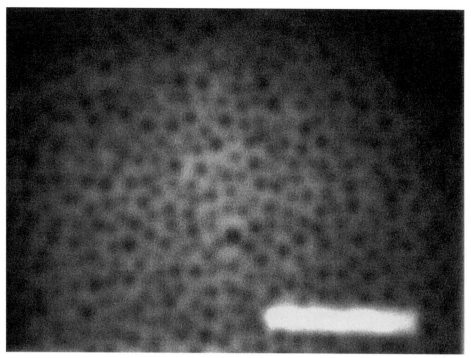

Figure 1.3.2.

image of the scene background, a reference image is created from the original by low-pass filtering; **xconv** (Section 3.1), used with a Gaussian low-pass filter (Section 3.2), produced Fig. 1.3.2.

```
xconv domains.tif domREF.tif -g 65
REM Section 3.1
REM     Generate 2D Gaussian of size 65x65
REM     Produce reference image ("background") by low-pass filtering
REM     Store output in domREF.tif
REM Note: REM indicates comment line
```

Using the convolution operation just described, Fig. 1.3.2 GUI illustrates the deployment of programs from within the GUI. In accordance with program usage, input and output image files can be selected (by typing or browsing), and parameter settings can be specified. Images are provided in the window, as shown. The GUI also provides a listing of all available programs in several formats, for example, by chapter and section, as shown in the figure.

Background Correction   With reference image in hand, we proceed to implement the actual background correction by applying an operation known as flat fielding. This involves the division of the original by the reference image (Section 3.3) and the subsequent scaling of resulting intensities to enhance contrast; **bc** (Section 3.3) produced Fig. 1.3.3.

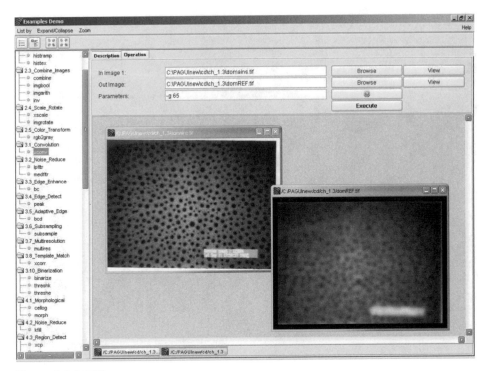

Figure 1.3.2 GUI.

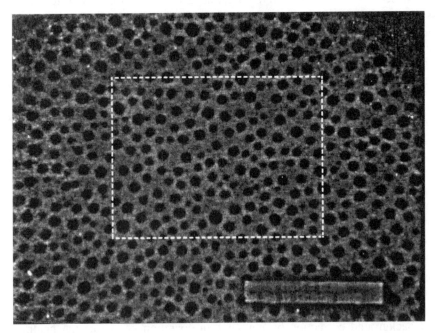

Figure 1.3.3.

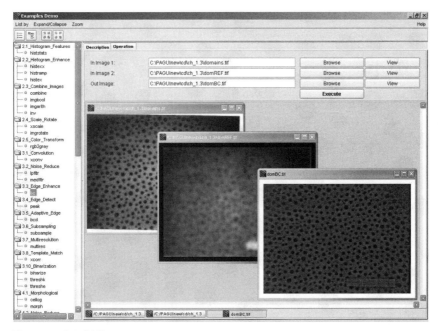

Figure 1.3.3 GUI.

```
bc domains.tif domREF.tif domBC.tif
REM Section 2.3
REM   Correct background nonuniformities: divide by reference image;
REM   Scale intensities
REM   Store output in domBC.tif
```

Fig. 1.3.3 GUI once again illustrates the execution of this program from within the GUI, requiring specification of two input image file names in the appropriate fields, as shown. The two input and the output images are displayed in the window.

Binarization   Binarization represents the final step in processing the image to convert it into a form that is suitable for subsequent analysis of image content. **threshm**, an optimal thresholding routine (Section 3.10), was applied to a portion of the image, also referred to as area of interest (AOI); this is selected by supplying AOI boundaries as command line arguments. Binarization was followed by magnification (Section 2.4) by means of **xscale** to produce Fig. 1.3.4.

```
threshm domBC.tif domBIN.tif
REM Section 3.10
REM   Specify A(rea)O(f)I(nterest)?y
REM   Input upper left, lower right AOI coordinates 300 200 850
      600
REM   Calculated threshold (by moment method) = 63.
REM Note: only a portion of the input image, specified by AOI,
      is binarized
REM   the binarized image appears magnified
```

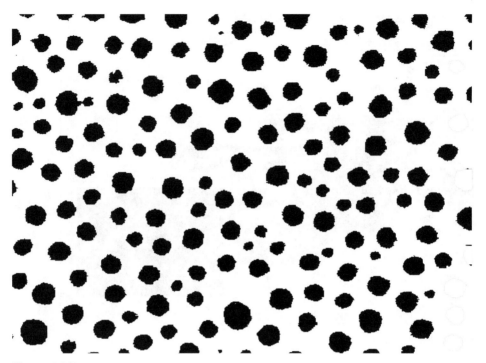

Figure 1.3.4.

Optional: Edge Detection As an optional step, domain edges in the binarized image are readily identified and marked by applying **xconv** in conjunction with a 3 × 3 Laplacian filter mask; the image is not displayed here. More typically, simple (Section 3.4) or advanced (Section 3.5) edge detection would be performed before binarization.

```
xconv domBIN.tif domLAP.tif -f laplac33.txt
REM Section 3.4
REM    Perform edge detection by using 3 × 3 Laplacian, provided in
       laplac33.txt
REM     0.00000  -1.00000   0.00000
REM    -1.00000   4.00000  -1.00000
REM     0.00000  -1.00000   0.00000
REM    Doing image convolution...
REM Note: for demonstration only
REM    edge detection usually applied to gray-scale image
```

Analysis A variety of analytical steps may now be applied to the binarized image to extract quantitative information regarding image content. The objective in analyzing this type of image is two-fold:

• analyze the shape of individual domains in order to measure deviations from circularity and to ascertain the balance of underlying forces,

- analyze positions of domains within the pattern to ascertain the presence or absence of spatial correlations: for example, is the pattern of domains completely random or does it reveal any regularity?

Domain Contours: Shape Descriptors   First, in pursuing the detailed analysis of individual domain shapes, we may wish to determine suitable shape descriptors to reveal the degree of distortion of domains away from a circular reference state (Sections 4.4 and 4.5).

We may accomplish this, for example, by invoking **xcp** (Section 4.3), which performs region encoding, or by invoking **xpm** (Section 4.4), which combines region encoding with contour analysis in terms of geometrical shape descriptors and moments; or, as shown here, by invoking **xbdy** (Section 4.5), which goes beyond **xpm** to provide a spectral representation of shape features in terms of Fourier descriptors.

```
xbdy domBIN.tif domBDY.tif
REM Sections 4.4 and 4.5
REM    Scan, link edge points, encode contour(s);
REM    Select domain(s) for shape analysis;
REM    Select various options (in response to prompting);
REM    Evaluate various shape descriptors;
REM    Create output image, containing:
REM    resampled contour/edge points; centroid, principal axes
REM Note: output image (domBDY.tif) not shown
REM    use XOR (see IMGBOOL, sect. 2.3) to overlay onto
           original
```

Domain Pattern: Region Filling and Centroids   An alternative mode of analysis would be to focus on the spatial configuration of domains within the pattern rather than on individual domain shapes. Accordingly, to locate domains in the pattern, an AOI within the image is scanned. The AOI is identified by means of coordinates supplied as command line arguments to **xah** (Section 4.3), a routine that scans the pattern, fills regions (Section 4.3), and produces a point pattern composed of domain centroids; this is shown superimposed on a binarized domain pattern in Fig. 1.3.5, a figure produced by addition (Section 2.3) of the point pattern and the binarized images.

```
xah domBIN.tif domAH.tif
REM Sections 4.3 and 4.5
REM    Scan, fill regions, and generate point pattern of centroids;
REM    Produce area histogram;
```

```
combine domBIN.tif domAH.tif domBV.tif
REM Section 2.3
REM    Superimpose (arithm addition) point pattern onto binary
           domain pattern
```

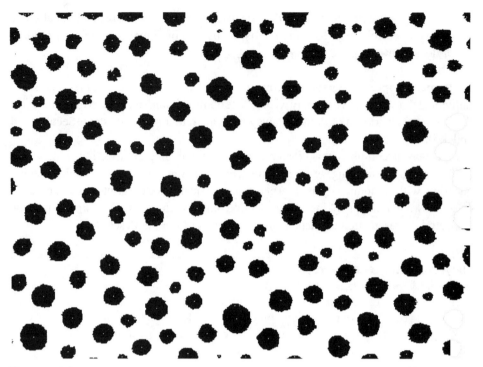

Figure 1.3.5.

Point Pattern: Voronoi Analysis A detailed quantitative analysis of the degree of order present in the pattern of centroids may be based on the Voronoi diagram, a graphical representation of the pattern generated by **xvor**. The output created by **xah** (Section 4.3) in the format of .vin files serves as input for **xvor**. Alternatively, coordinates of points in a given image may be scanned by **spp** to produce a .vin file for subsequent Voronoi analysis. In Fig. 1.3.6, the Voronoi diagram is shown superimposed on the point pattern produced by **xah**. The Voronoi output may be further analyzed by **xvora** and other programs introduced in Sections 6.2–6.4.

```
spp domAH.tif -w domAH.vin
REM Section 6.1
REM     Extract point coordinates
REM     Write data to file domAH.vin (for input to XVOR)
```

```
xvor domVOR.vin domVOR.tif -w domVOR.vdt
REM Section 6.1
REM     Perform Voronoi analysis, create output image domVOR.tif
REM     Write data to file domVOR.vdt (for analysis by XVORA)
```

Other multi-step procedures are feasible: as with the Guided Tour, these will generally follow the same progression from lower to higher chapters, with alternative or

Table 1.3.1. List of Programs and Input/Output Interdependencies

SECTION	PROGRAM	INPUT FILE SUFFIX	OUTPUT FILE SUFFIX	OUTPUT PROCESSED BY
4.3	xah		.gdt	spp; xsgt
	xcp		.zdt	xph
	xrg		.vin	
5.1	pcc		.pcc	pccde, etc.
	pccde	.pcc		
	pccfeat	.pcc		
5.2	linerid	.pcc	.pcc	
	linefeat	.pcc		
	linexy	.pcc		
	structrid	.pcc		
	structfeat	.pcc		
5.3	fitpolyg	.pcc		
5.4	fitcrit	.pcc		
5.5	fitline	.pcc		
5.6	fitspline	.pcc		
5.7	fitpolyg	.pcc	.seg	eh_seg, xsgll
	eh_seg	.seg	.hdt	
	xsgll	.seg	.sgl	eh_sgl
	eh_sgl	.sgl	.hdt	
6.1	spp		.vin	xvor, xptstats, xrg
	xvor	.vin	.vdt	xvora
6.2	xptstats	.vin		
	xvora	.vdt	.std	xsgt
6.3	xsgt	.gdt, .std		

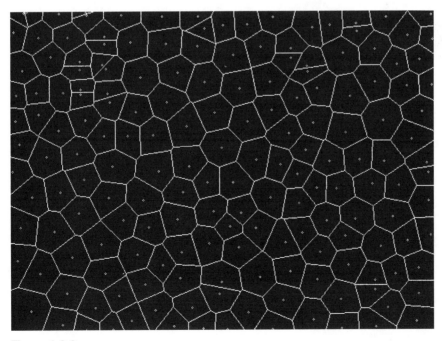

Figure 1.3.6.

multiple analytical paths opening subsequent to binarization. For example, we would have analyzed a raw image containing line patterns primarily by subjecting it to thinning (Section 4.7), encoding by means of **pcc** (Section 5.1), feature analysis (Section 5.2), polygonalization (Section 5.3), line and critical point analysis (Sections 5.4–5.6), and perhaps segment cluster analysis (Section 5.7), as well as Fourier transformation (Section 7.1).

Programs performing quantitative image analysis frequently provide for the option, either on the command line or during run time, to save output data to a file that may be required input for other programs performing further analysis. We use a variety of file suffixes to facilitate the interaction between such programs. The input/output file interdependencies in Table 1.3.1 identify programs calling for data input and lists programs producing requisite data output.

2 Global Image Analysis

In this chapter, we describe global operations for image analysis. A global operation is performed in an equivalent manner to all pixels in the image; it may be independent of the pixel values in an image, or it may reflect statistics calculated for all pixels, but not a local subset of pixels. In contrast, a local operation is performed on a pixel with respect to only its own value and those of its neighbors. In this sense, a local operation can be adaptive because results depend on the particular pixel values encountered in each image region. Local operations will be treated in Chap. 3. A typical application of a global operation is to increase overall image contrast or brightness; a typical application of a local operation is to enhance preferentially certain regions of low contrast or low brightness without affecting other regions.

Although a particular task can often be addressed by either a global or a local operation – for instance, contrast can be enhanced globally or locally – there is an overriding trade-off between the two approaches. When both global and local operations are available, the user must make a choice based on experimentation that is guided by an understanding of the general differences in applicability mentioned above.

Requiring no location-dependent computation, global operations are generally simpler to implement but are also less flexible than local operations, which provide adaptive image modification. Because global operations treat all pixels equivalently, it may be impossible to improve parts of the image without deleteriously affecting other parts. Thus a global increase in brightness will cause dark regions to be more visible, but may cause very bright regions to clip at the top range of brightness. Reflecting the statistics of the entire image, global operations effectively average over local features. Noise effects are thus reduced, but features associated with a desired signal also are averaged. This is acceptable if only a single type of signal is distributed uniformly throughout the image. However, it is a disadvantage in the more frequent case in which the signal reflects objects of various sizes in nonuniform distribution, as, for example, in microscope images of biological cells.

Although this book does not deal with computer architectures, we note that global operations can be implemented by using special-purpose hardware such as single-input–multiple-data (SIMD) computers. The design of SIMD computers is optimized to enhance the speed of performing global operations; most special-purpose boards will not perform general, more complex, image analysis operations.

Section Overview

Section 2.1 discusses the intensity histogram as a tool for the examination of statistical characteristics of the image intensity distribution, including the average intensity, the range of pixel intensities represented in the image, and the predominant intensities.

Section 2.2 describes the use of the histogram in transforming the intensity distribution of an image. The purpose of such transformations is to enhance the visible contrast of an intensity range about an object of interest, sometimes at the expense of intensity regions that are not of interest. We provide several methods of histogram transformation for automatic or interactive enhancement.

Section 2.3 introduces elementary operations for combining images including arithmetic operations (addition, subtraction, and multiplication) and logical operations (AND, OR, XOR). These operations address such tasks as comparing images and eliminating known background noise.

Section 2.4 describes several important geometric image transformations involving a change of size (scaling) and orientation (rotation). These transformations are used in image analysis to reposition the image for better visual inspection and for reduction of distortions introduced by cameras or other image recording devices. The process of interpolation between neighboring pixels plays an essential role in these transformations and is also described here, although it is not properly a global operation.

Section 2.5 presents a discussion of image color. Color can be represented in a computer image by the combination of three different color components, usually red, green, and blue. It is shown here how transformation to different three-component representations can aid analysis. The most popular and useful transformation is to intensity, hue, and saturation components. Very often it is sufficient to perform analysis just on the intensity image. All operations described elsewhere can be applied separately to one, two, or three of the three image components constituting a color image. Accordingly, color-specific methods are usually not needed.

We defer the discussion of additional global operations to later sections to provide appropriate context. Specifically, the following topics and sections are of interest:

Binarization – The process of reducing a gray-scale image to a binary, two-level image, usually for the purpose of separating foreground objects from a background. Binarization can be performed by global or local methods. Both of these are described in Section 3.10.

Image Profiling – The image profile is a histogram of ON-valued pixels summed across rows or columns of a binary image. The profile serves to examine the density and distribution of pixels in a binary image as well as to determine whether a rowwise or a columnwise pattern exists in the image; an example of the latter is text on a page, which is written in horizontal text lines. Image profiling is described in Section 4.9.

Frequency Analysis – Images are often characterized not by particular objects, but by the global configuration of *all* image objects and resulting patterns or texture. These global features can be effectively analyzed on the basis of a spatial frequency representation of the image. The frequency domain also permits spatial filtering and statistical analysis. Chapter 7 introduces the Fourier transform, or the fast Fourier transform (FFT), as the tool to interconvert between spatial domain and spatial frequency representations and discusses spatial frequency analysis.

2.1 Intensity Histogram: Global Features

Typical Application(s) – image quality test, object location.

Key Words – intensity histogram, global image features.

Related Topics – global enhancement (Section 2.2); binarization (Section 3.10)

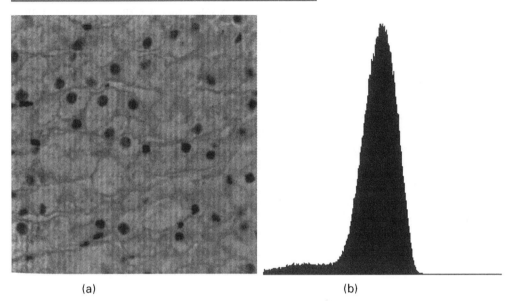

(a)   (b)

Figure 2.1.1. Pictorial Example. Illustration of intensity histogram: (a) image of a liver tissue biopsy, (b) intensity histogram computed from the image in (a). Note the two peaks: the larger peak represents the large area of gray background, and the much smaller peak (almost a plateau) to the right of the large peak represents the small, black regions. (Reprinted with Permission from [O'Gorman and Sanderson 85], Copyright © 1986 by IEEE.)

A useful initial characterization of an image may be based on statistics derived from the entire set of constituent pixels. The so-obtained descriptors, usually constructed from an intensity histogram of the image, are referred to as global features.

The most basic feature is the average pixel value $\bar{I}$ in an image. This represents a measure of image brightness. In particular, $\bar{I} = 0$ identifies an empty image. A second feature is the spread, or variance, of pixel values about the average. This represents a measure of image contrast: the narrower the spread, the lower the contrast. A convenient approach to determine such global features is to extract them from an analysis of an image intensity histogram.

The image intensity histogram is a plot showing the number of image pixels that display each of the possible discrete intensity values (see Fig. 2.1.1). We construct the intensity histogram by examining the intensity value of each pixel in the image and counting the number of pixels displaying each of the possible values. On the histogram plot, each intensity value is represented by a histogram bin whose height represents the number of image pixels displaying that intensity. Therefore the process of constructing the histogram corresponds to the filling of histogram bins.

Histograms of other image characteristics, such as pixel color or object size, etc., may be compiled in corresponding fashion; in each case, the histogram represents the number of instances of each feature versus the discrete set of possible values this feature can assume.

The intensity histogram offers a concise representation of the global intensity characteristics of an image and facilitates the determination of global features. In statistical terms, the histogram is a distribution of sample values for a population of intensities. Just as for any distribution, a statistical analysis may be performed for a histogram to extract what we refer to as global image features. In the remainder of this section, we discuss some common histogram features describing global image characteristics.

2.1.1 HISTOGRAM FEATURES

The simplest of histogram features is the very presence of occupied bins. If only one bin is occupied, then the corresponding image is completely featureless (uniformly white, black, or gray). This may indicate that there is nothing of interest in this image or that the camera has found an empty portion of a larger scene (or that the lens cap of the camera was not removed!).

A very useful feature is the range of intensity levels, defined as the difference between maximum and minimum pixel values represented in the histogram (see Fig. 2.1.2). This often serves as a means for feedback in the image acquisition stage to attain the best image quality. Appropriate adjustments can be made to the camera or scanning device to optimize contrast by taking advantage of the full range of intensity values

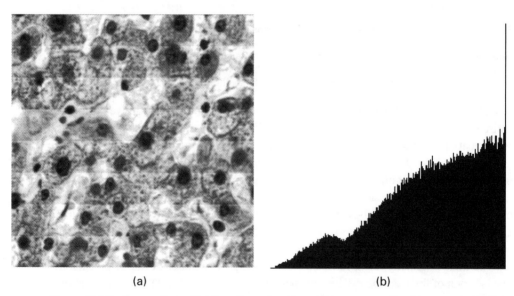

(a)   (b)

Figure 2.1.2. Comparison of (a) liver tissue image and (b) corresponding histogram with Fig. 2.1.1 reveals a wider range of contrast in the present image. The histogram range here is 0–255, whereas the range in Fig. 2.1.1 is 71–255. This image also appears lighter, and this can be seen by two measures, the peak intensity bin: 0 versus 130 for the image in Fig. 2.1.1 and the average intensity level; 82 versus 135 for the image in Fig. 2.1.1. (Reprinted with Permission from [O'Gorman and Sanderson 85], Copyright © 1986 by IEEE.)

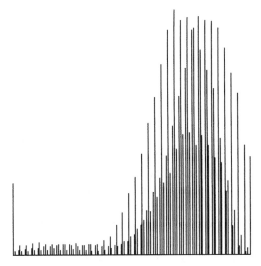

Figure 2.1.3. Histogram for same image as in Fig. 2.1.1, except originally digitized by 7 bits instead of 8 bits, as in Fig. 2.1.1. The histogram shows 128 occupied bins that are spread across the full 256 levels. Since each occupied bin is separated by an unoccupied bin, the histogram appears as a series of impulses rather than as a smooth plot. (Reprinted with Permission from [O'Gorman and Sanderson 85], Copyright © 1986 by IEEE.)

available for a given display device; specifically, n-bit intensity resolution translates into a dynamic range of 2^n levels. For example, 8-bit intensity resolution per pixel yields $2^8 = 256$ levels.

The number of image intensity levels is apparent from the histogram plot. In some cases, the digital camera by which the image was acquired has fewer intensity levels than the display device. For example, if the camera has 4 levels, then the image can have $2^4 = 16$ levels, and the histogram a maximum of 16 occupied bins. If, on the other hand, the display device has $2^8 = 256$ levels, then the image values will usually be multiplied as $16 \times 16 = 256$ so that the intensities cover the full display range. While this operation spreads the recorded bins over the entire available display intensity range (see Fig. 2.1.3), it leaves the number of bins unchanged.

Contrast enhancement (or histogram stretching) techniques of this type may be invoked in the processing stage should image reacquisition not be possible. These techniques are described in Section 2.2.

It is also useful to determine the location of the majority of occupied bins within the histogram. If the occupied set of histogram bins falls near the upper end of the dynamic range of the display, the image may appear too light; otherwise, the image may appear too dark (assuming high intensity is white and low is black). As before, either problem can be remedied by improved reacquisition or by histogram transformation (Section 2.2).

An examination of bin heights can reveal certain image features. The presence of a single bin with large occupancy indicates that the camera has captured a non-black, but featureless region within the image. In such a case, additional bins with low occupancy usually indicate slight nonuniformity or noise in the image.

The most commonly examined histogram feature is the number of peaks; a histogram peak usually comprises a group of adjacent bins whose respective counts exceed a baseline level. Histograms frequently display one predominant peak corresponding to either the lowest or the highest intensity. The large pixel count represents the image background and indicates that the fraction of the field of view occupied by background exceeds that occupied by foreground objects. For example, in a biomedical image of blood cells on a white background or in a document image of black text on white paper, the white background will predominate and will generate the maximally occupied histogram bin.

Aside from a peak representing background pixels, there will often be a distinct second peak representing foreground objects. In the example of the blood cell image, this peak will be located at an intensity level that may represent the level of dye staining used to make a particular cell type visible. Binarization is an important and common image processing operation that relies on histogram analysis: A valley location is found that best separates the two peaks representing foreground and background. Foreground objects are set to one value, for example, 1 or 255, and background pixels are set to another value, usually 0. Binarization is described in more detail in Section 3.10.

Histograms may have multiple peaks. For example, if a blood cell specimen has two types of cells that are stained with dyes of different colors, the histogram will have three peaks representing the background and each of the two cell types. More complex histograms, such as those corresponding to images of outdoor scenes, will contain overlapping peaks corresponding to multiple, possibly overlapping, objects that display a large variance in intensities. While it is generally impossible in such cases to match individual bin peaks to specific objects or object types, the histogram may still provide some image clues, such as lightness, darkness, or the presence of a single predominating object.

Another feature that can be derived from the histogram is entropy. The entropy measures the average, global information content of an image in terms of average bits per pixel. For an 8-bit image, an entropy approaching 8 indicates an information-rich image, that is, pixel intensities cover the full range and do so throughout the image. An entropy approaching 0 indicates the presence of predominating regions in the image in which there is little variation in pixel values.

References and Further Reading

[O'Gorman et al., 1985] L. O'Gorman, A. C. Sanderson and K. Preston, Jr., "A system for automated liver tissue damage analysis: methods and results," IEEE Trans. Biomedical Engineering Vol. BME-32(9), pp. 696–706 (Sept. 1985).

Program

```
histstats
            determines image intensity histogram, and
            yields some global image histogram statistics.
   USAGE:   histstats inimg outimg [-f HIST_FILE] [-L]
ARGUMENTS:   inimg: input image filename (TIF)
            outimg: output image filename (TIF)
 OPTIONS: -f file: write output data file holding histogram values
            -L: print Software License for this module
```

2.2 Histogram Transformations: Global Enhancement

Typical Application(s) – enhancement of image contrast by global histogram transformations.

Key Words – histogram transformations, expansion, equalization, mapping function.

Related Topics – global features (Section 2.1); local operations (Section 3.1).

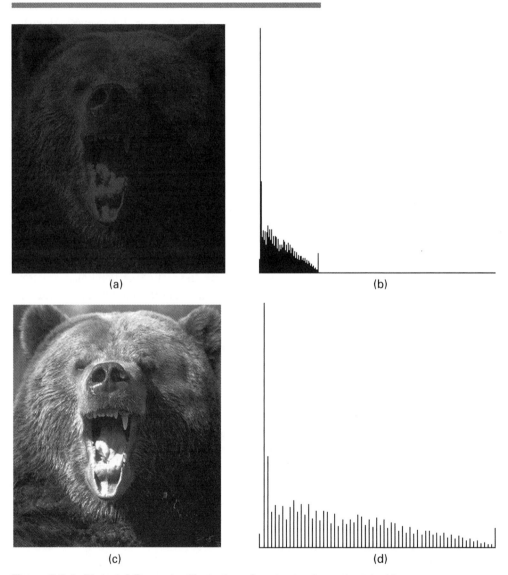

(a)

(b)

(c)

(d)

Figure 2.2.1. Pictorial Example. Illustration of contrast enhancement by histogram expansion: (a) original image with poor contrast; (b) histogram of (a), the narrow range of occupied histogram bins indicating poor contrast, the concentration of intensities at the high end indicating dark image; (c) result of enhancing contrast in (a); (d) histogram of (c). (This product/publication includes images from Corel Stock Photo Library, which are protected by the copyright laws of the U.S., Canada, and elsewhere. Used under license.)

Image enhancement is discussed here as a processing technique to increase the visual contrast of an image in a designated intensity range or ranges. Although the degree of enhancement may be subjective, procedures to perform a given type of enhancement can be directly related to the desired purpose. The methods described here are histogram transformations: These generate a new output histogram by modifying the shape of the input histogram according to a specific mapping function that is chosen to enhance contrast in the range of interest. The gray-scale values of the original image are modified to match the so-obtained output histogram. The result of this is that the gray-scale values of the original image are modified to improve its appearance or effectiveness for visual analysis.

Several definitions of contrast are in use in such different fields as psychology, optics, and photography. For present purposes, we define image contrast at a pixel as the difference between its intensity value $I(x, y)$ and the average background intensity $\bar{I}$ normalized by the full intensity range, i.e.,

$$C(x, y) = \frac{I(x, y) - \bar{I}}{I_{max} - 0}.$$

This equation implies that the most direct way to enhance image contrast is to stretch or expand the occupied intensity range, as shown in Fig. 2.2.1 and discussed below.

In this chapter, we have restricted ourselves to global image enhancement techniques. In contrast to local techniques (Section 3.3), these do not alter the information content of the image. That is, the histogram transformations discussed here conserve the number and heights of occupied intensity bins. In practice, certain transformations may lead to inadvertent merging of bins so that pixels of different input intensities are mapped to a single output intensity; in such a case, the information content of the original image is reduced. Conservation of information content may be important for some applications, such as medical diagnostics, in which it is imperative that the observed image be true and that no artifacts or information reduction be introduced in the course of processing. This can be checked by measurement of the entropy before and after transformation (as discussed in Subsection 2.1.1).

In contrast to global methods, local methods such as spatial filtering (Section 3.3) require the empirical choice of a size for the locally applied filter window. Because of significant variations among images, and even within a single image, local techniques warrant the choice of filter size. By definition, global techniques apply to the entire image and thus require no such choice. Three types of histogram transformations are distinguished here and introduced below.

2.2.1 HISTOGRAM EXPANSION (H_E)

Histogram expansion, denoted by H_E, is the most straightforward and conservative of the transformations introduced here. This is a linear transformation that entails stretching the nonzero input intensity range, $x \in [x_{min}, x_{max}]$, to an output intensity range, $y \in [0, y_{max}]$, to take advantage of the full dynamic range, typically 0 to $y_{max} = 255$.

As a result, the interval $[x_{min}, x_{max}]$ is stretched to cover the new, larger interval $[0, y_{max}]$. Each intensity value x is thus mapped to an output value y according to the following linear mapping function:

$$y = \frac{x - x_{min}}{x_{max} - x_{min}} y_{max}.$$

H_E increases contrast without modifying the shape of the original histogram. The transformation can be applied broadly without concern for deleteriously altering the image appearance. This is the histogram expansion transformation chosen for contrast enhancement shown in Fig. 2.2.1.

Some images may have low contrast even though their lowest and highest occupied intensity bins are close to the minimum and the maximum of the full range. These images are characterized by a histogram with a narrow peak and long tails to either side of the peak. The simple histogram expansion H_E will not be effective here in significantly enhancing image contrast. However, a suitable modification is readily made by introduction of a user-selected cutoff percentage p. This is chosen such that the mid-$p\%$ of intensities will be stretched to the maximum range, thus eliminating (or compressing to one bin each) the upper and the lower histogram tails comprising $(100 - p)/2\%$ of the pixels each. That is, p serves to determine new lower and upper bounds, x'_{min} and x'_{max}, to be placed on the input histogram: intensity values below x'_{min} and above x'_{max} are clipped, that is, they are set to the minimum and the maximum values, respectively. H'_E is a very useful transformation: While still fairly conservative, given that the shape of most of the histogram is left unchanged, H'_E accomplishes greater enhancement than does simple histogram expansion.

The lower x'_{min} and upper x'_{max} bounds of the input range are determined in relation to p as follows:

$$x'_{min}: \frac{(100 - p)}{2} = \frac{100}{T} \sum_{x=x_{min}}^{x'_{min}} h(x), \quad \text{where } T = \sum_{x_{min}}^{x_{max}} h(x);$$

$$x'_{max}: \frac{(100 - p)}{2} = \frac{100}{T} \sum_{x=x'_{max}}^{x_{max}} h(x).$$

Thus the modified expansion H'_E compresses the tails of the input histogram in order to expand the central histogram range, which is occupied by pixels contributing to the central intensities. The percentage p is selected interactively; common values for p range from $p = 98\%$ down to $p = 85\%$. The lower the chosen percentage, the more significant the effect on the image. The onus is on the user to select p so as to preserve salient information.

2.2.2 OUTPUT HISTOGRAM SPECIFIED (H_O)

A second class of histogram transformations is defined by specification of the shape of the output histogram and is designated as H_O. This produces a mapping to an output histogram that has the shape of a linear ramp. For a positive slope, the range of occupied

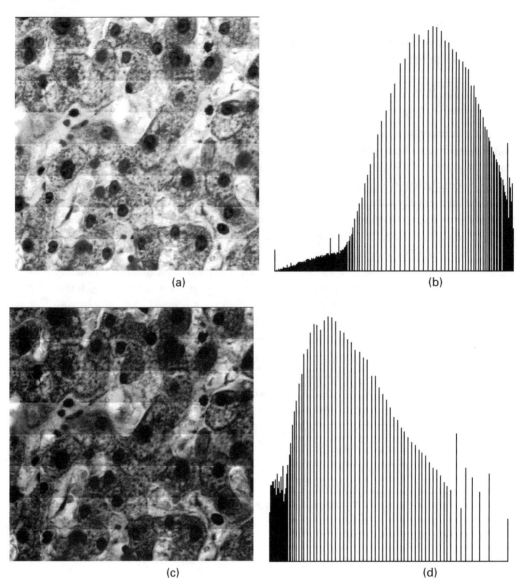

Figure 2.2.2. Histogram transformation, H_O, to a negatively sloped ramp function to enhance contrast in light intensity range. The Original and its histogram are shown in (a) and (b), the transformed image and its histogram are shown in (c) and (d), respectively. (Reprinted with Permission from [O'Gorman et al. 85], Copyright © 1986 by IEEE.)

bins in the low-intensity range is stretched and contrast is enhanced there (Fig. 2.2.2). For negative slope, the range of occupied bins in the high-intensity range is stretched and contrast is enhanced there (Fig. 2.2.3). The case of zero slope corresponds to histogram equalization, which is a popular transformation whose results lie between the two extremes of sloped ramps (Fig. 2.2.4).

H_O is particularly useful for automatic (noninteractive) enhancement, because the output shape is independent of input shape. For instance, H_O thus provides a convenient way to normalize the intensity distribution – to that specified output shape. This

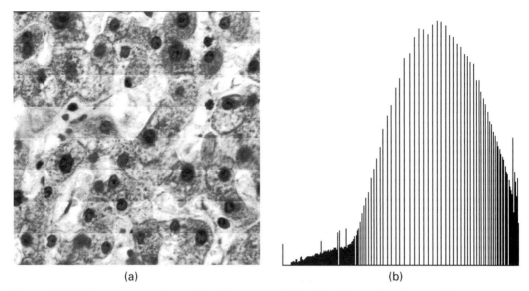

(a) (b)

Figure 2.2.3. Histogram transformation, H_O, to a positively sloped ramp function to enhance contrast in dark intensity range. The original and its histogram are shown in (a) and (b), the transformed image and its histogram are shown in (c) and (d), respectively. (Reprinted with Permission from [O'Gorman et al. 85], Copyright © 1986 by IEEE.)

normalization facilitates comparison among a number of different images. For this reason, it is often applied as a preprocessing step, when subsequent analysis is to be applied to images with normalized intensity statistics.

The function that maps the input histogram to a ramp-shaped histogram will depend on the slope m, of the ramp, on the range of output intensity $y \in [0, y_{\max}]$, and on the

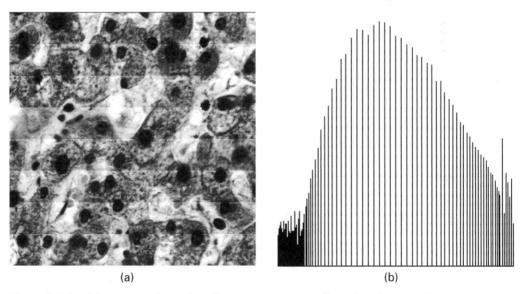

(a) (b)

Figure 2.2.4. Histogram transformation, H_O, to a zero-sloped, uniform, function to treat contrast equally over all intensity range. The original and its histogram are shown in (a) and (b), the transformed image and its histogram are shown in (c) and (d), respectively. (Reprinted with Permission from [O'Gorman et al. 85], Copyright © 1986 by IEEE.)

partial sum $T(x)$ of bins from $x = x_{min}$ to x in the input histogram:

$$T(x) = \sum_{x_{min}}^{x} h(x) - \frac{1}{2}[h(x_{min}) + h(x_{max})],$$

where $h(x)$ denotes bin heights and x_{min} and x_{max} denote the positions of the lowest and the highest occupied bins, respectively, in the input histogram, $T_{max} \equiv T(x_{max})$. For a given value of the slope m the requisite transformation has the form

$$y = \begin{cases} \frac{y_{max}}{2} - \frac{T_{max}}{m y_{max}} - \sqrt{\left(\frac{T_{max}}{m y_{max}} - \frac{y_{max}}{2}\right)^2 + \frac{2T(x)}{m}}, & m < 0 \\ \frac{y_{max}^2}{T_{max}} T(x), & m = 0, \\ \frac{y_{max}}{2} - \frac{T_{max}}{m y_{max}} + \sqrt{\left(\frac{T_{max}}{m y_{max}} - \frac{y_{max}}{2}\right)^2 + \frac{2T(x)}{m}}, & m > 0 \end{cases}$$

In our experience, transformations specified by three particular values for m cover most practical situations. Thus, to enhance

- high intensities, set $m = -2T/y_{max}^2$,
- low intensities, set $m = 2T/y_{max}^2$,
- all intensities equally, set $m = 0$ (histogram equalization).

These choices ensure transformations to an output histogram with maximum negative slope, maximum positive slope, and zero slope, respectively.

2.2.3 HISTOGRAM TRANSFORMATION SPECIFIED (H_T)

The third class of histogram transformations, H_T, requires the explicit specification of a mapping function. Applied to a range of input intensities, the mapping function directly produces the output histogram. This mode of histogram modification is most useful in a setting that permits visual inspection of the output histogram and interactive optimization of parameters specifying the mapping function.

The variable enhancement of intensities, or histogram stretching, around a particular value of the input intensity – with the goal of enhancing image contrast – calls for this type of histogram transformation.

Specifically, the requisite mapping function should stretch the given range of input intensities into a wider range of output intensities. A particular intensity value, or a range of input intensities, is chosen; let $x = x_c$ denote the chosen intensity value or the average intensity in a region of interest. Histogram stretching is achieved by specification of an additional parameter, namely the desired degree of stretching; this corresponds to setting the local slope of the mapping function to $m > 1$.

Given a range of input intensities, $x \in [x_{min}, x_{max}]$, and output intensities, $y \in [0, y_{max}]$, a third-order polynomial can be specified to meet these criteria, mapping input intensities x to output intensities y_S according to

$$y_S = Ax^3 + Bx^2 + Cx + D,$$

where

$$A = \frac{1 - m}{x_{\max}^2 - 3x_c x_{\max} + 3x_c^2},$$
$$B = -3Ax_c,$$
$$C = m + 3Ax_c^2,$$
$$D = 0.$$

By design, $y(x)$ will be limited to a range of output intensities between 0 and an upper bound of $y_{\max}$. However, when actual output intensity values produced by the mapping function exceed $y_{\max}$, the output range must be artificially clipped at $y_{\max}$, and this clipping often leads to an undesirable appearance. The severity of clipping can be reduced by a gradual tapering toward upper and lower bounds; this is achieved by a ramp function of the form

lower bound: $y_{\text{lower}}(x) = kx$,

upper bound: $y_{\text{upper}}(x) = k(x - x_{\max}) + y_{\max}$,

where k, the slope of the ramp, is an additional adjustable parameter. We choose the value of k by bearing in mind that, as k is increased from zero, the severity of clipping will be reduced, but that, as k approaches 1, the range of contrast enhancement is narrowed, and the overall effect of the transformation is diminished, eventually leaving the input image unaltered. A typical value of k is 0.2.

Taking into account the tapered bounds, we may thus specify the H_T mapping function in the following form:

$$y = \begin{cases} y_{\text{lower}}(x), & y_S(x) < y_{\text{lower}}(x) \\ y_{\text{upper}}(x), & y_S(x) > y_{\text{upper}}(x). \\ y_S(x), & \text{otherwise} \end{cases}$$

This mapping function is shown in Fig. 2.2.5 for low, medium, and high values of x_c. A typical value for the slope of the mapping function is $m = 2$; results of this transformation are shown in Fig. 2.2.6.

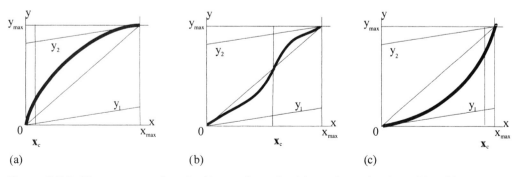

(a)   (b)   (c)

Figure 2.2.5. Histogram mappings for H_T transformation (a) to enhance low intensities, (b) to enhance middle intensities, (c) to enhance high intensities.

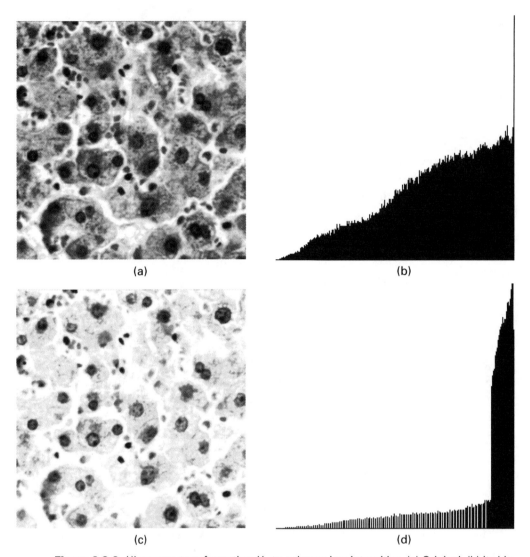

(a)

(b)

(c)

(d)

Figure 2.2.6. Histogram transformation H_T to enhance low intensities. (a) Original, (b) its histogram, (c) the transformed image, (d) its histogram. Note the greater contrast within cells, revealing structure within dark nuclei.

This transformation can also enhance the upper- and the lower-intensity ranges at the expense of the midrange. When the slope is less than unity, $m < 1$, the range around x_c loses contrast, while higher- and lower-intensity ranges gain contrast. In this case, for practical reasons, the requisite choice for x_c is $x_c = x_{max}/2$.

References and Further Reading

[O'Gorman et al., 1985] L. O'Gorman, A. C. Sanderson and K. Preston, Jr., "A system for automated liver tissue damage analysis: methods and results," IEEE Trans. Biomedical Engineering Vol. BME-32(9), Sept. 1985, 696–706.

Histogram modifications have been discussed extensively in the literature. Representative examples relevant to the techniques in this chapter are R. A. Hummel, "Histogram modification techniques," Comput. Graph. Image Process. **4**, 209–224 (1975).

V. T. Tom and G. J. Wolfe, in "Adaptive histogram equalization and its applications," in *Applications of Digital Image Processing IV*, A. G. Tescher, ed., Proc. SPIE **359**, 204–209 (1982).

J. S. Lim, "Image Enhancement," in *Digital Image Processing Techniques*, M. P. Ekstrom, ed. (Academic, Orlando, FL, 1984), pp. 1–51.

L. O'Gorman and L. Shapiro Brotman, "Entropy-constant image enhancement by histogram transformation," Proceedings of 29th Annual SPIE, San Diego, Aug. 1985, pp. 106–113.

Programs

histex (Histogram Expansion)

enhances image contrast by performing histogram expansion.

USAGE: histex inimg outimg [-p PCT_KEEP] [-m MAXVAL] [-L]

ARGUMENTS: inimg: input image filename (TIF)

outimg: output image filename (TIF)

OPTIONS: -p PCT_KEEP: percent of histogram to keep: tails at upper and lower edges of histogram are removed equally; default = 98%;

-m MAXVALUE: maximum value of image intensity (if different than 255); default = 255.

-L: print Software License for this module

histramp (Histogram Ramp)

transforms image intensity distribution to enhance contrast in dark, medium, or light intensities.

USAGE: histramp inimg outimg [-s SLOPE_SIGN] [-L]

ARGUMENTS: inimg: input image filename (TIF)

outimg: output image filename (TIF)

OPTIONS: -s SLOPE_SIGN: value of -1, 0, or 1, connoting the sign of the intensity histogram shape; corresponding increase in contrast in high- (-1), medium- (0), or low- (1) intensity ranges; the default SLOPE is 0, which yields uniform contrast distribution across all intensities (also known as histogram equalization)

-L: print Software License for this module

histexx

expands intensity range about a chosen value.

USAGE: histexx inimg outimg [-x X_CENTER][-s SLOPE][-t TAPER_SLOPE] [-L]

ARGUMENTS: inimg: input image filename (TIF)

outimg: output image filename (TIF)

OPTIONS: -x X_CENTER: intensity value about which to increase contrast (default = 128);

-s SLOPE: slope of enhancement transform; as the value is set above 1, contrast becomes

greater about X_CENTER, but contrast
is reduced at other intensities; if slope
is set below 1, contrast is reduced
at X_CENTER and increased at high and low
intensities. If slope is 1, intensities
remain unchanged (default = 2.0)

-t TAPER_SLOPE: slope of bounds of transformation;
values > 0 produce more tapering;
default = 0.2

-L: print Software License for this module

2.3 Combining Images

Typical Application(s) – image overlays, background subtraction; XOR binary operation to control composite image.

Key Words – addition, subtraction; AND, OR, XOR.

Related Topics – histogram operations (Section 2.1), color image transformation (Section 2.5); flat fielding (Section 3.3), binarization (Section 3.10); morphological and cellular processing (Section 4.1).

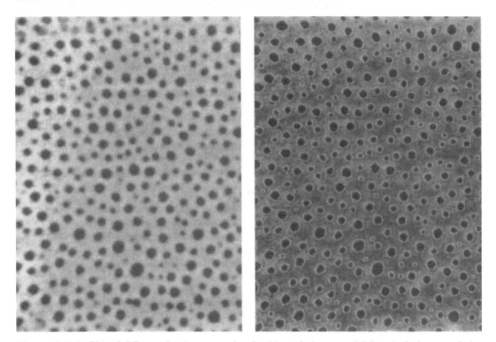

Figure 2.3.1. Pictorial Example. An example of arithmetic image addition. Left, image of circular domains observed in a two-dimensional monomolecular film by means of fluorescence microscopy; right, original with superimposed edge contour image. Note: methods for producing the domain contours overlaid on the original images are discussed in Sections 3.4 and 3.5. (Reprinted with Permission from [Morgan and Seul 95], Copyright © 1995 American Chemical Society.)

The elementary process of creating an image from a combination of two separate input images finds frequent use in the processing of gray-scale as well as binarized images. The former case calls for arithmetic operations, notably addition, subtraction, and, in rare cases, division (Section 3.3); the latter case requires the application of logical, or Boolean, operations, notably AND, OR, and (e)X(clusive)OR. While the majority of the image processing operations discussed in Chaps. 2 and 3 apply to groups of pixels (neighborhoods) in a single image, the binary operations described below are applied to pairs of corresponding individual pixels in separate images.

2.3.1 ARITHMETIC OPERATIONS

Addition and Subtraction   The most commonly required arithmetic operations for combining two separate input images are (pixel-by-pixel) addition and subtraction:

$$\text{Addition}\quad I(x, y) = \min[I_1(x, y) + I_2(x, y);\ I_{max}],$$

$$\text{Subtraction}\quad I(x, y) = \max[I_1(x, y) - I_2(x, y);\ I_{min}].$$

Given the finite range of gray values, limited by $I_{min} = 0$ and $I_{max} = 255$, the most important point to be aware of is the possibility of overflow and underflow, respectively, in connection with addition and subtraction. In either case, clipping of the image intensity will result. That is, in the first case, if the sum of two image intensities exceeds I_{max}, it will be set to I_{max}; in the second case, if the difference $I_1 - I_2$ falls below 0, it will be set to 0.

Clipping can be avoided if the range of intensity values encountered in both input images is rescaled before the images are combined. This entails scanning the two input images and computing all pixel sums and differences to identify maximal and minimal output intensity values, $(I_1 + I_2)_{max}$ and $(I_1 - I_2)_{min}$, respectively. These values are then applied as scale factors according to the following prescription:

$$\text{Modified Addition}\quad I(x, y) = [I_1(x, y) + I_2(x, y)] \times 255/(I_1 + I_2)_{max},$$

$$\text{Modified Subtraction}\quad I(x, y) = [I_1(x, y) - I_2(x, y) + (I_1 - I_2)_{min}]$$
$$\times 255/[255 + (I_1 - I_2)_{min}].$$

An example of addition is that of combining an original (gray-scale) image with a superimposed edge map as illustrated in Fig. 2.3.1. This facilitates the evaluation of the accuracy of edge detection procedures; these procedures are discussed in Sections 3.4 and 3.5. In the experimental sciences, subtraction of a reference image (background) from an incoming raw image (foreground) is a common operation to eliminate blemishes or inhomogeneities of the scene illumination, as discussed in more detail in Section 3.3. It is often possible to record a standard reference image, for example, by defocusing the imaging optics and to store this background image for subsequent use.

Division   While less commonly required than addition and subtraction, division of one image by another is often called for in video microscopy when images are recorded with cameras that exhibit nonlinear output characteristics (see also modified flat fielding, Subsection 3.3.1). In such a case, subtraction can introduce distortions in the image intensity distribution:

$$\text{Division}\quad I(x, y) = I_1(x, y)/I_2(x, y):\ I_2(x, y) \neq 0.$$

Implementation of the division of an image I_1 by a second image I_2 requires two modifications of this simple prescription. First, pixels of zero intensity are removed from I_2, adding a constant of unity to produce $I_2' \equiv I_2 + C$, $C = 1$. Second, division of two images with comparable intensity distribution leads to small (and generally noninteger) output values and therefore requires the scaling of the output image to the full dynamic range of the display device, say [0, 255], as follows:

$$\text{Modified Division} \quad I(x, y) = [I_1(x, y)/I_2'(x, y)] \times 255/(I_1/I_2')_{\text{max}}.$$

2.3.2 LOGICAL (BOOLEAN) IMAGE OPERATIONS

Referring to the two possible values of pixels in a binary image as ON and OFF, respectively, we may summarize the most commonly encountered Boolean operations as follows:

AND	a AND b = ON	if $a = b$ = ON,
OR	a OR b = ON	if a = ON or if b = ON,
XOR	a XOR b = ON	if a = ON and b = OFF or
		if a = OFF and b = ON.

The primary use of the AND operator is to find all ON pixels common to two input images. The OR operator is useful in combining all ON pixels of both images into a single output image, thereby producing the Boolean equivalent of the arithmetic addition operation. The XOR operator is used to highlight positions whose pixel values in the two images differ. Boolean operations also are encountered in connection with morphological and cellular logic operations, applied to pixel neighborhoods (Section 4.1).

As we have indicated, all operations are applied to individual pairs of pixels from input images I_1 and I_2 so that, for any of the Boolean operators, represented here by $\odot$, the output image I is composed as follows:

$$I(x, y) = I_1(x, y) \odot I_2(x, y) \tag{1}$$

for all positions (x, y) in the input images. Figures 2.3.2 and 2.3.3 illustrate the results.

These Boolean operators are readily combined (or concatenated) to achieve additional effects. Other Boolean operations, such as NOT, NAND, and NOR, are not commonly encountered in the context of image processing.

References and Further Reading

[Morgan and Seul, 1995] N. Y. Morgan and M.. Seul, "Structure of disordered droplet domain patterns in a monomolecular film," J. Phys. Chem. **99**, 2088–2095 (1995).

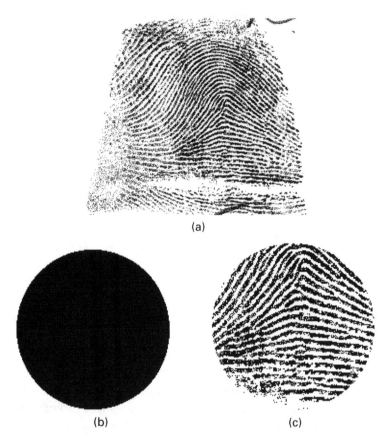

(a)

(b)   (c)

Figure 2.3.2. Illustration of Boolean AND operation: (a) the fingerprint pattern and (b) the circular disk (assumed here to be composed of ON pixels) were combined to produce (c) the cropped image.

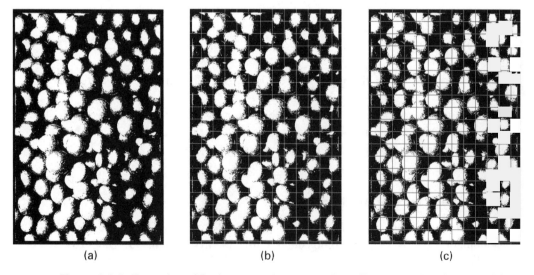

(a)   (b)   (c)

Figure 2.3.3. Illustration of Boolean OR and XOR operations: The pattern in (a) was combined with a square grid composed of ON pixels (not shown) to produce (b) by OR and (c) by XOR.

Programs

imgarith

performs arithmetic combination of two images: addition, subtraction, or multiplication

USAGE: imgarith in1img in2img outimg [-o OPERAND] [-x X] [-y Y] [-L]

ARGUMENTS: im1img, in2img: input image filenames (TIF)
outimg: output image filename (TIF)

OPTIONS: -o OPERAND: <+,-,x> for addition, subtraction, or multiplication; default is addition (+).
-x X: X-offset of second image relative to first
-y Y: Y-offset of second image relative to first; default = (0,0).
NOTE: for images of different sizes, the output image will have the size of the first specified image.

-L: print Software License for this module

imgbool

performs Boolean combination of two images: OR, AND, or XOR

USAGE: imgbool in1img in2img outimg [-o OPERAND] [-x X] [-y Y] [-L]

ARGUMENTS: im1img, in2img: input image filenames (TIF)
outimg: output image filename (TIF)

OPTIONS: -o OPERAND: <O,A,X> for OR, AND, or XOR; default is OR (O).
-x X: X-offset of second image relative to first
-y Y: Y-offset of second image relative to first; default = (0,0).
NOTE: for images of different sizes, the output image will have the size of the first specified image
-L: print Software License for this module

combine

adds two images and writes the resulting image to the specified output file.

USAGE: combine in1img in2img outimg [-c clip][-L]

ARGUMENTS:
in1img: first input image filename (TIF)
in2img: second input image filename (TIF)
outimg: output image filename (TIF)

OPTIONS:
-c clip: clipping factor for output image 0-255 (default=255)
-L: print Software License for this module

inv

inverts the input image and writes the resulting image to the specified output file.

USAGE: inv inimg outimg [-L]

ARGUMENTS:
inimg: input image filename (TIF)
outimg: ouput image filename (TIF)

OPTIONS:
-L: print Software License for this module

2.4 Geometric Image Transformations

Typical Application(s) – scaling: magnification and reduction; rotation.

Key Words – nearest-neighbor interpolation, bilinear interpolation; resampling; scaling, rotation.

Related Topics – convolution (Section 3.1), subsampling (Section 3.6), multiresolution analysis (Section 3.7), adaptive thresholding (Section 3.10); morphology and cellular processing (Section 4.1); sampling (Section A.4).

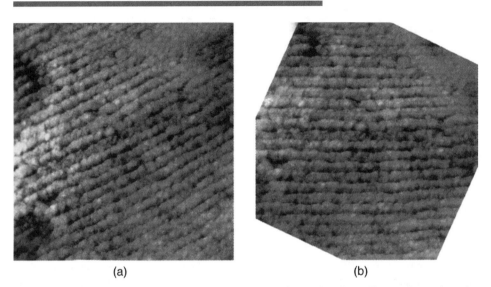

(a)   (b)

Figure 2.4.1. Pictorial Example. Image clipping introduced by direct (forward) rotation. As a result of rotation, portions of the original image (a) are placed out of bounds, while other regions of the field of view are left vacant (b). Remedies to this problem are discussed in the text. (Reprinted with Permission from [Woodward and Zasadzinski 96], Copyright © 1996 American Physical Society.)

This section addresses changes in image scale and orientation that are applied to facilitate processing (see, e.g., Section 3.7) or to correct misorientation of an image, such as a page of text on a scanner, before further analysis. Other corrective transformations include shearing and warping (not discussed here) that can be necessary to compensate for shape distortions introduced by recording devices.

2.4.1 GEOMETRIC INTERPOLATION

Operations such as scaling, rotation (see Fig. 2.4.1), and corrective warping rely for their proper implementation on interpolation and image resampling. Interpolation may be visualized as a process generating a continuous function $g(t)$ from a given discrete sample $g[m]$ by application of a suitable fitting function to a number n of nearest neighbors at each of the sample points. The interpolation function may then be evaluated at the desired target points to complete a process known as resampling. In this section,

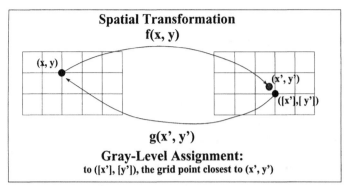

Figure 2.4.2. Interpolation is called for to assign approximate gray levels to pixels that are mapped to noninteger target coordinates under spatial transformations.

we describe the two most common interpolation schemes that have the benefit of speed and simplicity of implementation. Resampling will be discussed again in Sections 3.6 and 3.7, which also contain a more general presentation of interpolation based on linear filters.

Interpolation

An image transformation of the form

$$\mathbf{H} \equiv (\mathbf{h}_x, \mathbf{h}_y),$$

where $x' \equiv \mathbf{h}_x(x, y)$ and $y' \equiv \mathbf{h}_y(x, y)$ denote pixel coordinates in the target image, prescribes the mapping of each original pixel value, $I(x, y)$, into a new pixel value, $I(x', y')$. However, as illustrated in Fig. 2.4.2, the transformation $\mathbf{H}$ will generally yield noninteger target coordinates x' and y', a problem encountered in the specific cases of scaling and rotation. Geometric interpolation is invoked to provide gray-level values at valid – that is, at integer – pixel coordinates. We evaluate the requisite new pixel intensities by relying on the values at one or several actual (noninteger) target image locations.

The simplest form of interpolation relies on the inspection of nearest-neighbor locations. That is, each integer location in the target image is assigned the gray level of the transformed pixel (x', y') in its most immediate proximity. We obtain smoother images by considering the shell of (four) nearest neighbors and computing a new gray-level value on the basis of a weighted average of these four input values. The standard bilinear interpolation relies on weights derived by simple integer truncation in both x- and y directions: $a = x - [x]$ and $b = y - [y]$, respectively, where $[x]$ denotes the largest integer not exceeding x. Alternatively, weights can be based on the actual distance R between target location (x', y') and the (integer) pixel location of interest. These most commonly used nearest-neighbor (order $n = 1$) and bilinear ($n = 2$) interpolation schemes are illustrated in Fig. 2.4.3 and summarized in Fig. 2.4.4.

We perform higher-order interpolation by considering the shell of the (four) nearest, as well as the shell of the (twelve) next-nearest neighbors ($n = 3$). We may implement this scheme by invoking cubic spline functions [Hagan, 1991], but the computational effort involved diminishes its practical role and it is not included here. In general, the preferred choice of interpolation function depends on the nature of the requisite

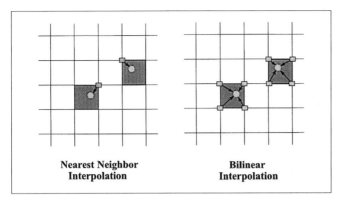

**Nearest Neighbor
Interpolation**

**Bilinear
Interpolation**

Figure 2.4.3. Graphical illustration of nearest-neighbor and bilinear interpolations.

transformation, keeping in mind the rapidly increasing computational cost of higher-order interpolation schemes.

Interpolation and Resampling

Resampling must be preceded by conversion of the (discrete) original to a continuous representation by means of convolution with a suitable interpolation function. While

Figure 2.4.4. Common Geometric Image Interpolation Procedures (input image: $I[]$; output image: $O[]$).

Nearest-Neighbor Interpolation (Simple Rounding): order $n = 1$

- *strategy*: assign gray value of closest pixel

- *comments*: image shifted by up to 1/2 pixel; gray values remain unchanged

- *expression*:

$$O[x', y'] \;=\; I[(int)(y + 0.5), (int)(x + 0.5)]$$

Bilinear Interpolation: $n = 2$

- *strategy*: select weighted sum of 4 nearest pixels in I
  - (a) weights proportional to distance
  - (b) weights proportional to square of distance

- *comments*: new gray values appear

- *expression*:

$$(a) : O[x', y'] \;=\; abI[y, x] + (1 - a)bI[y, x + 1] +$$
$$a(1 - b)I[y + 1, x] + (1 - a)(1 - b)I[y + 1, x + 1]$$

$$(b) : O[y', x'] \;=\; \sum_{i=1}^{i=4} \frac{I_i}{R_i^2} \Big/ \sum_{i=1}^{i=4} \frac{1}{R_i^2}$$

where, in (a), $a = x - floor(x)$, $b = y - floor(y)$.

conceptually distinct, the two operations may be combined to attain greater efficiency. This is accomplished by restriction of the evaluation of the interpolation function to the target points of interest. In cases such as image rotation, we determine the desired locations at which to resample the input image by invoking the inverse $\mathbf{H}^{-1}$ of the prescribed geometric transformation, as discussed below.

A more general perspective of this combined interpolation and resampling is possible but relies on the concept of convolution, introduced in Section 3.1. The convolution operator facilitates the modification of images by application of linear filter masks. Proper masks may be constructed to combine resampling and interpolation into a single step. Interpolation and resampling operations are discussed again in Section 3.6 [Hagan, 1991].

2.4.2 IMAGE SCALING

Magnification and Reduction

Adjustment of the scale in an image is often desirable either to zoom in on details by magnification or to zoom out to gain an overall overview by reduction. Magnification permits closer inspection of feature details, usually at the expense of excluding from view other parts of the image. Reduction facilitates displaying large images or multiple images such as a collection of thumbnail sketches on one screen.

Simple integer scaling is accomplished by expansion or reduction of the dimension of the displayed image by an integer factor k; for magnification, $k > 1$. Scaling may be performed separately for horizontal and vertical image dimensions with respective scale factors k_x and k_y; preservation of the aspect ratio requires that $k_x = k_y = k$.

Magnification requires an increase in the number of pixels. The simplest procedure to create the requisite additional pixels is to replicate each existing pixel k_x times and k_y times, respectively, to fill in the additional columns and rows of the magnified image. Conversely, reduction requires deletion of pixels from the original: $k - 1$ of every k pixels are removed to achieve a k-fold image reduction.

This simple single-pixel replication or deletion procedure often results in an objectionable, blocky appearance. In the case of image magnification, we often achieve an improvement by invoking interpolation. That is, rather than simply replicating existing pixels, we combine their values to compute intermediate values for the newly required pixels in the magnified image. The simplest and most common procedures for geometric interpolation are described in Subsection 2.4.1.

General Scaling: The Binary Code

The scaling of images by noninteger factors requires a generalization of the simple replication and deletion of pixels just described for scaling by an integer factor. While this can be achieved by interpolation, an approach based on the representation of lines and curves on pixel grids provides an alternative solution that yields very fast algorithms.

Consider the representation of a line with fractional slope on a pixel grid: The set of pixels selected to provide the closest discrete approximation to the ideal continuous line will form a staircase of varying step widths and heights, as illustrated in Fig. 2.4.5 for a line of slope $5/9$.

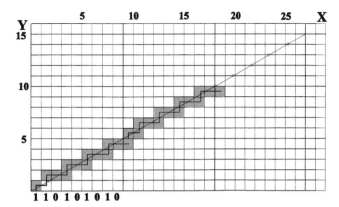

Figure 2.4.5. Binary code for a line of slope 5/9. Columns in which the continuous line intersects horizontal lines in the grid are indicated by ○. The corresponding binary code, a string of length 9, is indicated for two successive intervals.

This approximation is readily constructed by the marking of all columns in which the ideal line intersects a horizontal line in the pixel grid; once this threshold is reached, a new step is introduced in the staircase. This procedure thus defines a binary sequence by assigning 0 and 1, to unmarked and marked columns, respectively. For example, as illustrated in Fig. 2.4.5, a line of slope 5/9 generates the binary code 110101010.

This graphical construction corresponds to integer truncation, i.e., to the assignment $y = [y]$, where [ ] denotes the largest integer not exceeding y. That is, if an image of q columns is to be expanded to a width of p columns, the corresponding binary code would show $q < p$ set bits that are uniformly distributed throughout the string of p bits. As the target image is constructed column by column, each set bit in the code indicates that the corresponding column in the target image is to receive an original source image column, while each unset bit indicates that the corresponding column is to receive a duplicate of the most recently copied source image column. This procedure thus spreads p source image columns over q target image columns. Conversely, in a horizontal reduction, each source image column marked by 0 is eliminated.

Vertical scaling proceeds analogously, based on a binary code evaluated on the basis of vertical grid–line intersections, and the resulting binary code gives instructions for the handling of image rows.

Based on a generalization of this idea to two dimensions, the implementation provided here also avoids floating-point calculations by a simple trick and thus achieves very fast execution [Clark, 1997]. The images in Fig. 2.4.6 illustrate noninteger reduction and enlargement.

2.4.3 IMAGE ROTATION

Image rotation is a transformation to change the orientation of the original image on the screen. In general, rotation alters image size.

A special case is the rotation by multiples of 90° and the related reflection about the horizontal or the vertical center lines. These operations have the special property that the overall rectangular shape of the image is preserved. Implementation simply relies on appropriate translations of rows and columns. In particular, rotation by 180° can be

(a)

(b)

(c)

Figure 2.4.6. Illustration of noninteger scaling. The original in (a) was enlarged by a factor of 1.7 to produce (b) and was reduced by 1.7 to produce (c). (Reprinted with Permission from [BL-TJ 98] – Copyright © 1988 by Lucent Technologies Inc.)

performed in place, i.e., without creating additional copies of the image in memory; if the input image is square, in-place rotation is also possible for multiples of 90°.

Rotation by an arbitrary angle is more complex. The transformation of interest, $x' \equiv \mathbf{R}_x(x, y)$ and $y' \equiv \mathbf{R}_y(x, y)$ is explicitly specified in the form of an orthogonal transformation:

$$\mathbf{R}(x, y) = \begin{bmatrix} \mathbf{R}_x \\ \mathbf{R}_y \end{bmatrix} = \begin{bmatrix} \cos\phi & -\sin\phi \\ \sin\phi & \cos\phi \end{bmatrix}.$$

The naive way to proceed would be to use the transformation $\mathbf{R}$ to find target locations for all pixels in the original image. However, as illustrated in Figs. 2.4.1 and 2.4.7, this approach encounters two problems. First, target coordinates may fall on noninteger locations. Simply ignoring noninteger target coordinates would create holes in the rotated image, not a desirable result. The alternative is to invoke interpolation: Pixel values in the target image are computed on the basis of values assigned to nearby noninteger target coordinates. Second, rotation of a rectangular input image will generate a shape that is no longer congruent with the display: Some portions of the display region will be left unassigned and some portions of the rotated image will extend beyond the boundaries of the screen.

In fact, a preferable strategy is to work backwards. That is, rather than the use of $\mathbf{R}$ with the attendant problems discussed above, each screen pixel is regarded as an output pixel (x', y') whose source coordinates we evaluate by invoking $\mathbf{R}^{-1}$, the inverse of $\mathbf{R}$:

$$\mathbf{R}^{-1}(x, y) = \begin{bmatrix} \cos\phi & \sin\phi \\ -\sin\phi & \cos\phi \end{bmatrix}.$$

The assigned source coordinates (x, y) will generally not correspond to integer values,

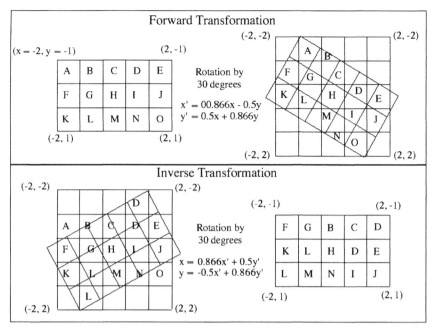

Figure 2.4.7. Illustration of difficulties encountered in implementing rotation of a rectangular input image (here of dimensions 3 × 5) by $\phi = 30°$ so that $x' = 0.866x - 0.5y$ and $y' = 0.5x + 0.866y$. The forward transformation (top) requires a 5 × 5 output array for displaying the rotated image. In contrast, a strategy relying on the inverse transformation, $x = 0.866x' + 0.5y'$ and $y = -0.5x' + 0.866y'$, evaluates source pixels for all pixels in the 3 × 5 display (bottom).

indicating, as before, a requirement for interpolation to assign a source value to each pixel in the output image. Suitable procedures are discussed in Subsection 2.4.1. As for those target pixels whose source locations are found to lie outside the original image boundaries, they are considered lost. In this case, the shape of the image is preserved. The alternative, not practical in many cases, is to provide for a sufficiently large target image size to accommodate all source image pixels rotated to new positions. In this case, the image size required to accommodate any rotation is a square with sidelengths equal to the original image diagonal length.

References and Further Reading

[BLTJ., 1998] Bell Labs Technical Journal, "The Transistor – 50th Anniversary: 1994–1997," Vol. 2(4), Autumn 1997.

[Clark, 1997] D. Clark, "A 2d DDA algorithm for fast image scaling," in *Dr. Dobb's Journal*, (April 1997) pp. 46–49; provides an alternative interpretation of the representation of lines on pixel grids and presents a simple, general and fast implementation of scaling.

[Hagan, 1991] G. T. Hagan, "Resampling methods for image manipulation," C Users J., 53–58 (August 1991); provides a concise summary of common interpolation schemes and gives an introduction to the concept of resampling by means of convolution.

[Woodward and Zasadzinski, 1996] J. T. Woodward and J. Zasadzinski, "Amplitude, Waveform and Temperature Dependence of Bilayer $P_{\beta'}$ Phase," Phys. Rev. E-Rapid Comm. **53**, R 3044–3047 (1996).

Reference points are further discussed in W. K. Pratt, *Digital Image Processing and Digital Image Processing*, 2nd ed. (Wiley-Interscience, New York, 1991); and in R. C. Gonzalez and R. E. Woods, *Digital Image Processing* (Addison-Wesley, Reading. MA, 1992), Chap. 5.

The results of applying bivariate interpolation (see also Section 2.4) are given in Subsection 8.4.1 of J. S. Lim, *Two-Dimensional Signal and Image Processing* (Prentice-Hall, Englewood Cliffs, NJ, 1990).

A discussion of common image distortions and their modeling is given by P. Mansbach, "Calibration of a camera and light source by fitting to a physical model," Comput. Vis. Graph. Image Process. **35**, 200–219 (1986).

Programs

xscale

scales the input image by fast integer interpolation and writes the scaled image to the specified output file

USAGE: xscale inimg outimg x_scale y_scale [-L]

ARGUMENTS: inimg: input image filename (TIF)
outimg: output image filename (TIF)

x_scale: scaling factor for x axis (float) >= 0.0
y_scale: scaling factor for y axis (float) >= 0.0

OPTIONS: -L: print Software License for this module

imgrotate

rotates image by specified angle;

USAGE: imgrotate in.img out.img angle [-o X_ORIGIN Y_ORIGIN]
[-q] [-L]

ARGUMENTS: imimg: input image filename (TIF)
outimg: output image filename (TIF)
angle: value (in degrees) of desired rotation with respect to the horizontal; counterclockwise is positive.

OPTIONS: -o X_ORIGIN Y_ORIGIN: coordinates about which to rotate; (0,0) is at top-left corner; default is center of image.

-q: if set, performs quicker rotation by using nearest-pixel instead of default bilinear interpolation.

-L: print Software License for this module

2.5 Color Image Transformations

Typical Application(s) – segmentation of color-specific image components, reduction of color image to gray-scale image.

Key Words – color, bases, primaries, transformations, intensity, brightness, value, hue, saturation, RGB, IHS, YIQ.

Related Topics – gray-scale analysis (Chap. 3).

(a)

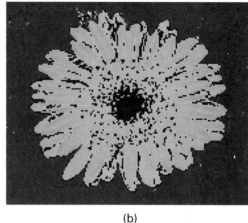

(b)

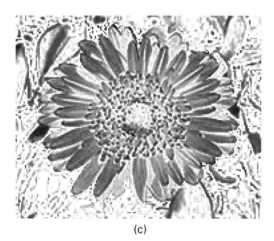

(c)

Figure 2.5.1. Pictorial Example. Illustration of the intensity (I), hue (H), and saturation (S) representation of a color image: (a) gray-scale image, essentially representing the intensity (I) component of an IHS-transformed image; (b) hue; (c) saturation. Together these represent a color image: I, its brightness, H, its color, and S, its deepness of color. (This product/publication includes images from Corel Stock Photo Library which are protected by the copyright laws of the U.S., Canada and elsewhere. Used under license.)

A common representation of color in an image relies on storing three color values at each pixel. The color values, usually red, green, and blue (RGB), are obtained by the capture of each image pixel through separate color filters. Combinations of the three basis colors, also referred to as bases or primaries, are used to describe different colors. For instance, secondary colors result from primary combinations: equal red and blue yield magenta, blue and green yield cyan, and red and green yield yellow. All other colors – nonprimary or secondary – can be produced from other combinations of the three primary colors. Because a color image may be thought of as a combination of three independent gray-scale images, most methods of color image analysis do not differ significantly from those applied to gray-scale images; they just entail application of the same methods as those used for a single gray-level image, but applied threefold to the different color images.

Partly because of the threefold increase of data compared with a gray-scale image, a useful first step in analyzing color images is to perform data reduction to a single gray-scale image by color recombination. The simplest procedure is the elimination of undesired portions of a given image by color filtering. For instance, if only magenta regions are of interest, a single image can be generated by a color combination of red and blue intensities for each pixel.

A more general, commonly used color transformation maps the representation in terms of three color bases (RGB) to a representation in terms of intensity, hue, and saturation (IHS) (see Fig. 2.5.1). These terms describe characteristics of the color image. Intensity (I) is a measure of brightness, just the gray-scale value. It is also commonly called brightness, value, or gray-level intensity. For example, the intensity values are all that one sees of a color picture on a black-and-white television screen. Hue (H) represents the color value and is roughly proportional to its corresponding color wavelength. Saturation(s) gives the depth of color by measuring lack of white. The red color of a stop sign is deep red, so is said to have a high saturation value, whereas the color pink has a low saturation value. The RGB $\rightarrow$ IHS transformation allows further processing to be concentrated on one image plane of interest. The most common basis for image analysis is the intensity plane. This is because, in most cases, color conveys much less picture information than the relative brightness values.

2.5.1 COLOR BASES AND TRANSFORMATIONS

A color image $I_C(x, y)$ has the RGB representation

$$I_C(x, y) = [I_R(x, y), I_G(x, y), I_B(x, y)],$$

where the color primaries are red (R), green (G), and blue (B), $0 \leq R, G, B < I_{\max}$. It is a simple matter to obtain, from this representation, an image of only red, green, or blue regions. It is also straightforward to generate any other color as a combination of these primaries. This can be visualized from the RGB color cube shown in Fig. 2.5.2. For instance, the image containing all cyan- (green–blue-)colored regions has the form

$$I_{\text{cyan}}(x, y) = [G(x, y) + B(x, y)]/2.$$

Most real-life images display nonuniformity of coloring, restricting the applicability of this method of separation to simple cases of synthetic images in computer graphics in which color values can be read and displayed by exact values.

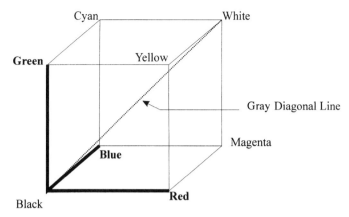

Figure 2.5.2. RGB color cube. From the color primaries, the secondaries are generated as yellow equals equal parts of red and green, cyan equals equal parts of blue and green, magenta equals equal parts of blue and red. Equal parts of red, green, and blue yield gray, shown as the diagonal from zero intensity (black) to maximal intensity (white).

The RGB → IHS Transformation

In practice, to facilitate further analysis, the most frequently used transformation is that to the IHS bases.

Intensity is a measure of brightness (also called value) and is the average of the color values:

$$I = (R + G + B)/3.$$

Saturation is a measure of color purity, that is, of the lack of whiteness of color and is found from the primaries as

$$S = 1 - \min(R, G, B)/I.$$

A saturation value of 0 indicates equal values of R, G, and B, and corresponds to a gray value. A saturation value of 1 indicates one or two color values of zero and corresponds to a pure color with no pastel characteristic.

Hue is proportional to the average wavelength of the color. It gives the gradation of color as an angle:

$$H = \cos^{-1}\{[(R - G) + (R - B)]/2[(R - G)^2 + (R - B)(G - B)]^{1/2}\}.$$

In this polar coordinate system, red corresponds to 0°, green to 120°, and blue to 240°. Figure 2.5.3 shows a diagrammatic representation of this color space. The IHS space provides a representation in terms of familiar color features and thus is a natural basis for image feature analysis.

Related to the above-mentioned problem of determining an image with objects of a single color, the more pertinent problem in practice is to generate an image with a single range of colors. For instance, to find all objects in the purple range, we would

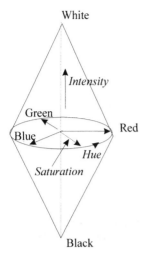

Figure 2.5.3. IHS color space. Intensity is shown on the vertical axis from black to white. Hue is represented angularly: red at 0°, green at 120°, and blue at 240°. Saturation is represented as the radial distance from the origin, 0 for no saturation (gray) and 1 for fully saturated color.

first define the purple range to lie somewhere between red and blue, say from 270° to 330°. Then the purple image would consist of the intensity values of all image pixels with hue within the specified range.

Conversely, it may be of interest to find all objects of a certain shape or size and to characterize them by a number of features, including color features. In this case, analysis to locate the object may be performed exclusively on the intensity image; color attributes of the object, once located, may then be extracted from the other two images as the average hue and saturation within the bounds of the object.

Other Transformations

Although the RGB $\rightarrow$ IHS transformation is the most useful for image analysis, different applications may call for other color bases. For instance, the National Television Systems Committee standard for television uses a YIQ basis set, where Y specifies the intensity or brightness in the same way as for IHS, I is a red–cyan range, and Q is a magenta–green range. This is advantageous because black-and-white television sets need extract only the black–white Y basis. In addition, for color video transmission, the bandwidth associated with I and Q can be limited to a degree without noticeable image degradation. This latter feature further reinforces the practice of using just the intensity information for locating an object and the two other color primaries only to obtain color feature information.

References and Further Reading

For a concise summary of color representations, see App. 1 of W. K. Pratt, *Digital Image Processing*, 2nd ed. (Wiley-Interscience, New York, 1991).

Program

rgb2gray

transforms RGB color image into gray-scale image
according to type specified with -t option.

USAGE: rgb2gray inimg outimg type [-L]

ARGUMENTS: inimg: input image filename (TIF)
           outimg: output image filename (TIF)
             type: r|R produces RED plane
                   g|G produces GREEN plane
                   b| B produces BLUE plane
                   i|I produces INTENSITY values
                   h|H produces HUE values
                   s|S produces SATURATION values

             NOTE: the HUE values are arbitrarily scaled
                   from 0 to 200, where Red=0, Green=67, and
                   Blue=133. The gray value, i.e., (R=G=B),
                   is arbitrarily set to 255.

OPTIONS: -L: print Software License for this module

Gray-Scale Image Analysis

This chapter introduces local operations for the processing and analysis of gray-scale images. This set of methods comprises the most general type of processing steps such as noise removal, flat fielding, filtering, and binarization, which usually precede more detailed analytical operations of the type discussed in subsequent chapters. For example, binarization must precede binary analysis, described in Chap. 4; edge detection is often performed as a preprocessing step to yield line images that are analyzed in Chaps. 4 and 5; the identification of points of peak intensity within gray-scale objects may be followed by point analysis, described in Chap. 6.

Gray-scale images represent the most commonly encountered type of image; for present purposes, a pixel in a gray-scale image has one of 256 possible intensity values, ranging from $0 = $ OFF to $255 = $ ON; that is, pixel intensities have an 8-bit representation. Generally, 0 corresponds to black, 255 to white, and intermediate values to various gray levels that span the range from black to white. In special cases, such as document images, the background (OFF) is white and the foreground (ON) is black, that is, the gray scale is reversed. In this book, we are consistent in the use of OFF $= 0$ for the background and ON $= 255$ for the foreground.

We discussed in Chap. 2 that a common procedure for color image analysis entails transformation to an intensity image on which the analysis is performed. This intensity image is effectively a gray-scale image. Thus analytical procedures discussed in this chapter apply equally to gray-scale images and to color images subjected to this transformation.

Section Overview

Section 3.1 introduces the fundamental operation of spatial domain convolution, which forms the basis of spatial filtering of images. Smoothing (low-pass filtering) and edge enhancement (high-pass filtering) are discussed in this section as introductory examples of filtering.

Section 3.2 describes noise reduction by various methods including Gaussian low-pass and median filtering. Noise is prevalent in images, and it is deleterious to subsequent analysis, making it prudent practice to apply noise reduction before most analysis. We discuss how to choose the noise reduction filter such that it reduces only the data considered by the user as noise and leaves intact the desired signal.

Section 3.3 presents a discussion of edge enhancement and flat fielding. These operations are performed to enhance the crispness of objects and features in the image and thus to improve the appearance of the image. In addition, the operations facilitate object detection and image segmentation, common analytical steps.

Section 3.4 deals with edge and point detection by convolution with suitable filters. Edge detection, a fundamental processing step, differs from enhancement in that the edges are actually found and isolated by the detection operation. Enhancement yields a crisper rendition of the entire image; edge detection preserves only edge features. Edges delineate objects, and their identification facilitates the process of detecting and distinguishing objects and features in an image. An alternative approach to this image segmentation task also is described that proceeds by determination of the location of an object's peak intensity (rather than delineating contours).

Section 3.5 addresses the rather typical complications encountered for edge detection in real images containing edge features of differing acuity that may also be corrupted by noise. This situation calls for advanced methods of edge detection.

Section 3.6 describes subsampling as an analytical operation. In general, subsampling is used to reduce image size by discarding every nth pixel. However, in image analysis the purpose is more constrained. If the image is larger than necessary, it contains more data than are needed for the analysis at hand. When the image is subsampled, noise is commonly reduced, thus facilitating subsequent analytical steps. By reducing the number of pixels in an image, subsampling also reduces the computation time for analysis.

Section 3.7 generalizes the concept of subsampling to the methodology of multi-resolution analysis. An image of interest may not be available at the best resolution (or size) for analysis of particular objects. In general, the best size may not be known *a priori*. Multiresolution analysis serves to transform a single image to many images, each of different resolution. Analysis may then be performed on all or only some of the resulting images to optimize an operation such as object detection.

Section 3.8 describes template matching, a method used to identify objects of known size and shape in the presence of noisy background. Among the earliest practitioners of template matching were military engineers who used the technique to identify targets in aerial and satellite photos. While template matching is straightforward, its use is limited. Whereas actual shape and size of the object may not vary, the image will vary because of changes in perspective, scale, lighting, etc. If these variables are known and can be kept constant, template matching is a powerful technique. If not, then identification by means of object features, such as shape (Section 4.4) and critical points of the outline (Section 5.4), is more appropriate.

Section 3.9 describes Gabor wavelets and their use for analyzing images, especially images containing textures and patterns. Wavelet analysis has one property akin to frequency domain analysis (Chapter 7), in that it can be used to detect spatial frequencies and their orientations in images. However, since frequency domain analysis is a global operation, it can only determine that all or a portion of an image contains texture or pattern of determined frequencies. Wavelet analysis, being a local operation, retains in addition the location of image sub-regions containing these textures or patterns. Wavelet analysis can be considered as a generalized combination of three other operations discussed in this book: multi-resolution analysis, template matching, and frequency domain analysis. As such, this general tool has become popular for those analyzing images containing textures and patterns.

Section 3.10 introduces the binarization operation that is used to transform the gray-scale image to a binary two-level image composed of ON and OFF pixels. One purpose of such a transformation is to reduce an image that contains only two levels of *information*, but 256 levels of *data*, down to two levels of data that correspond to the information. For instance, although the text on this page has only black-on-white information, there are many more gradations due to nonuniform inking, shadows, uneven lighting, etc. The purpose in intensity level reduction is to simplify the image for future analysis, described in Chaps. 4 and 5, which deal with binary object images and binary line images, respectively.

3.1 Local Image Operations: Convolution

Typical Application(s) – image contrast manipulation by means of filtering, required in many of the applications discussed in subsequent sections.

Key Words – filter mask, kernel; smoothing, sharpening; edge and point detection; transfer function; cyclic convolution.

Related Topics – smoothing/noise reduction (Section 3.2), feature enhancement (Section 3.3), edge and point detection (Section 3.4), filter mask design (Section 7.2); correlation (Section A.1), serial product (Section A.2).

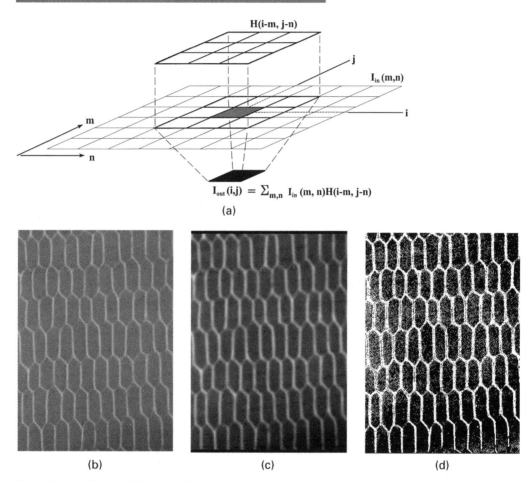

(a)

(b)   (c)   (d)

Figure 3.1.1. Pictorial Example. Two-dimensional discrete convolution operation: (a) graphical representation (Adapted with permission from [Castleman 96], Copyright © 1996 by Prentice-Hall, Inc.); (b), (c), (d) illustration of the effect of convolution. In these examples, a Gaussian filter (Section 3.2) of size 11 × 11 and a high-pass filter (Section 3.3) of size 3 × 3 were applied to respectively blur, (c), and sharpen, (d), the regular foreground features in a micrograph (b) of adult bovine lens fiber cells. Border strips of corrupted image pixels are apparent; these are further discussed in the text. (Panel (b) Reprinted with Permission from [Costello 97] – Copyright © 1997 by Dr. J. Costello.)

The two-dimensional (2D) convolution operation is fundamental to the analysis of images. It ascribes a new value to a given pixel based on the evaluation of a weighted average of pixel values in a $k \times k$ neighborhood of the central pixel. The weights are supplied in a square matrix, usually referred to as the filter mask or the convolution kernel; in general, the same mask is applied to each pixel in an image. By selection of different masks, diverse operations may be performed. These notably include the application of low-pass filters for smoothing, high-pass filters for sharpening, and gradient operators for edge and point detection. These topics are covered in detail in Sections 3.2–3.5. More specialized filter masks are encountered in subsequent sections of this chapter.

A brief summary of the essential concepts and basic properties of the convolution operation is given in the appendix (Section A.2). Here we focus on the discrete formulation of the 2D convolution operator and on image filtering, the most important application of the convolution operator in the context of image processing: This is introduced by way of the example of neighborhood averaging. We generalize this example by providing a description of the performance characteristics of ideal low-pass filters (ILPFs) and ideal high-pass filters (IHPFs), designed to effect image smoothing and sharpening, respectively. This serves as preparation for the actual application of these concepts in subsequent sections. Finally, we also discuss the selection of filter length on a heuristic basis.

General methods for the construction of mask coefficients to encode a desired filter operation are described in connection with Fourier transformation in Section 7.2, in which we also examine more closely the trade-offs between direct-space (pixel domain) and Fourier space (frequency domain) implementations of the convolution operator.

3.1.1 CONVOLUTION AND IMAGE ANALYSIS

Discrete Convolution in Two Dimensions

Image analysis invokes the discrete form of the 2D convolution operator. This is defined by the following relationship between the elements $g_i(x, y)$ of the input image, the elements $h(\xi, \eta)$ of the convolution kernel **H**, and the elements $g_o(x, y)$ of the output image:

$$g_o(x, y) = \sum_{\xi=-\infty}^{\infty} \sum_{\eta=-\infty}^{\infty} g_i(\xi, \eta) h(x - \xi, y - \eta),$$

where $x, y, \xi,$ and η are integers. As discussed in Sections 3.2–3.4 and in Section 7.2, the coefficients of the kernel **H** represent a discrete approximation of the analytical form of the response function characterizing the desired filter (Section A.2).

While rectangular kernels are possible, $k_x \neq k_y$, the kernel, in the great majority of practical cases, is a square array, $k_x = k_y \equiv k$, where k is odd and much smaller than the linear image dimension N. The general expression may be rewritten to reflect this simplification:

$$g_o(x, y) = \sum_{\xi=-(k-1)/2}^{(k-1)/2} \sum_{\eta=-(k-1)/2}^{(k-1)/2} g_i(\xi, \eta) h(x - \xi, y - \eta). \tag{3.1}$$

We can perform any 2D convolution by explicitly invoking this master formula, Eq. (3.1). As illustrated in Fig. 3.1.1, this entails, for each pixel $P \equiv (x, y)$ in the image, the following steps:

- placement of (the center element of) **H** on P,
- multiplication of each pixel in the $k \times k$ neighborhood by the appropriate filter mask coefficient superimposed on it,
- summation of all products, and
- placement of the suitably normalized sum into position P of the output image.

We make frequent use of this direct-space convolution in subsequent sections and chapters.

Special Considerations

Border Effects   Implementation of the convolution according to the master formula, Eq. (3.1), inevitably corrupts a strip of pixels adjacent to each image border. When applied to pixels within a strip of width $(k - 1)/2$ along image borders, a $k \times k$ filter mask will straddle the border so some of the mask coefficients will not be superimposed on valid pixels. This corrupts the convolution output. While this is not a serious problem in practice as long as k is small, it must be kept in mind when further processing steps are applied to a filtered image. The example in Fig. 3.1.1 serves as a reminder of this loss of border pixels; particularly when k is large, the effect can be substantial. The analogous effect is encountered in one-dimensional (1D) convolution and is illustrated in Section A.2.

Common procedures to handle this situation are the following:

- Pixels within the four border strips are set to zero, and the convolution is performed on only the remainder of the image of size $N - (k - 1)/2$; or
- Rows and columns are considered to wrap around, as illustrated in Table 3.1.1, for a 5×5 input image that is convolved with a 3×3 uniform filter: Only the central element is left intact. This cyclic convolution has the advantage of ease of implementation. In contrast to the zeroing of border strips, no preprocessing is required while memory management can take into account the fact that dimensions of input and output images are the same. However, pixels in the border strip are nonetheless corrupted.

Separability   Significant computational advantages may be realized in cases involving a separable kernel:

$$h(x - \xi, y - \eta) = h(x - \xi)h(y - \eta).$$

Separable kernels permit the replacement of the 2D operation by two successive 1D operations. This situation is frequently encountered in practice. For example, the Gaussian (low-pass) filter is separable, as are other common low-pass, high-pass filters and edge detection filters (see, e.g., Tables 3.1.1, 3.2.1, and 3.4.1). The topic thus warrants more detailed consideration.

Table 3.1.1. Neighborhood Averaging as Image Convolution

Averaging Over a 3×3 Neighborhood is Achieved by Application of the Convolution Kernel (Note the Decomposition of This Separable 2D Filter into a Dyadic Product of Two 1D Filters):

$$H(u, v)_{\text{NNA}} = \frac{1}{9}\begin{bmatrix} 1 & 1 & 1 \\ 1 & 1 & 1 \\ 1 & 1 & 1 \end{bmatrix} = \frac{1}{3}\begin{bmatrix} 1 \\ 1 \\ 1 \end{bmatrix}\frac{1}{3}\begin{bmatrix} 1 & 1 & 1 \end{bmatrix}.$$

Illustration of Convolution: $H_{\text{NNA}}G_I = G_O.$

$$G_I = \begin{bmatrix} 1 & 1 & 1 & 1 & 1 \\ 1 & 9 & 27 & 9 & 1 \\ 1 & 27 & 99 & 27 & 1 \\ 1 & 9 & 27 & 9 & 1 \\ 1 & 1 & 1 & 1 & 1 \end{bmatrix}.$$

Zeroing of Border Strip

$$\frac{1}{9}\begin{bmatrix} 1 & 1 & 1 \\ 1 & 1 & 1 \\ 1 & 1 & 1 \end{bmatrix}\begin{bmatrix} 0 & 0 & 0 & 0 & 0 \\ 0 & 9 & 27 & 9 & 0 \\ 0 & 27 & 99 & 27 & 0 \\ 0 & 9 & 27 & 9 & 0 \\ 0 & 0 & 0 & 0 & 0 \end{bmatrix} = \frac{1}{9}\begin{bmatrix} 0 & 0 & 0 & 0 & 0 \\ 0 & 162 & 198 & 162 & 0 \\ 0 & 198 & 243 & 198 & 0 \\ 0 & 162 & 198 & 162 & 0 \\ 0 & 0 & 0 & 0 & 0 \end{bmatrix}.$$

Cyclic Convolution

$$\frac{1}{9}\begin{bmatrix} 1 & 1 & 1 \\ 1 & 1 & 1 \\ 1 & 1 & 1 \end{bmatrix}\begin{bmatrix} 1 & 1 & 1 & 1 & 1 \\ 1 & 9 & 27 & 9 & 1 \\ 1 & 27 & 99 & 27 & 1 \\ 1 & 9 & 27 & 9 & 1 \\ 1 & 1 & 1 & 1 & 1 \end{bmatrix} = \frac{1}{9}\begin{bmatrix} 17 & 43 & 51 & 43 & 17 \\ 43 & 167 & 201 & 167 & 43 \\ 51 & 201 & 243 & 201 & 51 \\ 43 & 167 & 201 & 167 & 43 \\ 17 & 43 & 51 & 43 & 17 \end{bmatrix}.$$

As with the 1D case of the convolution, discussed in detail in Section A.2, a close connection exists between convolution and matrix multiplication and pertinent notation may be used. Separability then implies the existence of a factorized representation of the convolution kernel in the form of a dyadic,

$$\mathbf{H} \equiv \mathbf{h}_C \odot \mathbf{h}_R,$$

where $\mathbf{H}_{ij} \equiv \mathbf{h}_{C_i}\mathbf{h}_{R_j}$ and subscripts C and R indicate column and row vectors, respectively. In most cases of practical interest, $\mathbf{H}$ is symmetric so that $\mathbf{h}_R = \mathbf{h}_C{}^T$. For the example of the Gaussian, this implies that every Gaussian filter mask of size $n \times n$ can be constructed, row by row, by multiplying each element in a 1D Gaussian filter of length n, represented by an n-element row, by the elements an n-element column of the same elements. Thus, (1 2 1) would be multiplied first by 1 to give (1 2 1), next by 2 to give (2 4 2), and last by 1 to give (1 2 1) as the three rows of a 3×3 Gaussian, as shown in Table 3.2.1.

The 2D convolution for separable kernels may be performed as a sequence of two matrix multiplications involving matrices $\mathbf{F}_C$ and $\mathbf{F}_R$, respectively composed of columns of unit-shifted copies of $\mathbf{h}_C$ and of unit-shifted rows of $\mathbf{h}_R$, so that

$$\mathbf{G}_o = \mathbf{F}_C \mathbf{G}_i \mathbf{F}_R.$$

This reduction of the 2D convolution to two successive 1D convolutions entails multiplication of the column matrix with the image and the subsequent multiplication of the resulting intermediate image with the row matrix. The result is normalized by the sum of all filter coefficients. This procedure has the advantage of increased computational efficiency. The number of requisite multiplications decreases from $k^2 N^2$ for the general 2D operation to $2k N^2$ for the separable case.

3.1.2 FILTERING IN THE SPATIAL DOMAIN

The convolution operator is indispensable to image processing because it facilitates the implementation of filtering operations in the spatial domain. We illustrate the application of the master formula, Eq. (3.1), by describing one of the simplest filtering operations, namely neighborhood averaging. We then introduce the two most basic convolution operations, namely image smoothing and image sharpening, facilitated by application of low-pass and high-pass filters, respectively. The term filtering refers to the effect of the convolution operation on the spectrum of spatial frequencies, as apparent from the description of ideal smoothing and sharpening filters below; the topic is discussed in greater detail in Section 7.1.

Smoothing by Means of Neighborhood Averaging

The simplest smoothing operation is that of neighborhood averaging, invoked to reduce the prominence of sharply localized image features that often are associated with noise. The prescription is simple: Each pixel in a given image is replaced by the average value of pixels in a square neighborhood, of size $k \times k$, of that pixel. As a result, abrupt variations in image intensity appear more gradual and less pronounced.

We can easily see the connection of neighborhood averaging with 2D convolution by referring back to the master formula, Eq. (3.1): if all coefficients of the convolution kernel $\mathbf{H}$ are set to $1/k^2$, i.e., $h(x - \xi, y - \eta) = 1/k^2$ for $-(k-1)/2 \le \xi, \eta \le (k-1)/2$, the elements of the output image are precisely in the form of an average of input elements over a $k \times k$ neighborhood. That is, neighborhood averaging is equivalent to a 2D convolution of the image with a uniform filter, also known as a box filter: All coefficients in the corresponding $k \times k$ mask are equal. As with virtually all masks encountered in practice, k is chosen to be odd.

Application of the general prescription according to the master formula, Eq. (3.1), then dictates the procedure of applying the averaging filter: All pixel values in a $k \times k$ square surrounding the center pixel are summed, and the sum is divided by $k \times k = k^2$. The resulting value is the average of the $k \times k$ neighborhood, and it replaces the pixel at the center location of the neighborhood. The uniform filter is separable, and it can therefore be applied by two successive convolutions with 1D masks of length k.

Neighborhood averaging represents a simple linear operation on a group of pixels, and it is this linearity that ensures the direct equivalence of the operation to a matrix multiplication. As we have discussed, this equivalence has the advantage of efficiency of implementation, especially in the frequently encountered situations involving separable masks. In later sections, we will encounter a number of nonlinear operations; these cannot be implemented in the form of a convolution and consequently exact greater computational cost.

Smoothing and Sharpening: Ideal Low-Pass and High-Pass Filters

An ideal smoothing filter would remove (or reduce, as discussed in Section 3.2) all sharply localized features, such as random dots or blobs, whose characteristic dimensions fall below a preset cutoff length L_0 while leaving unaffected those features with dimensions exceeding L_0. An ideal sharpening filter would remove all features whose characteristic dimensions exceed L_0, while leaving unaffected sharply localized features with characteristic dimensions below L_0. In this way, it preserves (or actually enhances, as discussed in Section 3.3) localized image features, such as edges, and suppresses other, less well-defined features, such as slow variations in the background intensity.

The standard way to specify the performance characteristics of a filter is to define its transfer function. The transfer function encodes the desired filter performance in terms of the corresponding (spatial) frequency domain characteristics (see Section A.2). As depicted in Fig. 3.1.2, the ideal filters are characterized by a step-function cutoff at a characteristic frequency $D_0 \sim 1/L_0$. Ideal smoothing corresponds to the elimination of features with frequency content above D_0, while ideal sharpening corresponds to the elimination of features with frequency content below D_0 (see also Section 7.2).

The transfer functions $H(u, v)$ of the ILPF and the IHPF are simply step functions of azimuthal symmetry, i.e,

$$H(u, v)_{\text{ILPF}} = \begin{cases} 1 & \text{if } D(u, v) \leq D_0 \\ 0 & \text{if } D(u, v) > D_0 \end{cases},$$

$$H(u, v)_{\text{IHPF}} = \begin{cases} 0 & \text{if } D(u, v) \leq D_0 \\ 1 & \text{if } D(u, v) > D_0 \end{cases},$$

where u and v denote (spatial) frequencies, $D(u, v) = (u^2 + v^2)^{1/2}$ gives the distance between the origin and the point (u, v) in the frequency plane, and D_0 denotes the cutoff frequency, introduced above. As may be ascertained from Fig. 3.1.2, these transfer functions specify passage of all spatial frequencies below (ILPF) and above (IHPF) the cutoff and complete attenuation of all other spatial frequencies.

Unfortunately, the infinitely sharp cutoff of the ideal filters brings with it some undesirable aspects in the filter performance. Foremost among these is a ringing distortion. This phenomenon, further discussed in Section 7.2, manifests itself as a doubling or a tripling of sharp edges due to ghost edges; it may be reduced if the infinitely sharp cutoff in the filter transfer function is replaced with a finite roll-off. Specifically, the family of Butterworth filters offers a practical approach to approximate the ideal filters by filters with finite, tunable roll-off. Butterworth low-pass filters (BLPFs) and Butterworth high-pass filters (BHPFs) are illustrated in Fig. 3.1.3 and are defined by

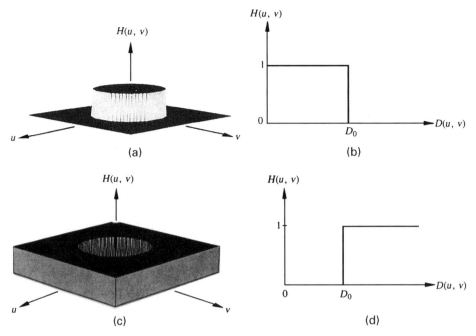

Figure 3.1.2. Transfer functions for ILPFs (top; (a), (b)) and IHPFs (bottom; (c), (d)); perspective views (left; (a), (c)) and radial cross sections (right; (b), (d)) are shown. (Reprinted with Permission from [Gonzalez & Woods 92], Copyright © 1998 by Addison-Wesley Publishing Company.)

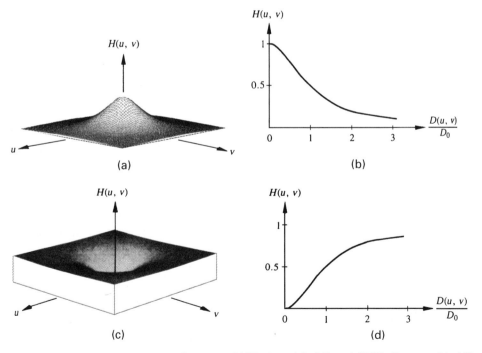

Figure 3.1.3. Transfer functions for $n = 1$ BLPFs (top; (a), (b)) and BLPFs (bottom; (c), (d)); perspective views (left; (a), (c)) and radial cross sections (right; (b), (d)) are shown. (Reprinted with Permission from [Gonzalez & Woods 92], Copyright © 1998 by Addison-Wesley Publishing Company.)

transfer functions of the form

$$H(u, v)_{\mathrm{BLPF}} = \frac{1}{1 + c[D(u, v)/D_0]^{2n}},$$

$$H(u, v)_{\mathrm{BHPF}} = \frac{1}{1 + c[D_0/D(u, v)]^{2n}},$$

where $D(u, v)$ is as defined above. The cutoff frequency D_0 is customarily defined by the condition $H(u, v)|_{D_0} = 1/2$; with $c = 1$ in the expression for the transfer function, this is seen to be equivalent to the condition $D(u, v) = D_0$. The parameter n in the transfer function denotes the order of the filter: the larger the n, the steeper the roll-off at D_0. High-order Butterworth filters thus approximate the step-function form of the ideal filter transfer function.

In most common use are low-order Butterworth filters, and their preferred application entails evaluation of the convolution in the frequency domain. We return to this topic in Section 7.2, in which we also discuss the relative merits of convolution in the spatial and the frequency domains.

Selection of Filter Length

An important question in connection with the use of filters is the selection of the proper filter size, $k \times k$. Qualitatively, the proper k, also known as the filter length, is determined by the characteristic scale, or spatial extent, of intensity variations associated with actual features of interest versus those attributed to noise or background. Thus, to smooth slowly varying features, of characteristic scale L_0, one must make the effective filter length comparable; features whose dimensions exceed L_0 remain largely unaffected.

The choice of k sets the effective cutoff frequency, $D_0 \sim 1/k$: the longer a smoothing filter, the lower the corresponding cutoff frequency, with the result that only large-scale features remain unaffected. Conversely, the shorter a sharpening filter, the higher its cutoff frequency, with the result that all large-scale features will be suppressed.

More generally, as with the Butterworth filters introduced in this section, two parameters, filter length and roll-off, are available to tune filter performance. The requisite choice of desirable values for k and n must reflect the fact that given a filter length k, the choice of a smaller value for n implies a more gradual roll-off and hence less severe ringing, but the more gradual roll-off implies in turn a more narrowly constricted range of frequencies in which flat transmission characteristics are ensured. That is, the effect of the rounding of the filter transfer function naturally extends to above and below the actual roll-off frequency and thus impairs the flat transmission characteristics.

Specific rules of thumb may be given once a particular filter has been selected, and we return to this question in Section 3.2. In practice, best performance is often ensured by trial and error.

References and Further Reading

[Castleman, 1996] K. Castleman, "Digital Image Processing," (Prentice-Hall, Inc. Englewood Cliffs, NJ, 1996).

[Costello, 1997], J. Costello, University of North Carolina, unpublished.

[Gonzalez and Woods, 1992] R. C. Gonzalez and R. E. Woods, *Digital Image Processing*, 2nd ed. (Addison-Wesley, Reading, MA, 1992) Chap. 4.

R. W. Hamming, *Digital Filters*, 3rd ed. (Dover, Mineola, NY 1997).

The analogies between the analysis of 1D audio and other signals and 2D images has been noted by many authors. An in-depth discussion is available in J. S. Lim, *Two-Dimensional Signal and Image Processing* (Prentice-Hall, Englewood Cliffs, NJ, 1990).

Lim also discusses the merits of separable kernels (see [Lim, 1990], Chap. 1.2). When multiplication and addition as well as edge effects are taken into account, the number of arithmetic operations decreases from $(k + N - 1)^2 k^2$ for the general 2D operation to $(k + N - 1)kN + k(k + N - 1)^2$ for the separable case. We may further improve efficiency by taking explicit advantage of the fact that in the overwhelming majority of cases of practical interest the matrix **H**, when expanded to the full size of an image, is extremely sparse. By optimizing the algorithmic implementation, we can eliminate multiplication with vanishing mask elements.

Program

xconv

convolves an image with a kernel; kernels may be
read from a file; Gaussian kernels of any
length can be generated automatically.

```
     USAGE:  xconv inimg outimg [-f filter_file] [-g len] [-L]
 ARGUMENTS:   inimg: input image filename (TIF)
              outimg: output image filename (TIF)
   OPTIONS:  -f filter_file: name of file containing filter
                             coefficients (ASCII)
                 -g len: use 2D Gaussian filter of length len
                         NOTE: either -f OR -g option MUST be
                         specified
                 -L: print Software License for this module
```

3.2 Noise Reduction

Typical Application(s) – reduction of extraneous image features; reduction of noise introduced by imaging system; removal of small extraneous objects.

Key Words – noise removal, filtering, low-pass filter, median filter, speckle noise, smoothing, blurring.

Related Topics – subsampling (Section 3.6); binary noise removal (Section 4.2); noise reduction (Section 5.2); frequency domain processing (Chap. 7).

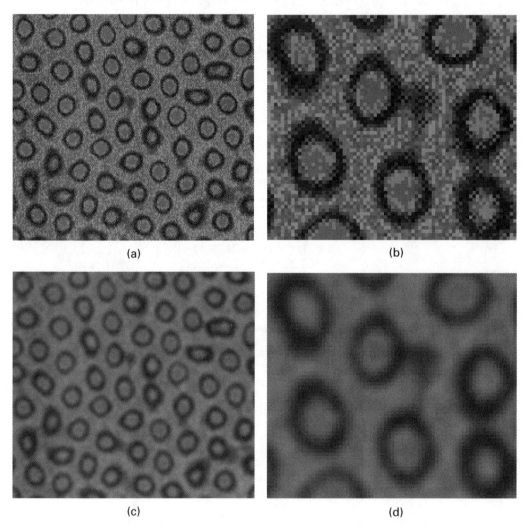

(a)   (b)

(c)   (d)

Figure 3.2.1. Pictorial Example. The results of noise reduction by low-pass filtering: (a) original noisy image, (b) magnified portion of (a) to show noise, (c) low-pass-filtered result of (a) to reduce noise, (d) magnified portion of (c) to show smoothed results compared with (b). (Reprinted with Permission from [LT-BL], Copyright © 1998, Lucent Technologies Inc.)

Image noise can be introduced by several sources, including electronic noise in transmission or simply dust in the optics or blemishes in the camera. The purpose of noise reduction is to rid the image of components considered to be extraneous. The definition of noise, that which is extraneous, is "in the eyes of the beholder" and may differ for different applications and images. Conversely, the definition of signal is that which is of actual interest to the observer. A common component of image noise is referred to as speckle noise or salt-and-pepper noise [although the latter is usually reserved for noise in binary images (Section 4.2)]. This noise manifests itself as unevenness in background and foreground regions and imparts a bumpy or jagged appearance to otherwise smooth regions of intensity. These noise characteristics are illustrated in the magnified image in Fig. 3.2.1(b). In what follows, we discuss procedures to reduce this type of noise to produce a cleaner picture and to facilitate subsequent image analysis.

We speak of noise reduction rather than noise removal, because it is usually impossible to remove all noise without simultaneously reducing or distorting some of the signal as well. There is this general trade-off: the greater the amount of noise to be reduced, the greater the amount of reduction of the desired signal. Equivalently, the more important it is to retain the signal without distortion, the more conservative must be the noise reduction.

A general approach to noise reduction is smoothing and, more specifically, the convolution of the image with an appropriate low-pass filter. The idea is simple: if localized (sharp) nonuniformities and jagged edges can be made to appear less severe if these features are smoothed, then this source of noise will be reduced. To accomplish the desired noise reduction, each pixel is examined with respect to its neighboring pixels and, if its intensity value is found to differ sharply from those of its neighbors, this difference may be reduced by an adjustment of the value of the pixel.

This is the domain of low-pass filtering. In low-pass filtering, the center pixel is replaced by a weighted average of the neighboring pixels. Alternatively, in median filtering and in maximum and minimum filtering, the center pixel is replaced by a nonlinear combination of its neighbors (as suggested by the names of the examples).

3.2.1 THE UNIFORM FILTER

The simplest smoothing filter is the uniform filter or averaging filter (Section 3.1), which results in the smoothing of sharp features and thus in the reduction of intensity disparity of pixels whose intensities are much above or below those of the neighbors.

The size of the uniform filter is chosen according to the characteristic scales of desired image and noise features. The rule of thumb for practical uniform filters is to set $k = 2w_N + 1$ pixels to blur objects with diameter w_N or smaller. For example, if speckle noise features are estimated not to exceed 2 pixels in diameter, then a uniform filter length, $k \geq 2 \times 2 + 1 = 5$, is recommended to reduce this noise. Conversely, if one wishes to retain all image features of diameter w_s or above, a suitable choice would be $k \leq 2 \times w_s - 1$. For example, if image features with a diameter of 4

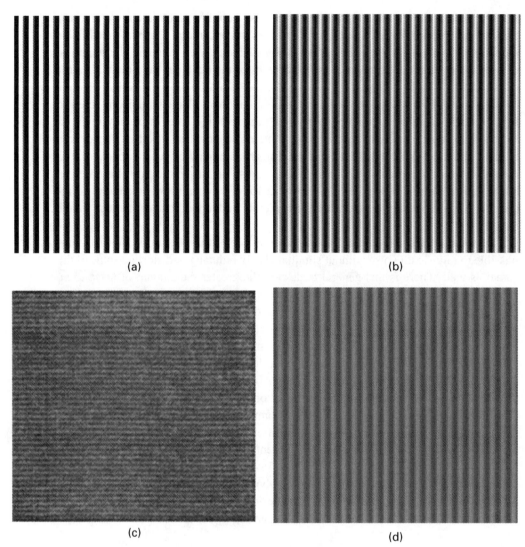

(a)   (b)

(c)   (d)

Figure 3.2.2. Results of low-pass filtering with uniform filters of different sizes: (a) original stripe image; (b) result for 3 × 3 filter, in which signal is retained; (c) result for 7 × 7 filter, in which signal is corrupted (number of stripes is reduced) because of too large a filter size; (d) result for 9 × 9 filter, in which signal is also corrupted.

or greater are to be retained, the maximum recommended filter length would be $k \leq 2 \times 4 - 1 = 7$.

The filter performance is illustrated in Figs 3.2.1 and 3.2.2. Close examination of Figs. 3.2.1(c) and 3.2.1(d) reveals that noise spots, while far less noticeable because of smoothing, are nonetheless still visible. As we stated at the outset, application of this filter only reduces, and does not eliminate, noise.

Figure 3.2.2 contains an example of the effect of applying a uniform filter to an image of vertical stripes, 3 pixels in width. It is apparent that the relevant characteristic scale of image features to be retained is in fact the stripe width. A filter of length $k = 2w + 1$ will reduce any feature of characteristic size smaller than w. Figure 3.2.2 demonstrates the result of applying filters of side lengths 3, 7, and 9, to the stripes of width 3.5. The smallest filter retains the signal because its length falls below $2 \times 3.5 + 1$; in contrast, the two larger filters reduce the signal because their widths equal or exceed $2 \times 3.5 + 1$. Since the 7×7 filter is a multiple of the stripe period (3.5 ON and 3.5 OFF), it results in a uniform intensity output over the image, losing the stripe pattern completely.

The fact that the uniform filter is limited to the adjustment of its length to control performance has some drawbacks. First, while speckle noise may be reduced substantially, important image features such as edges and textures may be equally affected and thus blurred if features and noise exhibit similar degrees of acuity. This is a fundamental problem of all low-pass filters, but it is particularly severe for the uniform filter. Closely related is a second problem of the uniform filter, namely its propensity to introduce a ringing distortion, mentioned in Subsection 3.1.2. These problems can be prevented by use of a filter whose coefficients, rather than exhibiting a sharp drop, taper to zero at the mask edges (see Table 3.2.1). We discuss these next.

3.2.2 GAUSSIAN FILTERS

A Gaussian filter mask has the form of a bell-shaped curve, with a high point in the center and symmetrically tapering sections to either side (Fig. 3.2.3). Application of the Gaussian filter produces, for each pixel in the image, a weighted average such that central pixels contribute more significantly to the result than pixels at the mask edges. The taper of the Gaussian filter reduces ringing, although it also reduces the degree of smoothing. That is, for a given filter length, the degree of smoothing achieved by application of the Gaussian filter falls short of that achieved by application of the uniform filter.

For a filter mask of size $k \times k$, Gaussian filter coefficients are obtained in the form $h_G(i, j) = \exp[-1/2(d/\sigma)^2]$, where $d = (i^2 + j^2)^{1/2}$ and $-(k-1)/2 \leq i, j \leq (k-1)/2$. Only one parameter, the standard deviation σ, is under the control of the filter designer. That is, both filter cutoff and roll-off are determined when σ is fixed: the lower the value

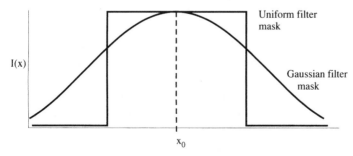

Figure 3.2.3. Cross-sectional shapes of a uniform filter and a Gaussian filter.

of σ, the greater the amount of filtering, but also the more pronounced the tendency toward ringing. For $\sigma = \infty$, the Gaussian filter is the same as the uniform filter.

This parameter is chosen in accordance with the scale w of the features to be smoothed. The rule of thumb is to set $\sigma = (2w + 1)/2$ and to choose $k > 2w + 1$. Since k must be odd, $k_{min} = 2w + 3$; in practice, either $k = 2w + 3$ or $k = 2w + 5$ is common. A larger value for k ensures less ringing, but requires more computation.

The Gaussian filter requires more computation time than a uniform filter to accomplish the same amount of smoothing, given that the Gaussian filter exceeds the size of the equivalent uniform filter and contains coefficients of several values and hence requires multiplication (in contrast to simple addition for the uniform filter) for performing the convolution operation.

3.2.3 THE MEDIAN FILTER

The median filter is used to reduce speckle noise while retaining sharp edges of image features. This is achieved when the intensity of the center pixel in a $k \times k$ neighborhood is replaced with the median value of the neighboring intensities. The straightforward way to determine a median value for a given $k \times k$ neighborhood is to sort pixels in the order of lowest- to highest-intensity values and then to find the middle value. There are fast sorting programs available for this task.

When the median filter is centered on the noise pixel, its intensity is replaced by the median of the neighborhood intensities. The effect of the median filter is to replace pixel values that differ significantly from those of the other pixels in the $k \times k$ neighborhood by the middle neighborhood value, and this eliminates isolated noise very efficiently. This operation preserves edge discontinuities because replacement involves a pixel intensity value that already is present in the neighborhood rather than a new value (such as the average, as in the case of the uniform filter). However, the median filter typically requires more extensive computation than do the uniform and the Gaussian filters.

There is only one median filter parameter, the filter length k. The rule of thumb by which to choose k is as follows. If we want the median of a neighborhood to be the value of the background (as opposed to the value of a noise region), then the filter mask must contain a greater number of background pixels than noise pixels. Thus, given a noise feature of width w, the filter length must be at least $k = 2w + 1$.

Figure 3.2.4 illustrates the removal of isolated noise dots. Whether a noise pixel has higher or much higher intensity than the rest of the neighborhood, the result of median filtering is the same, that is, the result is set to the same neighborhood median value. This is a manifestation of the nonlinearity of the median filter, and it is in distinct contrast to the behavior of the uniform or the Gaussian filters.

The median filter is especially effective in eliminating isolated pixel noise. However, if the task at hand calls for the smoothing of texture or dense fields of noise, one of the linear smoothing filters is a better choice. This reflects the fact that, while the median filter always produces a value already present in the neighborhood, the other two filters can produce a new value better matched to effect smoothing.

Table 3.2.1. Examples of Smoothing Filters

Examples of 1D Binomial Filters

$$b_1 \qquad \tfrac{1}{2}(1 \ 1)$$

$$b_2 \qquad \tfrac{1}{4}(1 \ 2 \ 1)$$

$$b_3 \qquad \tfrac{1}{8}(1 \ 3 \ 3 \ 1)$$

$$b_4 \qquad \tfrac{1}{16}(1 \ 4 \ 6 \ 4 \ 1)$$

$$b_5 \qquad \tfrac{1}{32}(1 \ 5 \ 10 \ 10 \ 5 \ 1)$$

$$b_6 \qquad \tfrac{1}{64}(1 \ 6 \ 15 \ 20 \ 15 \ 6 \ 1)$$

$$b_7 \qquad \tfrac{1}{128}(1 \ 7 \ 21 \ 35 \ 35 \ 21 \ 7 \ 1)$$

$$b_8 \qquad \tfrac{1}{256}(1 \ 8 \ 28 \ 56 \ 70 \ 56 \ 28 \ 8 \ 1)$$

Examples of Gaussian Filters

$$B_2 = \frac{1}{16}\begin{bmatrix} 1 & 2 & 1 \\ 2 & 4 & 2 \\ 1 & 2 & 1 \end{bmatrix} = \frac{1}{4}\begin{bmatrix} 1 \\ 2 \\ 1 \end{bmatrix}\frac{1}{4}\begin{bmatrix} 1 & 2 & 1 \end{bmatrix}, \qquad B_4 = \frac{1}{256}\begin{bmatrix} 1 & 4 & 6 & 4 & 1 \\ 4 & 16 & 24 & 16 & 4 \\ 6 & 24 & 36 & 24 & 6 \\ 4 & 16 & 24 & 16 & 4 \\ 1 & 4 & 6 & 4 & 1 \end{bmatrix}.$$

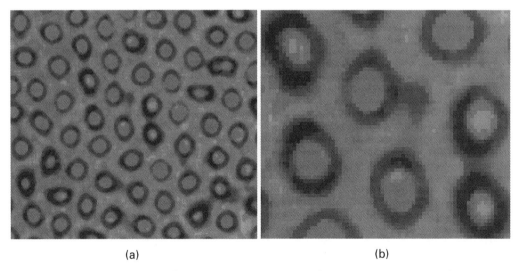

(a)   (b)

Figure 3.2.4. Results of median filtering on same image as in Figure 3.2.1(a): (a) median filter result for 3 × 3 sized filter – compare with Fig. 3.2.1(c); (b) magnified portion of (a) to show noise reduction – compare with Fig. 3.2.1(d). (Reprinted with Permission from [LT-BL], Copyright © 1998, Lucent Technologies Inc.)

Reference

[LT-BL] Lucent Technologies – Bell Laboratories Physical Sciences Division Image Library.

Programs

`lpfltr`

> performs low-pass filtering on image by using filter window of chosen size; the default is to use a Gaussian-shaped filter

USAGE:	lpfltr inimg outimg [-f CUTOFF <0.50>] [-s FLTR_SIZE <odd>] [-q] [-L]
ARGUMENTS:	inimg: input image filename (TIF)
	outimg: output image filename (TIF)
OPTIONS:	-f CUTOFF: <0.0 to 0.5> is the fraction of the passband; the default is 0.25, that is, half of original passband.
	-s FLTR_SIZE: size of filter mask; this will override the size that would otherwise be automatically set based on the cutoff frequency; size value must be odd.
	-q: quick (and dirty) rectangular filtering. NOTE: specifying both cutoff frequency and filter size is redundant for the rectangular filter; the cutoff is used if both are specified.
	-L: print Software License for this module

`medfltr`

> performs median filtering on image by using filter window of chosen size.

USAGE:	medfltr inimg outimg [-s FLTRSIZE] [-L]
ARGUMENTS:	inimg: input image filename (TIF)
	outimg: output image filename (TIF)
OPTIONS:	-s FLTR_SIZE: filter size; the larger the size, the greater the degree of smoothing; size value must be odd; default = 3
	-L: print Software License for this module

3.3 Edge Enhancement and Flat Fielding

Typical Application(s) – emphasis of localized features such as contours and object borders; elimination of inhomogeneities in scene illumination.

Key Words – high-pass filter, sharpening; unsharp mask, flat fielding; homomorphic filter.

Related Topics – convolution (Section 3.1), edge detection (Section 3.4); shape features (Section 4.4), Hough transform (Section 4.10); filtering in frequency domain (Section 7.2).

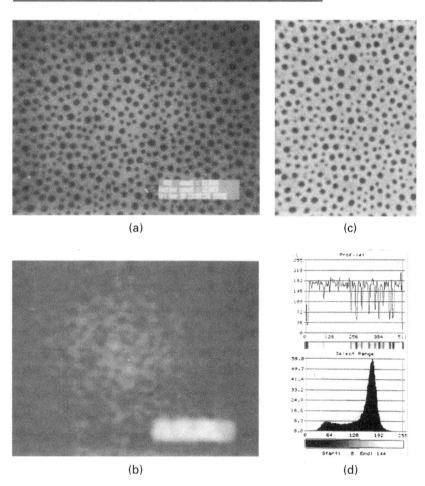

(a)

(c)

(b)

(d)

Figure 3.3.1. Pictorial Example. Flat fielding by means of division of the original (a) by a low-pass-filtered version (b). The original image recorded by a fluorescence microscope equipped with an imaging camera shows circular domains in a monomolecular film of amphiphilic molecules on a water surface. The uneven illumination is clearly visible. The left portion of the flat-field-corrected image in (c) illustrates the dramatic improvement in uniformity. A horizontal line profile through the center of (c) displays a constant level of "bright" background in (d), top; the histogram in (d), bottom, permits unambiguous binarization. (Reprinted with Permission from [Morgan and Seul 95], Copyright © 1995 American Chemical Society.)

Among the techniques that may be used to improve the appearance of an image, sharpening by enhancement of edges and contours plays an important role. In contrast to global image enhancement strategies involving the manipulation of image intensities by means of gray-scale mapping and histogram transformations (Chap. 2), the methods discussed here rely on the manipulation of the spatial frequency content in local areas throughout an image. The primary objective in enhancing edges and other localized features is to give an image a crisper appearance.

An edge or contour in an image is a feature marking a local variation in image intensity: an intensity profile crossing an edge will exhibit a transition from low to high or vice versa. As we will discuss in Section 7.2, such a variation in turn implies high spatial frequencies. Edge enhancement may thus be accomplished by amplification of these high-frequency components relative to the image background and other slowly varying intensity features that correspond to low-frequency content.

This is the domain of high-pass filtering and techniques discussed in this section may in fact be viewed as a direct analog to the low-pass-filtering techniques introduced in Section 3.2. Related in their aim, but entirely different in approach, are image restoration algorithms designed to reduce or eliminate blurring: These are generally based on a model of the process underlying the observed loss in acuity and are performed by deconvolution.

3.3.1 UNSHARP MASKING AND FLAT FIELDING

One approach to enhance sharply localized features is simply to remove or diminish slowly varying features, thereby producing a constant (flat) background intensity distribution. This is the essence of the procedure known as unsharp masking. A common application is the correction of nonuniformities in scene illumination that can be introduced by imperfections in the illumination source or by misalignment of optical componenents in microscopes or other imaging instruments. Typical in many experimental situations is the gradual, slow variation of the illumination profile across the field of view, corresponding to low spatial frequency components. If the spatial frequency components of interest in the actual image are well separated from the low-frequency components, unsharp masking represents an effective procedure to eliminate them. Long practiced in photography, this technique entails low-pass filtering, or unsharpening, of a given image and subtraction of this filtered version, suitably scaled, from the original.

If $f(x, y)$ represents the original image and $f_{LP}(x, y)$ the low-pass-filtered version, unsharp masking yields an output image $g(x, y)$, as follows:

$$g(x, y) = af(x, y) - bf_{LP}(x, y), \tag{3.2}$$

or, when the high-frequency component $f_{HP}(x, y) \equiv f(x, y) - f_{LP}(x, y)$ is introduced,

$$g(x, y) = (a - b)f_{LP}(x, y) + af_{HP}(x, y). \tag{3.3}$$

Here, a and $b < a$ are positive constants. Thus the latter expression makes explicit that this masking operation does indeed emphasize high spatial frequencies over low spatial frequencies.

Unsharp masking may be implemented by application of a Gaussian low-pass filter of suitable size to obtain a blurred version of the original image (Subsection 3.2.2); this is followed by scaling and subtraction (Section 2.3) of this low-pass filtered version from the original image, as prescribed by Eqs. (3.2) and (3.3) above.

The unsharp masking procedure rests on the assumption that image features display uniform intensities that are additively superimposed on the scene background. However, the response of an optical system is frequently nonlinear, and the intensity associated with a feature is in fact proportional to the background intensity in that part of the image. In such cases, a modified flat-fielding procedure is appropriate. Rather than subtracting the low-pass-filtered image from the original, the latter is divided by a scaled version of the former, with all pixel values of 0 replaced with 1 in the filtered image (Section 2.3). A suitable alternative is to record a reference image showing only the illuminated background field – without features or objects of interest – and to divide original images by this separate reference image to achieve flat fielding. Figure 3.3.1 shows an example of this method.

3.3.2 ENHANCEMENT BY MEANS OF HIGH-PASS FILTERING

The obvious alternative to the flat-fielding technique is high-pass filtering with masks such as those listed in Table 3.3.1. In contrast to the case of the ideal high-pass filter introduced in Section 3.1, the coefficients of each of the masks in the table sum to unity, and this ensures that the average intensity (dc component) of the original image remains unaltered. An example of this procedure is shown in Fig. 3.3.2.

The enhancement of background noise is a common side effect of high-pass filtering. It may thus be desirable to perform noise reduction as a prior step (Section 3.2).

Table 3.3.1. Masks for Image Enhancement

$$
\begin{bmatrix} 0 & -1 & 0 \\ -1 & 5 & -1 \\ 0 & -1 & 0 \end{bmatrix}, \quad
\frac{1}{3}\begin{bmatrix} 0 & 1 & -1 & 1 & 0 \\ 1 & -2 & 4 & -2 & 1 \\ 1 & 4 & -13 & 4 & 1 \\ 1 & -2 & 4 & -2 & 1 \\ 0 & 1 & -1 & 1 & 0 \end{bmatrix},
$$

$$
\begin{bmatrix} 1 & -2 & 1 \\ -2 & 5 & -2 \\ 1 & -2 & 1 \end{bmatrix}, \quad
\frac{1}{7}\begin{bmatrix} -1 & -2 & -1 \\ -2 & 19 & -2 \\ -1 & -2 & -1 \end{bmatrix}.
$$

3.3.3 RELATED METHODS

Homomorphic Filtering is a nonlinear operation that may be viewed as high-pass filtering not of image intensity (I) but rather of log I. Discussed in textbooks on image processing, homomorphic filtering is predicated on a model in which an image of a scene is assumed to be formed by the recording of the light reflected from an object that

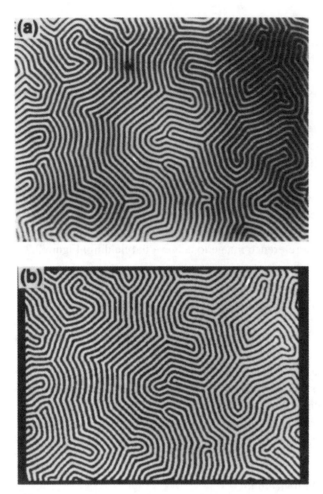

Figure 3.3.2. Example of high-pass filtering applied to a magnetic domain pattern of labyrinthine stripes [Seul et al., 1992]. The poor illumination in the original image (top) represents a slowly varying intensity component that is removed by application of the high-pass filter (bottom). (Reprinted with Permission from [Seul et al., 92], Copyright © 1992 Taylor & Francis.)

is illuminated by a source. Accordingly, the image intensity $I(x, y)$ is decomposed into illumination $i(x, y)$ and reflectance $r(x, y)$ components so that $I(x, y) = i(x, y)r(x, y)$. Slowly varying illumination and rapidly varying reflectance may be separated by logarithms. Homomorphic filtering is of limited practical importance and is not implemented here.

   Laplacian Sharpening involves the convolution of an image with an operator known as the Laplacian. As discussed in Section 3.4, the Laplacian is a linear difference operator that is commonly used in the context of detecting edges that of course represent high-frequency features in an image. The effect on an image of invoking the Laplacian is essentially equivalent to subtracting a blurred version of the image from the original image and is thus essentially equivalent to unsharp masking. However, the Laplacian is very sensitive to noise and tends to amplify it; this limits its practical utility.

References and Further Reading

[Lim, 90] J. S. Lim, *Two-Dimensional Signal and Image Processing* (Prentice-Hall, Englewood Cliffs, NJ, 1990), provides a discussion of unsharp masking (Chap. 8) as well as motion deblurring (Chap. 9).

[Morgan and Seul, 1995] N. Y. Morgan and M. Seul, "Structure of Disordered Droplet Domain Patterns in a Monomolecular Film" J. Phys. Chem. **99**, 2088–2095 (1995).

[Seul et al., 1992] M. Seul, L. R. Monar, and L. O'Gorman (see p. 279) "Pattern Analysis of Magnetic Stripe Domains: Morphology and Topological Defects in the Disordered State," Philos. Mag. **B 66**, 471–506 (1992).

Programs

xconv

```
            performs high-pass filtering by reading suitable filter
            mask coefficients, supplied in file

   USAGE:  xconv inimg outimg [-f filter_file] [-L]

ARGUMENTS:  inimg: input image filename (TIF)
            outimg:  output image filename (TIF)

 OPTIONS:  -f filter_file: name of file containing filter
                           coefficients (ASCII)

                     -L: print Software License for this module
```

bc

```
            divides the input image by a reference image to correct
            for nonuniform background illumination and/or shading;
            scales the result and writes the image to the specified
            output file

   USAGE:  bc inimg refimg outimg [-c clip] [-b] [-L]

ARGUMENTS:  inimg: input image filename (TIF)
            refimg: reference image filename (TIF)
            outimg: ouput image filename (TIF)

 OPTIONS:    -c: clipping factor for normalized image 0-255
                 (default=1)
             -b: generate binarized output by conversion to
                 integer

             -L: print Software License for this module
```

3.4 Edge and Peak Point Detection

Typical Application(s) – detect lines and points; determine object location by finding intensity peaks of extended pointlike objects such as atoms in scanning probe microscopy images, cells, vesicles, or colloidal spheres in optical micrographs; image segmentation.

Key Words – gradient, Laplacian; peak detection, converging squares.

Related Topics – edge enhancement (Section 3.3), advanced edge detection (Section 3.5), multiresolution analysis (Section 3.7); region detection (Section 4.3), shape analysis (Section 4.4), Hough transform (Section 4.10).

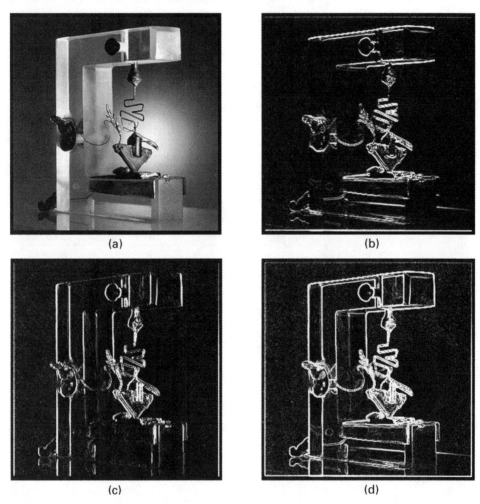

(a) (b)

(c) (d)

Figure 3.4.1. Pictorial Example. Illustration of edge detection, performed on (a) (see also: Fig. 2.4.6) the images in (b) and (c) were produced by invoking **xconv** with the - *f* option set to read text files containing coefficients for 3×3 Sobel filter masks (see Table 3.4.1): (b) horizontal Sobel filter (for filter coefficients, see sobel1.txt;); (c) vertical Sobel filter (for filter coefficients, see sobel2.txt); (d) combined image, produced by addition of (b) and (c).

This section introduces a set of filters for the detection of lines and isolated points in an image. The detection of contours, or edges, that delineate regions within an image is an important step in the segmentation of an image by identifying its principal features.

Edge enhancement, discussed in Section 3.3, invokes high-pass filtering to accentuate edges so as to produce a crisper appearance of an otherwise preserved image. In contrast, edge detection comprises detection and localization of edge points, followed by suitable linking of points into a contiguous contour that delineates each region. The result is an image containing only edge features, which provide the basis for subsequent further analysis.

In many instances, it suffices to rely for detection on the evaluation of derivatives, notably gradient and Laplacian, suitably approximated by a set of filter coefficients and applied by convolution. Popular edge detection filters in this category bear the names of Prewitt and Sobel.

3.4.1 SIMPLE EDGE DETECTION: DIFFERENCE OPERATORS

The most basic definition of an edge is that it marks a change in image intensity separating distinct regions. An edge will therefore manifest itself in the derivative of the image intensity $i(x, y)$. Consequently, the most direct approach to the problem of detecting an edge relies on the evaluation of local spatial variations of $i(x, y)$ (Figs. 3.4.1 and 3.4.2). Of particular interest in this context is the gradient ∇i, a vector quantity that yields the directional difference in intensity between adjacent pixels:

$$\nabla i(x, y) \equiv [\partial_x i(x, y), \partial_y i(x, y)] = |\nabla i(x, y)| \tan^{-1}[\partial_y i(x, y)/\partial_x i(x, y)].$$

Most commonly used are nondirectional gradient edge detectors that rely on only the magnitude of $\nabla i(x, y)$, commonly approximated as follows:

$$|\nabla i(x, y)| \equiv \sqrt{[\partial_x i(x, y)]^2 + [\partial_y i(x, y)]^2} \simeq |\partial_x i(x, y)| + |\partial_y i(x, y)|.$$

If desired, the direction of an edge may be determined by evaluation of the orientation of the gradient vector, as discussed below in this section.

The Laplacian is a scalar operator that yields the rate of change of intensity variations:

$$\Delta i(x, y) \equiv \nabla [\nabla i(x, y)] = \partial_{xx}^2 i(x, y) + \partial_{yy}^2 i(x, y).$$

The sign of the Laplacian determines the location of a given pixel relative to an edge. That is, the sign of the Laplacian is positive for pixels located on the darker side of a nearby edge, while it is negative for pixels located on the brighter side. For pixels marking an edge, the Laplacian evaluates to zero.

Such zero crossings can in principle serve to define edge positions, although this approach is of limited value in the presence of noise because application of the Laplacian tends to enhance the corresponding high-frequency features [Lim, 1990, Subsection 8.3.2]. This is a point to which we return below in connection with optimal edge detection.

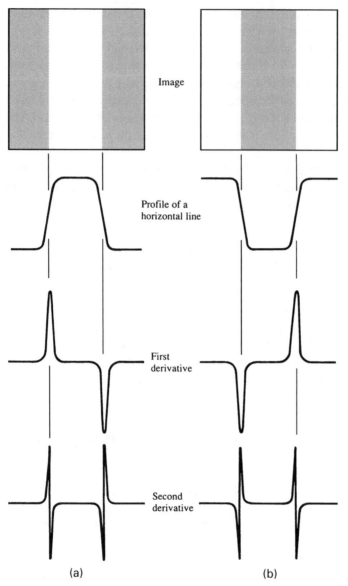

Image

Profile of a
horizontal line

First
derivative

Second
derivative

(a)   (b)

Figure 3.4.2. Detection of edges associated with a bright stripe on a dark background (left) and with a dark stripe on a bright background (right). The graphical illustrations show the intensity variations registered while crossing the edges in the direction perpendicular to the stripes; also shown are corresponding first and second derivatives, illustrating the fact that information on amplitude, sign, and location of an edge may be extracted from difference operators. (Reprinted with Permission from [Gonzalez & Woods 92], Copyright © 1998 by Addison-Wesley Publishing Company.)

The Laplacian vanishes when evaluated in the interior of a uniform region, and it may thus be invoked to eliminate all pixels comprising the interior of such regions to enhance features in subsequent histogram construction (Section 2.2). There is thus a natural connection to edge enhancement and sharpening (Section 3.3), but in contrast to the high-pass filters discussed in that context, the Laplacian does not preserve the average intensity value of the original image.

The evaluation of Laplacian $|\Delta i(x, y)|$ and gradient $\nabla i(x, y)$ on a discrete pixel grid relies on one of a variety of finite-difference approximations for the partial derivatives $\partial_x i(x, y)$ and $\partial_y i(x, y)$. As with other linear operators, the resulting filter masks are applied to the image by convolution.

Most important in practice is a symmetric central-difference approximation that corresponds to a nondirectional gradient operator and eliminates any displacement of the computed difference relative to the pixel (x, y) of interest. The simplest such approximation for partial derivatives along x and y are

$$\partial_x i(x, y) \simeq [i(x + 1, y) - i(x - 1, y)],$$
$$\partial_y i(x, y) \simeq [i(x, y + 1) - i(x, y - 1)],$$

with corresponding 1D masks $(-1, 0, 1)$ and $(-1, 0, 1)^T$.

3.4.2 SIMPLE EDGE DETECTORS: CONVOLUTION

As with sharpening filters, gradient operators tend to enhance the high-frequency features often associated with pixel noise. Accordingly, the image produced by the central-difference operator is improved by the subsequent application of a smoothing filter (Section 3.2). Better yet, given the linearity of the convolution operation (Section A.2), differentiation and smoothing may be combined into a single operation by convolution with suitable filter masks.

This is the idea behind the popular Prewitt and Sobel filters listed in Table 3.4.1. The Prewitt filters represent the combination of differentiation with subsequent neighborhood averaging by application of a box filter (Section 3.1), that is, application of the Prewitt filters is equivalent to averaging central differences of adjacent rows [for $\partial_x i(x, y)$] and columns [for $\partial_y i(x, y)$]:

$$\partial_x i(x, y) \simeq [i(x + 1, y) - i(x - 1, y)] + [i(x + 1, y + 1)$$
$$- i(x - 1, y + 1)] + [i(x + 1, y - 1) - i(x - 1, y - 1)],$$
$$\partial_y i(x, y) \simeq [i(x, y + 1) - i(x, y - 1)] + [i(x + 1, y + 1)$$
$$- i(x + 1, y - 1)] + [i(x - 1, y + 1) - i(x - 1, y - 1)].$$

The corresponding filter masks (Table 3.4.1) serve to detect vertical and horizontal edges. A combination of the filter masks representing $\partial_x i(x, y)$ and $\partial_y i(x, y)$ provides a simple nondirectional gradient edge detector, $|\nabla i| \simeq |\partial_x i(x, y)| + |\partial_y i(x, y)|$. That is, a given image is handled by a modified convolution routine that handles both directional masks: at each pixel position, both masks are applied, and the sum of the absolute values, computed for each filter output, is placed into the current pixel position to create the edge map.

The Sobel filters listed in Table 3.4.1 represent the combination of differentiation with subsequent Gaussian smoothing by application of the simple Gaussian mask (1 2 1).

Table 3.4.1. Commonly Used Different Operators

Gradient Operators

Prewitt

$$\partial_x i(x, y) = \begin{bmatrix} -1 & -1 & -1 \\ 0 & 0 & 0 \\ 1 & 1 & 1 \end{bmatrix} = \begin{bmatrix} -1 \\ 0 \\ 1 \end{bmatrix} [1 \quad 1 \quad 1],$$

$$\partial_y i(x, y) = \begin{bmatrix} 1 & 0 & -1 \\ 1 & 0 & -1 \\ 1 & 0 & -1 \end{bmatrix} = \begin{bmatrix} 1 \\ 1 \\ 1 \end{bmatrix} [-1 \quad 0 \quad 1].$$

Sobel

$$\partial_x i(x, y) = \begin{bmatrix} -1 & -2 & -1 \\ 0 & 0 & 0 \\ 1 & 2 & 1 \end{bmatrix} = \begin{bmatrix} -1 \\ 0 \\ 1 \end{bmatrix} [1 \quad 2 \quad 1],$$

$$\partial_y i(x, y) = \begin{bmatrix} -1 & 0 & 1 \\ -2 & 0 & 2 \\ -1 & 0 & 1 \end{bmatrix} = \begin{bmatrix} 1 \\ 2 \\ 1 \end{bmatrix} [-1 \quad 0 \quad 1].$$

Laplacian (3×3 approximation)

$$\Delta i(x, y) = \begin{bmatrix} 0 & -1 & 0 \\ -1 & 4 & -1 \\ 0 & -1 & 0 \end{bmatrix}.$$

By reducing ringing, Gaussian filters provide superior performance in image smoothing when compared with box filters of equal length (Section 3.2). The example in Fig. 3.4.1 illustrates the performance of 3×3 Sobel filters.

The separability of the masks in Table 3.4.1 implies that we may perform the 2D convolution operation with greater efficiency by performing two consecutive passes of a 1D convolution operation (Section 3.1).

3.4.3 EDGE DIRECTION

We may evaluate the direction of an edge by invoking the expression for the angle α, subtended by gradient and x axis,

$$\alpha \equiv \tan^{-1} \partial_y i(x, y) / \partial_x i(x, y).$$

Alternatively, this determination may be based on the application of the gradient operator in several orientations. The comparison of corresponding magnitudes identifies the edge orientation: The mask oriented normal to the local edge tangent produces the maximal magnitude.

Table 3.4.2. Edge Orientation by Rotated Gradient Operators

Kirsch

$$
\begin{bmatrix} 5 & -3 & -3 \\ 5 & 0 & -3 \\ 5 & -3 & -3 \end{bmatrix},
\begin{bmatrix} 5 & 5 & -3 \\ 5 & 0 & -3 \\ -3 & -3 & -3 \end{bmatrix},
\begin{bmatrix} 5 & 5 & 5 \\ -3 & 0 & -3 \\ -3 & -3 & -3 \end{bmatrix},
\begin{bmatrix} -3 & 5 & 5 \\ -3 & 0 & 5 \\ -3 & -3 & -3 \end{bmatrix},
$$

$$
\begin{bmatrix} -3 & -3 & 5 \\ -3 & 0 & 5 \\ -3 & -3 & 5 \end{bmatrix},
\begin{bmatrix} -3 & -3 & -3 \\ -3 & 0 & 5 \\ -3 & 5 & 5 \end{bmatrix},
\begin{bmatrix} -3 & -3 & -3 \\ -3 & 0 & -3 \\ 5 & 5 & 5 \end{bmatrix},
\begin{bmatrix} -3 & -3 & -3 \\ 5 & 0 & -3 \\ 5 & 5 & 3 \end{bmatrix}.
$$

Simple rotated gradient operators for four of the eight possible edge configurations in a 3×3 neighborhood (four directions, each with bright and dark regions on either side of the edge) may in principle be obtained from those given in Table 3.4.1 by consecutive left rotations of the outer mask elements about the stationary central element.

An optimized set of eight masks for the determination of edge orientation was introduced by Kirsch. Masks representing these Kirsch filters, listed in Table 3.4.2, give positive weights to five consecutive neighbors and negative weights to the remaining three consecutive neighbors of the central element, which is set to zero. As with the filter masks listed in Table 3.4.1, all coefficients sum to zero, guaranteeing a vanishing filter response in regions of constant gray level.

3.4.4 EDGE DETECTORS OF INCREASED SIZE

Another generalization of the form of gradient operators is to increase the size of the corresponding filter mask. The use of such filters provides greater smoothing, but comes at the expense of increased computational cost.

Implementation of the Laplacian relies on a finite-difference approximation of the second-order partial derivatives ∂_{xx}^2 and ∂_{yy}^2 to construct suitable filter masks whose specific form reflects the specific choice of approximation. A general requirement is that the central element be positive and that mask coefficients sum to zero. A popular 3×3 mask is that in Table 3.4.1.

The application of the gradient operator discussed in Subsection 3.4.2 facilitates the transformation of the original gray-scale image into the gradient image $g(x, y)$. To construct the actual edge map defining $g(x, y)$, the simplest approach involves a test of $G(x, y) \equiv |\mathrm{grad}\, i(x, y)|$ against a preset threshold T to suppress spurious edge points. This leads to an edge map of the form

$$
g(x, y) = \begin{cases} G(x, y) & \text{if } G(x, y) \geq T \\ L_B & \text{otherwise} \end{cases},
$$

where L_B denotes a gray scale assigned to the background. If all points at which $G(x, y)$ exceeds T are admitted as edge points, the resulting edge map will consist of strips of finite and varying widths. These strips constitute the boundaries of uniform regions

Table 3.4.3. Laplacian Peak Point Detector

$$\Delta i(x, y) = \begin{bmatrix} -1 & -1 & -1 \\ -1 & 8 & -1 \\ -1 & -1 & -1 \end{bmatrix}.$$

within the image. To obtain uniformly narrow edge contours (of width 1 pixel), it may be desirable to follow edge detection by edge thinning (Section 4.7).

3.4.5 DETECTION OF PEAK POINTS

The use of filters for the detection of points, that is, isolated bright spots in an image, may be considered a special case of edge detection. A peak point is simply a pixel whose intensity value exceeds that of its surroundings.

In practice, this is not a common task. Isolated (single-pixel) bright spots are typically indicative of noise and are preferably removed (see Section 3.2). More generally, we may encounter peak regions: these are pointlike regions, extended over more than a single pixel, approximately circular in shape and with a central-intensity peak; a method for peak region detection in gray-scale images is described below. Peak region detection in binary images is discussed in Section 4.3.

The simplest detection technique is based on the application of a high-pass filter to the image. Filter coefficients are chosen to yield a high filter output when the mask is positioned over a peak point and a low output otherwise. For the simple case of an image containing isolated points on a uniform background, convolution with a circularly symmetric difference operator may suffice to identify the image intensity discontinuities associated with these points. This operation is identical to that invoked for edge detection by a Laplacian filter mask. The Laplacian filter mask for peak point detection (Table 3.4.3) is simple and suitable for this purpose.

However, the simple Laplacian peak point detector detects point noise indiscriminately from peak regions whose support area is larger than a single pixel. One could use a larger filter mask for larger peaks, but this requires greater computation.

3.4.6 CONVERGING-SQUARES ALGORITHM

For a point region composed of more than a few pixels, the converging-squares algorithm affords an efficient means to determine the peak location. This method is applicable to any region whose spatial bounds are convex and whose intensity cross section is approximately monotonic toward a single peak. Approximate monotonicity implies that the algorithm deals well with noise (local nonmonotonicity), avoiding the detection of false, noisy peaks. Objects whose images fit this description include spherical and spheroidal shapes such as those of many biological cells.

The strategy of the converging-squares algorithm is to use an adjustable square window that is first placed at the outer boundaries of the image area of interest. On each iteration, the square is contracted by one pixel on each side length and approaches the region of greatest intensity. Iterations continue until the square is of size 2×2, so that the final iteration requires a comparison of only four pixel values to find the peak.

A common way to use this algorithm when there are multiple peak regions in the image and when the approximate (or maximum) size of the regions is known is the following. The image is examined in raster order until a pixel $I(x, y)_T$ is found whose intensity exceeds a preselected threshold. The raster search is suspended, and a converging-squares sequence is initiated whose initial square is centered on $(x, y)_T$, whose side length is set equal to the (known) maximum region diameter. After the peak within that region is found, pixels in the initial square area are set to zero to ensure that no further peak search will be undertaken for this same region. The raster scan is then continued for further peaks.

The converging-squares algorithm represents an instance of a multiresolution algorithm (Section 3.7) that performs processing successively at low, then at higher resolutions. The lowest resolution is used on the first iteration when the (entire) initial area of interest, of size $k \times k$, is divided into four subsquares of size $(k - 1) \times (k - 1)$ each. These four subsquares can be thought of as pixels of a low-resolution image of size 2×2, with each of the pixels representing a suitable average over the actual pixels of the original, constructed by low-pass filtering with a uniform filter (Section 3.2). In this way, each iteration in fact involves a comparison of four pixels at the current image resolution. Each following iteration then proceeds to the square corresponding to the highest-intensity pixel of the 2×2 field. As the square shrinks with each iteration, the resolution concomitantly increases. The final comparison is among four pixels of the initial image, whereupon the peak is located at the highest resolution. Because of this low-to-high-resolution sequence by a succession of low-pass-filtering operations, the converging-squares algorithm is maximally tolerant to noise.

Approaches Combining Other Methods

A variety of additional methods are available for the task of detecting peak regions. One method proceeds by application to the original image of as large a smoothing filter as possible, which is chosen to reduce false peak intensities while avoiding the merging of adjacent true peaks. We find peaks by performing a raster search to find pixels whose intensities exceed a preselected threshold, then choosing the maximum intensity in a window of chosen size. However, unless the filtered result is very smooth, the presence of noise often results in smaller, local peaks' being erroneously found.

Morphological processing is commonly used to establish the number of (disjoint) point regions. The image is converted to binary by thresholding at some intensity (see Section 3.9) chosen to isolate individual peak regions. These thresholded regions are shrunk or eroded by removal of a single layer of their outer pixel boundary on successive image iterations by means of morphological processing (see Section 4.1) until the region is reduced to a single pixel. The entire image now consists of single ON-valued pixels, and each of these indicates one of the initial peak regions. If the initial peak region is symmetric both with respect to geometric shape and to intensity cross section, the locations of these remaining ON pixels also indicate the true peaks of the initial regions. Under this latter condition, the peak will coincide with the centroid of the binary region for any single intensity level.

An efficient approach to peak detection in the case just described is first to generate binary regions corresponding to the peak regions by means of binarization. The methods of Sections 4.3 and 4.4 may be applied to the binarized image.

References and Further Reading

[Lim, 1990] J. S. Lim, *Two-Dimensional Signal and Image Processing* (Prentice-Hall, Englewood Cliffs, NJ, 1990), provides a concise derivation of various approximations for gradient and Laplacian operators and discusses edge thinning.

[Gonzalez and Woods, 1992] R. C. Gonzalez and R. E. Woods, *Digital Image Processing*, 2nd ed., (Addison-Wesley, Reading., MA, 1992) contains a useful discussion of edge detection.

Programs

xedgefltr

> performs simple edge detection by reading suitable gradient filter masks coefficients, supplied in file, and invoking xconv

USAGE: xedgefltr inimg outimg [-f filter_file] [-g len] [-L]

ARGUMENTS: inimg: input image filename (TIF)
              outimg: output image filename (TIF)

OPTIONS: -f filter_file: name of file containing filter coefficients (ASCII)

         -g len: use 2D Gaussian filter of length len
                 NOTE: either -f OR -g option MUST be specified

         -L: print Software License for this module

peak

> finds peak within a square region of an image; the area of interest is specified by supplying its top-left coordinates and its size
> NOTE:
> - program finds ONE peak based on iteratively reducing the region size; therefore the input specifying and size strongly affects which peak will be region location identified;
> - program works best for convex shapes with smoothly changing slopes; it may give strange results for geometric shapes, especially those with edges aligned to horizontal or vertical axes.

USAGE: peak inimg outimg -x X0 -y Y0 -s SIZE_INITIAL
                         [-f SIZE_FINAL] [-d] [-L]

```
ARGUMENTS:              inimg:  input image filename (TIF)
                       outimg:  output image filename (TIF)
                       -x X0:  top left x-coord of region
                       -y Y0:  top left y-coord of region
             -s SIZE_INITIAL:  square size of region
OPTIONS:  -f SIZE_FINAL:  final square size, after which peak
                                pixel
                          -d:  to display peak location
                          -L:  print Software License for this module
```

3.5 Advanced Edge Detection

Typical Application(s) – image segmentation in the presence of noise.

Key Words – optimal detection, matched filter, Wiener filter.

Related Topics – edge enhancement (Section 3.3), edge and point detection (Section 3.4), multiscale analysis (Section 3.7), template matching (Section 3.8); shape analysis (Section 4.4), Hough transform (Section 4.10).

 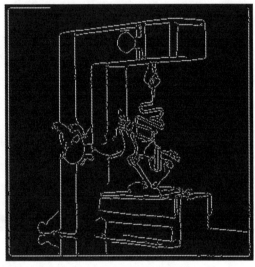

Figure 3.5.1. Pictorial Example. Optimal edge detection, followed by edge linking, illustrated by application of the Boie–Cox edge detector to the original in (a) (see also Fig. 2.4.6); to produce (b), **bcd** was run with a Gaussian filter width value of 1.5 and with -T set to 20; for comparison, see also Fig. 3.4.1.

The tasks of detecting and precisely locating edge features in real images generally pose challenging problems and have inspired the development of a variety of advanced edge detection algorithms. In this section, we discuss a systematic approach to edge detection. While making reference to several advanced concepts encountered in later sections of the book, we emphasize the basic concepts incorporated in the robust optimal edge detector we introduce here.

A significant early contribution to this topic is that of Marr and Hildreth (1980). Their approach, further developed by Canny (1986), relies on the inspection of intensity changes on different scales. The more general strategy, developed by Boie and Cox (1986) and presented here, uses matched filters to accomplish both detection and localization of an edge and a Wiener filter to minimize the effect of additive noise (see Fig. 3.5.1). Additional information on these filters can be found in Section A.3.

3.5.1 "OPTIMAL" EDGE DETECTION

Multiscale Edge Detection

Real images may contain edges of widely varying width or acuity. The presence of sharp lines along with blurred boundaries in the same image may in fact indicate different physical origins. Accordingly, several length scales will generally be of interest when an image is examined for intensity variations (see Section 3.7). In such a situation, the reliable detection of significant intensity changes calls for requisite adjustments in filter length. That is, construction of several edge maps from a given image, each representing intensity changes on one of several pertinent length scales, may be envisioned. The judicious choice of gradient operators, relying on the adjustment of filter length and mask coefficients, provides a degree of control over edge detection in the presence of noise, but this process is primarily guided by experimentation.

Edge Detection in the Presence of Noise: An Optimization Problem

A systematic approach requires a general formulation of a statistical nature in terms of a predetermined optimality criterion. Canny introduced the following three useful performance criteria for reliable edge detection in the presence of noise:

- *Good Detection*: The probability of missing real edge points and that of falsely marking spurious nonedge points should be minimal; this requires optimization of the signal-to-noise ratio.
- *Good Localization*: The points marked as edge points should be as close as possible to the center of the true edge.
- *Unique Response*: Multiple responses to a single edge feature must be avoided. This is a condition on the choice of the functional form of the edge detector: Multiple responses can arise if the transfer function of the chosen filter exhibits several local maxima, an aspect of the ringing problem encountered in Section 3.2.

Figure 3.5.2 illustrates the nature of the problem and underscores the relevance of Canny's three criteria. As illustrated in the figure, convolving the image with a Gaussian low-pass filter has distinct advantages in this context.

Edge Detection by Matched Filters: Boie–Cox Algorithm

The edge detection algorithm developed by Boie and Cox (1986) is predicated on the assumption of a specific functional form for the intensity variation associated with edges. Given an edge template, the strategy is to construct an optimal filter to detect and locate matching features in the image.

The approach entails the optimization of a specified attribute associated with an edge while contributions due to noise are minimized. Within this context, it is important to distinguish between the detection, or recognition, of an edge and its localization. The objective in detection is the verification of the presence of a region of changing image intensity. In contrast, the location of an edge may be identified as a particular position within a transition region.

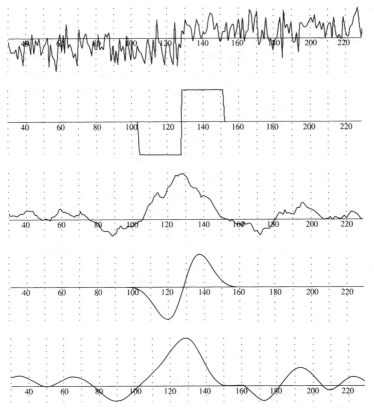

Figure 3.5.2. Edge detection in the presence of noise. From top to bottom, the respective panels show: a scan through a noisy edge; a difference operator constructed from two box functions; output produced by application of the "difference of boxes" operator to the edge; first derivative of Gaussian operator; and output produced by application of the Gaussian difference operator to the edge. (Reprinted with Permission from [Canny 86], Copyright © 1986 by IEEE.)

Detection   An optimal detection filter is one that, in response to the signal attribute of interest, provides the best estimate of the signal while minimizing the contribution of noise. Assuming the presence of stationary, additive noise of given spectral density, we may specify the transfer function of the desired optimal filter explicitly by requiring that its application maximize the ratio of signal power to noise power. This condition defines the transfer function of the matched detection filter, as elaborated in Section A.3. Application of a matched filter corresponds to evaluation of the cross correlation of the given signal with the predefined template.

Localization   The location of an edge is determined by the position, within an intensity transition region, at which the signal reaches a preset threshold. The accuracy of such a determination depends on both signal amplitude and on slope. An explicit form of the desired optimal location filter is in fact given by the first derivative of the matched detection filter. This suggests an implementation of the optimal location measurement: Differentiate the matched detection filter output, then determine a zero crossing of this first derivative. The situation is summarized in Fig. 3.5.3. Subpixel localization may be achieved by suitable interpolation of the localization filter output.

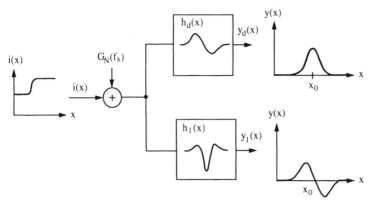

Figure 3.5.3. Schematic representation of matched detection and location filtering of a (1D) input signal $i(x)$ contaminated by additive noise of spectral density $G_N(f_x)$. The filter response functions $h_d(x)$ and $h_l(x)$ are indicated, along with the respective filter output. (Reprinted with Permission from [Boie et al., 86], Copyright © 1986 by IEEE.)

Treatment of Noise   Explicit contour extraction from an arbitrary 2D shape requires additional considerations to obtain a continuous edge. If a specific profile is assumed for the edge gradient in the direction normal to the contour of interest, the foregoing discussion implies that optimal detection and localization of individual edge points may be achieved by application of a matched filter in the edge-normal direction.

The common corruption of contours by noise, in the form of transverse displacements from the ideal contour position, presents an additional complication that further compromises the accuracy of edge localization. Systematic treatment of this type of edge corruption aims to restore the ideal contour, a task limited to a statistically optimal estimate of the uncorrupted form of the signal attribute associated with the contour. If the contour degradation may be represented as noise that is stationary and not correlated with the edge signal, optimal restoration is achieved by application of the Wiener filter, which is discussed in Section A.3.

Edge Orientation   Optimal performance of matched and Wiener filters is ensured only if the filters are aligned in the local edge-normal direction. Given its sensitivity to signal slope, localization in particular depends critically on proper filter orientation. As Fig. 3.5.4 illustrates, filter misalignment diminishes and broadens the signal; accordingly, the relative contribution of noise to the filter output increases. In the limit of filter alignment parallel to the contour, the output will in fact be composed entirely of noise. To ensure optimal performance, outputs from a set of filters probing different orientations must be compared. For the commonly encountered Gaussian edge model (see below), a set of four filters oriented at 45° from one another yields acceptable performance. Edge maps are constructed from a combination of the output of all filters by means of hysteresis thresholding which is further discussed below.

Edge Profile   Explicit specification of the transfer functions for matched and Wiener filters requires, first, the identification of a signal attribute characterizing an edge and, second, explicit functional forms of both the expected edge signal and the noise. The gradient represents an obvious choice for the requisite signal attribute. Optimal edge detection will then yield the best estimate of the signal gradient amplitude.

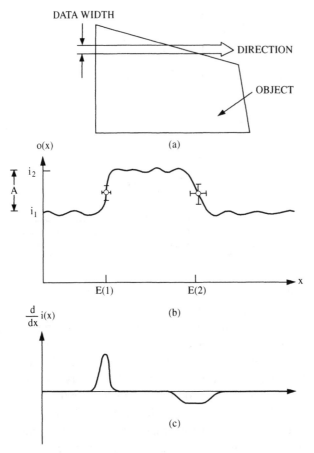

Figure 3.5.4. Adaptation of detection and localization filters to orientation of contours corrupted by noise. A horizontal scan through an imaged object (a) produces an intensity profile in which edge acuity reflects the relative orientation of edge and scan direction (b); accordingly, detection filters are matched to the corresponding gradient signal di/dx (c). (Reprinted with Permission from [Boie et al., 86], Copyright © 1986 by IEEE.)

In a Gaussian model, the edge signal is described by the integral of a Gaussian added to a fixed background [see the input function $i(x)$ in Fig. 3.5.3]. For the signal gradient of an edge aligned normal to the x direction, this implies a functional form $As(x) = (A/\sigma)\exp(-x^2/2\sigma^2)$. Here, the amplitude $A = i_1 - i_2$ (Fig. 3.5.3) describes the magnitude and the sign of the transition and hence measures contrast; $s(x)$ represents the normalized functional form of the gradient; and the variance σ^2 indicates the signal slope and hence edge acuity. As discussed, edges of widely differing acuity may exist in a given image, and the variation of σ provides the means to model them. Proper handling of the different length scales implied by varying edge widths requires adjustment of the filter length (see also Section 3.7). In practice, this requires the adjustment of σ as a parameter of the edge detector.

In an early systematic investigation of edge detection, Marr and Hildreth (1980) proposed that the noise reduction achieved by Gaussian low-pass filtering would make it feasible to exploit the zero-crossing property of the Laplacian to locate edge points. As with the nondirectional gradient operators discussed in Section 3.4, low-pass filtering and application of the Laplacian both involve convolution and may thus be combined. The result is the application of a bandpass filter to the original image. In fact, this bandpass filter represents the optimal location filter for the Gaussian edge.

Noise Power Spectrum Specification of the noise power spectrum is the last ingredient for the filter transfer functions. If a Poisson distribution is assumed to model the camera and illumination noise, the corresponding power spectrum has a Lorentzian form that is included in the transfer function of the matched detection filter. Edge noise, to be suppressed by application of the Wiener filter (Section A.3), is modeled as white noise.

Edge Linking by Means of Local Analysis

To generate a reliable edge map that provides a truthful representation of edges in the original, additional steps beyond detection and localization of edge points may be required. Some form of thresholding is generally used to eliminate spurious filter responses to noise. However, the imposition of such a single threshold in the presence of noise invariably leads to omission of some actual edge points, thus leaving breaks in the contour (see Section 3.4). To remedy this defect, edge detection is generally followed by edge linking to construct actual boundary contours by suitably augmenting the set of candidate edge points originally detected. Local analysis relies on additional criteria that are applied to the pixels connected to certified edge points.

A simple approach that illustrates this strategy relies on the comparison of gradient direction, as well as magnitude, for a pair of potential edge points. Given an edge point (x_0, y_0), characterized by a gradient grad $i(x_0, y_0)$ of magnitude $G(x_0, y_0) \equiv |\text{grad } i(x_0, y_0)|$ and direction $\alpha(x_0, y_0) \equiv \tan^{-1}[\partial_y i(x, y)/\partial_x i(x, y)]$, other points (x, y) in an $N \times N$ neighborhood of (x_0, y_0) may be classified as edge points and linked to the central point if $|G(x_0, y_0) - G(x, y)| \leq L_{\text{mag}}$ and $|\alpha(x_0, y_0) - \alpha(x, y)| \leq L_{\text{dir}}$, where L_{mag} and L_{dir} represent preset limits for the acceptable deviation of gradient magnitude and direction, respectively.

The implementation of the Boie–Cox algorithm invokes a clever hysteresis thresholding scheme, first suggested by Canny, that uses a pair of thresholds instead of a single threshold. Edge points yielding a filter response above the higher threshold are immediately accepted; in addition, all points are accepted that produce a response above a second lower threshold, provided they are connected to one of the primary edge points.

References and Further Reading

[Boie and Cox, 1986] R. A. Boie, I. Cox, and P. Rehak, "On optimum edge recognition using matched filters," in *Proceedings of the IEEE Conference on Computer Vision and Pattern Recognition* (IEEE, New York, 1986), pp. 100–108; R. A. Boie and I. Cox, "Two dimensional optimum edge recognition using matched and wiener filters for machine vision," in *Proceedings of the IEEE First International Conference on Computer Vision* (IEEE, New York, 1987), pp. 450–456.

[Canny, 1986] J. Canny, "A computational approach to edge detection," IEEE Trans. Pattern Anal. Mach. Intell. PAMI-8, 679–698.

[Marr and Hildreth, 1980] D. Marr and E. Hildreth, "Theory of edge detection," Proc. R. Soc. London **B 207**, 187–217 (1980).

For additional discussion, see Chap. 8 in J. S. Lim, *Two-Dimensional Signal and Image Processing* (Prentice-Hall, Englewood Cliffs, NJ, 1990); a useful discussion of edge linking is in R. C. Gonzalez and R. E. Woods, *Digital Image Processing*, 2nd ed. (Addison-Wesley, Reading, MA, 1992).

Program

bcd

```
            detects and marks contours of regions in a gray-scale
            image by invoking Boie—Cox algorithm
   USAGE:   bcd inimg outimg sigma [-h] [-t] [-T thld] [-L]
ARGUMENTS:  inimg:   input image filename (TIF)
            outimg:  output image filename (TIF)
            sigma:   width of Gaussian filter (float)
 OPTIONS:     -h:  do not apply hysteresis
              -t:  do not thin image
         -T thld:  threshold value (int)

              -L:  print Software License for this module
```

3.6 Subsampling

Typical Application(s) – scale reduction of images to be subjected to object or feature detection.

Key Words – subsampling, image size reduction, resolution adjustment.

Related Topics – geometric interpolation (Section 2.4); noise reduction and smoothing (Section 3.2), multiresolution analysis (Section 3.7); frequency domain filtering (Section 7.2).

(a)

(b)

Figure 3.6.1. Pictorial Example. Illustration of subsampling: (a) original image, (b) smaller image subsampled at a rate of 4 from the original. (This product/publication includes images from Corel Stock Photo Library, which are protected by the copyright laws of the U.S., Canada, and elsewhere. Used under license.)

It is often more convenient to perform analysis on a smaller version of an image while retaining all essential information of interest. For instance, for the task of locating an object or feature in an image, one option is to perform a search by means of template matching or feature detection over the original image. However, if the object or feature is just as evident in a smaller image, then great gains in efficiency can be realized by first reducing image size before proceeding with object or feature detection. This yields computational savings in two ways: first, the smaller image contains fewer pixels to be visited; second, the template is also smaller, and this implies a reduction, for every pixel site, in the time required for evaluating the match. Reducing image size requires a selection of pixels that are to be retained, and this selection is made by subsampling (see Fig. 3.6.1).

If an image is to be reduced to one-half size, subsampling entails retaining only every second pixel of each row and column. Analogously, reduction to one-third size implies retention of every third pixel. In the general case, reduction by a factor of r implies retention of every rth pixel. This r is referred to as the subsampling rate; analogously, $1/r$ is the reduction rate.

Note that size refers to linear image dimension. Thus, reduction to one-half size ($r = 2$) implies reduction in both the x and the y dimensions of the image so that the number of pixels is actually reduced by a factor of $2 \times 2 = 4$. In the general case, subsampling by r reduces the number of pixels N to N/r^2. Consequently, algorithms with execution times proportional to a power of N will benefit substantially from subsampling with even modest rates.

3.6.1 SUBSAMPLING WITH FILTERING

Subsampling without accompanying processing may introduce interference effects due to aliasing, a phenomenon more fully discussed in Section 7.1. Briefly, aliasing arises if the reduced pixel density is no longer sufficiently high to represent acute image features faithfully. Rather than just disappearing completely at lower resolution, such acute features introduce unwanted interference. This effect is minimized by low-pass filtering to smooth such features before subsampling.

The appropriate size of the smoothing filter is determined with respect to the desired subsampling rate. (The sizes of low-pass filters are also discussed in more detail in Section 3.2.) Thus, if an image is to be reduced by $r = 2$ in both the x and the y dimensions, the maximum frequency content of the image should likewise be reduced by a factor of 2: this choice will ensure the maximal degree of smoothing. In practice, the application of low-pass filtering to reduce aliasing will also smooth all other image features. Therefore a compromise must often be made in selecting the proper filter length to attain a degree of smoothing somewhere between maximal effect and no effect at all. The most suitable choice depends on the type of image in question. If acute features such as lines and edges are present, aliasing effects can be large and very undesirable and the maximum degree of filtering should be applied. In contrast, for a softer image, filtering may not be required at all.

An intuitive way to choose the filter size for subsampling is the following. For a subsampling rate r, one pixel for every r pixels is retained. Since we want all original pixels to exert some influence on the final result, the filter should be at least of size $r \times r$. However, the filter should also be symmetric about the retained pixel, implying a filter side length that is odd. Therefore, for the case of r even, the minimum filter size is $(r + 1) \times (r + 1)$, and for r odd, the minimum filter size is $r \times r$.

It is important to realize that the low-pass filter need not be applied to every pixel in a given image if filtering is to be followed by subsampling: filtering need be applied only to those pixels that are to be retained in the subsampled image. Note, however, that this is not equivalent to the sequence of first subsampling the image and then applying a filter to all subsampled pixels. This is because, although each filter window is centered at the location of only the pixels to be retained, even pixels that will not be retained contribute to the filter output in the original image.

3.6.2 RELATED ISSUES

Noninteger Subsampling

In its most direct implementation, subsampling of discrete images is restricted to integer subsampling rates. However, noninteger subsampling is possible when filtering precedes subsampling, and this provides greater flexibility in reducing an image to any specified size. For example, for a subsample rate of $r = 1.5$, yielding an image reduction of $1/1.5 = 0.66$, the output corresponds to a sequence of pixel locations at $1.5, 3, 4.5, 6, \ldots$. Output values for the noninteger pixel locations are determined by application of a filter mask that is centered at these locations. The only complication is that when a discrete filter mask is centered at a noninteger location in the image, the filter coefficients will not correspond to image locations. This complication is overcome by interpolation between given neighboring filter coefficients to generate new filter coefficients corresponding to the noninteger locations.

Binary Subsampling

Subsampling of binary images differs from gray-scale subsampling in one important way. When a binary image is low-pass filtered before subsampling, the result at each pixel is not binary. For example, for a $k \times k$ sized averaging filter, there are 0 to k^2 possible results. We can simply binarize each filter result to yield a binary subsampled image, but we have the flexibility of two other options. The better of these options, as far as image quality, is to retain the filtered results in the subsampled image, effectively transforming a binary image to a smaller, gray-scale image. Here, the gray-scale intensity gradations retain spatial information from the original binary image.

The second option is to perform binarization, but to do so by means of a user-selected contrast control parameter c. For the example of a uniform filter, where $I'(x, y)$ is the result at a pixel, then the result is contrast enhanced by $I''(x, y) = \text{ON if } 255/k^2 > 128$ or OFF otherwise. The value of c is chosen to be higher or lower than 1 to yield thicker or thinner ON regions, respectively.

Supersampling

The inverse operation to subsampling is supersampling. This entails an increase in image size and hence in the number of pixels. It is used to view a larger rendition of the image and is also known as zooming (Section 2.4). Supersampling is often described in connection with subsampling; however, it is not an operation that has much benefit for image analysis. This is because supersampling *does not* increase feature resolution but only enlarges the image without adding information. Other than to aid visual analysis, it would be counterproductive to enlarge an image by supersampling and then to perform further processing or analysis operations on it.

References and Further Reading

Further applications of subsampling can be found in A. Rosenfeld, ed., *Multiresolution Image Processing and Analysis* (Springer-Verlag, Berlin, 1984). Foundations of filtering as required for subsampling and supersampling are detailed in R. E. Crochiere and L. R. Rabiner, "Optimum FIR digital filter implementations for decimation, interpolation, and narrow-band filtering," IEEE Trans. Acoust. Speech Signal Process. **24**, 444–464 (1975).

Program

```
subsample
          performs subsampling after low-pass filtering; the
          default low-pass filter is a Gaussian.
   USAGE: subsample inimg outimg [-r SAMPLE_RATE <2>]
                                  [-d DEG_FILTERING <0.5>][-q ][-L]
ARGUMENTS:          inimg: input image filename (TIF)
                   outimg: output image filename (TIF)
 OPTIONS: -r SAMPLE_RATE: rate of image reduction; the default
                          setting, -r 2, reduces image by 2 in x
                          and y
        -d DEG_FILTERING: value, set between 0.0 and 1.0,
                          controlling the degree of low-pass
                          filtering; a value of 1.0 produces
                          maximum filtering to reduce aliasing,
                          but may lead to blurring; a value of 0.0
                          produces no filtering at the risk of
                          aliasing; the default is 0.50.
                     -q : when set, performs quick low-pass
                          filtering with rectangular filter rather
                          than Gaussian filter; the former is
                          faster but does not produce equally good
                          results as does the latter.
                     -L: print Software License for this module
```

3.7 Multiresolution Analysis

Typical Application(s) – object detection.

Key Words – scale-space processing, multiscale analysis, multiresolution pyramids, wavelets.

Related Topics – noise reduction (Section 3.2), edge enhancement (Section 3.3), subsampling (Section 3.6), template matching (Section 3.8).

Figure 3.7.1. Pictorial Example. A multiresolution image pyramid. At each level in the pyramid, the original image is reduced by 1/2 scale from the previous level; the original image is shown in the background with each multiresolution level superimposed. (This product/publication includes images from Corel Stock Photo Library, which are protected by the copyright laws of the U.S., Canada, and elsewhere. Used under license.)

Most image analysis tasks are undertaken with a knowledge of the typical dimensions of the features of interest. However, there are instances in which the size of an object may lie within a wide range. This poses no difficulty for human observers; for instance we have no problem identifying the same person in a small photograph and in a large poster, even though the characteristic scale may differ by orders of magnitude.

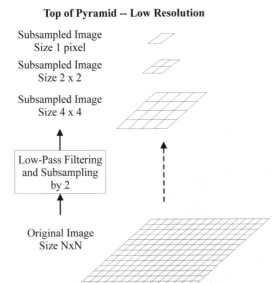

Top of Pyramid -- Low Resolution

Subsampled Image
Size 1 pixel

Subsampled Image
Size 2 x 2

Subsampled Image
Size 4 x 4

Low-Pass Filtering
and Subsampling
by 2

Original Image
Size NxN

Bottom of Pyramid -- High Resolution

Figure 3.7.2. Pyramid aspect of multi-resolution processing.

There are two approaches in image analysis to handle a range of scales. The most direct strategy is to perform template matching with a set of templates of varying sizes. The converse approach is to fix the template size and to change instead the size of the image. This multiresolution approach offers an advantage in that matching operations performed on an image of reduced size are more efficient (see Fig. 3.7.1).

Even when the size of an object is known, multiresolution analysis may expedite the task of locating the object in the image. If the image resolution is first reduced, then a search on this smaller image will be more efficient. Because of the reduced resolution, the accuracy of location is also reduced. However, given an approximate location of an object found at a lower resolution, a subsequent search can be made in the high-resolution image, restricted to the approximate location already found. A similar approach is described for the method to detect circular objects by a coarse raster-scan search in Section 4.3.

The concept of multiresolution analysis is straightforward. The original image is transformed to a hierarchy of images of successively smaller sizes and correspondingly lower resolutions. This representation is often referred to as a multiresolution pyramid (Fig. 3.7.2). The detection tool of interest such as a template or feature detector – of a single size – is applied to each image of the pyramid. Detection is performed on the image where detection mask and image object match in size and shape.

3.7.1 CONSTRUCTION OF MULTIRESOLUTION PYRAMIDS

The original full-resolution image resides at the lowest level of the multiresolution pyramid. On the next level resides the image generated by subsampling of the original by a given subsampling rate. On the next level above resides the image produced by subsampling the first subsampled image by the same rate; this image reduction is usually continued so long as the object of interest remains discernible.

Analysis on this multiresolution pyramid usually begins with the image of lowest resolution because this smallest number of pixels expedites the detection operation. Analysis then proceeds to successively higher resolutions, as necessary, until the object of interest is detected.

For the sake of simplicity, we assume that the original is square, with side lengths L; L is assumed to be a power of 2. With a subsampling rate of 2, an image of original dimension $L \times L$ yields a complete pyramid of $\log_2 L$ levels, with the lowest level comprising an image of size 2×2.

The direct approach to generating a single subsampled image was described in Section 3.6. Given a subsampling rate of 2, apply a low-pass filter to every second pixel in the original and retain the results in the final $L/2 \times L/2$ subsampled image. Having reduced the original image to a subsampled version of size $L/2 \times L/2$, one may construct the pyramid iteratively, that is, one generates the $L/4 \times L/4$ image by applying the same low-pass filter to the $L/2 \times L/2$ image, and so on, until the lowest-resolution stage is reached.

3.7.2   APPLICATIONS OF MULTIRESOLUTION ANALYSIS

Multiresolution analysis serves to render analytical procedures such as template matching more efficient. For instance, Section 3.8 describes how to perform template matching with the multiresolution pyramid to account for multiple sizes of the same shape. The approach is to apply the same template successively to multiple images of successively higher resolution. This multiresolution approach is more efficient than the converse approach of successively applying templates of adjusted size to the original image (which usually substantially exceeds the template in size). In practice, a size range is often known for the object of interest, so that only a limited set of levels in the resolution pyramid need be examined.

Multiresolution techniques also are very useful in finding multiple objects of different sizes. Multiresolution analysis ensures that the image is methodically analyzed at regular levels of resolution and that desired features are detected at their optimal resolution level of recognition.

If only the approximate location of an object or feature is of interest, application of object detectors at lower resolution offers substantial gains in efficiency. Although precise location may not be required, further analysis can be undertaken at a higher resolution to determine better features for more reliable recognition. An example of this task is the identification of facial features. In the first step, faces can be detected in a low-resolution image merely on the basis of overall (oval) shape and general features of a human face. To make a positive identification, facial features may be examined at higher resolution. This is analogous to our visual process: A shape can be identified as human at long distances, but only at shorter distances does it become possible to identify a particular person.

An added benefit of operating on the pyramid of images from the top down is that low-pass-filtered images are less noisy, simply as a result of repeated low-pass filtering. If the image of appropriate resolution is obtained in the pyramid, analysis is restricted to the level that matches object sizes of interest and omits features, textures, and noise at other scales.

3.7.3   MULTIPLE PASSBAND ANALYSIS

Texture is an image feature that is usually associated with a particular spatial frequency component (Section 7.2). For instance, an image of sand, with a fine grain texture, will have a high-frequency component; an image of a brick wall will display a coarser pattern, corresponding to a lower-frequency component. One way to detect and examine texture is to extract from the image all components that have the same range of spatial frequencies. Multiresolution analysis is useful when this range of frequencies is unknown beforehand. That is, if the particular spatial frequency is not known, a multiple passband transformation can be performed and examined at each level of the pyramid.

For this purpose, we construct a pyramid not by low-pass filtering, but by bandpass filtering. The original image is subjected to low-pass filtering to reduce the frequency cutoff by half, but it is not subsampled. The filtered version is subtracted from the original image to obtain an image containing all image frequencies in the upper half of the frequency spectrum. The low-pass image (before subtraction) can now be subsampled by a rate of 2 and low-pass filtered, then subtracted from the first subsampled low-pass-filtered image to obtain the second image. This contains a frequency between $1/4$ and $1/2$ of the original frequency spectrum. This process of subsampling, low-pass filtering, and subtracting images can be continued to obtain successively lower and narrower bandpass images. As with the low-pass multiresolution pyramid, a particular texture can be examined from low to high resolution.

References and Further Reading

Various aspects of multiresolution processing are discussed in
A. Rosenfeld, ed., *Multiresolution Image Processing and Analysis* (Springer-Verlag, Berlin, 1984); M. D. Kelly, "Edge detection in pictures by computer using planning," in *Machine Intelligence*, B. Meltzer and D. Michie, eds. (Edinburgh U. Press, Edinburgh, 1971), Vol. VI, pp. 397–409; P. J. Burt, "Fast filter transforms for image processing," Comput. Graph. Image Process. **16**, 20–51 (1981); and L. O'Gorman and A. C. Sanderson, "A comparison of methods and computation for multi-resolution low- and band-pass transforms for image processing," Comput. Vis. Graphics Image Process. **35**, 276–292 (1986).

Program

multires

        performs multiresolution processing.

    USAGE: multires inimg outimg [<-c> || <-m> || <-l LEVEL>] [-b]
                            [-d DEG_FILTERING] [-q] [-L]

ARGUMENTS: inimg: input image filename (TIF)

          outimg: output image filename (TIF)

  OPTIONS: -c: composite image output of all multiresolution
                levels arranged in a single image; this is the
                default mode;

           -m: multiple images of each multiresolution level in
                separate output image files; the output files
                are: out1.tif out2.tif, ... for each level 1, 2,
                etc.

     -l LEVEL: output is single image of the multiresolution
                level specified; level 1 is half size of
                original, level 2 is half size of level 1, etc.
                Filename is output filename specified, with 1, 2,
                etc. appended.
                NOTE: only one of -c, -m, or -l should be chosen.

           -b: for composite image <-c>; if flag is set, then
                background to composite images is original image;
                otherwise, the default is a blank background of
                the original image size

-d DEG_FILTERING: value, set between 0.0 and 1.0, controlling the
                degree of low-pass filtering; a value of 1.0
                produces maximum filtering to reduce aliasing,
                 but may lead to blurring; a value of 0.0 produces
                no filtering at the risk of aliasing; the default
                is 0.50.

           -q: when set, performs quick (and dirty) low-pass
                filtering with rectangular filter rather than
                Gaussian filter; the former is faster but does
                not produce equally good results as does the
                latter.

           -L: print Software License for this module

3.8 Template Matching

Typical Application(s) – detection of objects of known shape in noisy environment.

Key Words – matched filter, template matching, cross correlation.

Related Topics – region peak detection (Section 3.4); binary region detection (Section 4.3), shape analysis (Section 4.4), moments (Section 4.6), Hough transform (Section 4.10); critical point detection (Section 5.4).

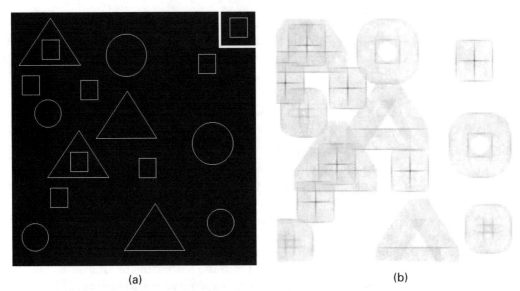

(a)   (b)

Figure 3.8.1. Pictorial Example. Illustration of template matching: (a) image containing geometrical objects, inset: template image of object to be matched, (b) results of matched filtering produced by **xcorr** (gray-scale values have been inverted for easier viewing) show peaks, indicated by crossed lines, at the image locations where the mask best matched the image.

Template matching, or matched filtering, is used to detect objects in an image of known shape, size, and orientation (Fig. 3.8.1). The method is straightforward and effective. A filter mask whose coefficients constitute a subimage that matches the desired object is convolved across the image. When this mask is located at the image object, the convolution result is an intensity peak. Therefore the presence and the location of desired objects are determined by peak detection (Section 3.4). In general, matched filtering is applicable only to objects of well-defined shapes. If the object shape is not fixed, this method is inappropriate. Furthermore, if object size or orientation is unknown, or the object is imaged from a different angle, or lighting introduces unexpected shadows, or the object is partially occluded, then matched filtering may yield unreliable results. In what follows, we will further discuss the advantages and the disadvantages of matched filtering and we introduce modifications to the basic strategy to circumvent some of the disadvantages.

3.8.1 IMPLEMENTATION

Given a gray-scale image of the usual form $I \equiv \{I(x, y); 0 \le x < X, 0 \le y < Y\}$, define a template or filter mask $F \equiv \{F(m, n); 0 \le m < M, 0 \le n < N\}$ containing a replica of an object to be located in the image. The mask size $M \times N$ corresponds to the bounding rectangular size of the object of interest.

The basic matched filtering technique is that of performing a cross correlation of the image with a template, i.e.,

$$I'(x, y) = \sum_{m,n} \sum F(m, n) I(m - x, n - y)$$

for all pixels in the input image.

Cross correlation with a large template is computationally expensive because the number of operations is proportional to the size of the mask. Above a certain size, implementation of the correlation operation by multiplication in the frequency domain is preferable to convolution in the spatial domain. This spatial domain versus frequency domain trade-off is further discussed in Section 7.2.

In practice, nonuniform illumination may cause an uneven image background intensity (Section 3.3), and it may be necessary to first normalize the image at each pixel with respect to its local intensity level and intensity range,

$$I'(x, y) = \sum_{m,n} \sum F(m, n)\{[I(m - x, n - y) - d(x, y)]r(x, y)\}$$

where $d(x, y) = [1/MN] \sum \sum_{m,n} I(m - x, n - y)$ represents the average of the local region and $r(x, y) = \max I(m - x, n - y) - \min I(m - x, n - y)$ represents the range of intensities in the local region.

The resulting image $I'(x, y)$ contains peaks at the locations of matches between the template and the underlying image objects. These peaks found by peak detection (Section 3.4) must be located to detect the presence of the object and determine their central locations.

3.8.2 MODIFIED IMPLEMENTATIONS

High-Pass Filtering of Nonuniform Images   As an alternative to performing normalization based on local averaging at each pixel, we can remove low-frequency information from the entire image by using a high-pass filter (Section 3.3), then performing a cross correlation of the modified image with a template that has been subjected to a similar high-pass filter. This is particularly useful when the salient features of the object of interest contain sharp edges. It is important here that a visual test be made of the filtered template to ascertain that all important template features have been retained; if not, the high-pass filter cutoff frequency should be lowered.

Edge Detection   Similarly to the high-pass filter approach, object edges will often serve to represent object shape faithfully. In such a situation, both the mask and the image are subjected to an edge detection filter (Section 3.4). Applying the matched filter to the resulting edge contour images (see Fig. 3.8.1) will yield a better correlation since fewer pixels – the important edge pixels – will form the basis of the match. This produces sharper peaks, which in turn facilitates subsequent peak detection.

Binarization   We can achieve a sharper correlation result by correlating binary images. If a given image contains objects that can be reliably thresholded to achieve a binary image (Section 3.9), this will produce a good correlation result. If the objects have good edges, then edge detection followed by thresholding will produce even better results.

Subsampling and Multiresolution   An approach to reduce the amount of computation incurred in the application of large templates is the reduction of mask and image sizes by subsampling (Section 3.6). For instance, if the side lengths of both image and template are halved, the raw number of multiplications for the requisite correlation operation is reduced by 16. This is so because, for a $k \times k$ template, correlation at a pixel requires $(k \times k) \times (k \times k) = k^4$ multiplications. Subsampling by 2 reduces the template size to $k/2 \times k/2$. Since each pixel correlation requires multiplication of a template with a portion of the image of same size, this requires $(k/2 \times k/2) \times (k/2 \times k/2) = k^4/16$ multiplications.

Size reduction is recommended, provided that template and object retain their characteristic shape – excessive size reduction will distort small objects into small (ill-defined) blobs or into single pixels.

If exact size and orientation of the object of interest are not available, it may be possible to search for a match within a range of sizes and orientations by supplying multiple masks to represent each possibility. Given the significant computational cost, this approach is practical only in situations in which only a small range of sizes and rotations may occur.

3.8.3 ALTERNATIVES TO MATCHED FILTERING

The utility of matched filtering is subject to a number of limitations. These are due to the fact that a complete object match against an image is computationally expensive. It is much more expensive when object rotation, scaling, perspective, lighting, occlusion, etc., require multiple correlations for all these possibilities.

Partial object matching, or subpart matching, is an alternative to complete object matching. For example, instead of trying to identify a particular make of automobile, the first step might be to identify a pair of dark regions close to one another (tires) to find the location of a car and to follow this partial identification with a more detailed template matching to determine the particular type of automobile of interest.

For the same example of detecting the make of a car, matched filtering may still be inappropriate if many possible orientations of the car are expected. In this case, it may be preferable to determine features of the car, such as its height, wheelbase, color, etc., and these features may form the basis for making a determination of the particular model. This strategy, referred to as feature detection, is a popular and effective alternative to matched filtering and is in fact the more general method of locating and identifying objects.

References and Further Reading

An introduction to the concept of matched filters is in G. L. Turin, "An introduction to matched filtering," IRE Trans. Inf. Theory, IT-6, No. 3, June 1960, 311–329 (1960); an additional discussion of matched filtering in image analysis is in W. K. Pratt, *Digital Image Processing*, 2nd ed. (Wiley-Interscience, New York, 1991); see also Section A.4.

Program

xcorr

finds locations of a pattern within an image by convolving the image with the pattern ("template"); the pattern is defined in a template file

USAGE: xcorr inimg templateimg outimg [-L]

ARGUMENTS: inimg: input image filename (TIF)

templateimg: name of image file containing template (TIF)

outimg: output image filename (TIF)

OPTIONS: -L: print Software License for this module

3.9 Gabor Wavelet Analysis

Typical Application(s) – texture and pattern detection

Key Words – texture, pattern, wavelets

Related Topics – multiresolution analysis (3.7), template matching (3.8), Fourier Transform (7.1), Frequency Domain Filtering (7.2)

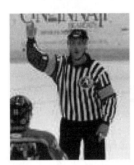 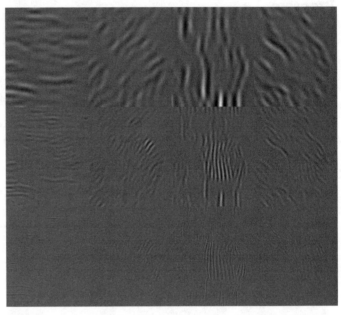

Figure 3.9.1. Pictorial Example. Illustration of Gabor filtering shows input image containing a stripe pattern (the referee's uniform) and output wavelet filtered results. The output shows a grid of 4 by 3 output results corresponding to different filter scales (vertical axis) and orientations (horizontal axis). The filter responsible for the image at grid location (3,2) produces a result that matches the input pattern most closely in scale and orientation.

Many types of image signals, especially textures and patterns, are made up of a repeating component (a wave) that has a repeated component length (wavelength) and one or more predominant orientations. For instance, Figure 3.9.2 contains four images that contain waves. Figure 3.9.2a contains a bull's eye pattern, 1b contains a brick wall texture, 1c contains a weave texture, and 1d contains a stripe pattern of a hockey referee's jersey. The wavelengths from largest to smallest are: brick, bull's eye, referee, and weave. There is one predominant orientation in the brick texture, one predominant and two less predominant orientations in the weave texture, all orientations in the bull's eye pattern (because it is circular), and the referee's jersey contains four regions of predominant orientations in the torso, shoulder, upper arm, and lower arm. To extract features of these waveforms in the image, there is a function that combines waveform and orientation matching, the wavelet. Furthermore, the wavelet localizes features in the spatial domain (unlike frequency domain analysis (Chapter 7), which also extracts frequency and orientation features, but conveys no location information).

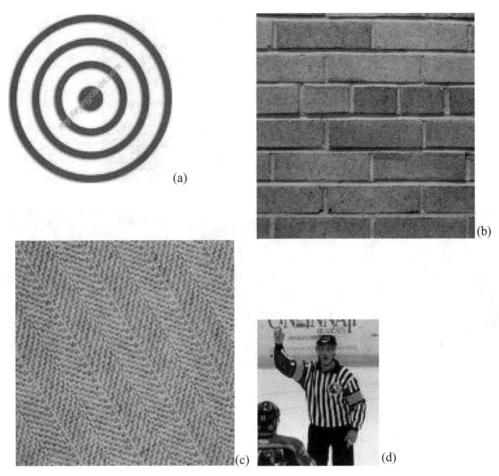

Figure 3.9.2. Images with patterns and textures: (a) bull's eye pattern, (b) brick wall texture, (c) weave texture, (d) referee's striped jersey.

Consider an image in Figure 3.9.3a that contains a sinusoidal waveform that travels from left to right throughout the whole image. If this is convolved (Sect. 2.1) with the bull's eye in Figure 3.9.2a, it will produce a signal showing that the same wavelength and orientation in Figure 3.9.3a is also in Figure 3.9.2a (the left and right vertical portions of the concentric circles). However, it will not show *where* it is because the full-image wave will match for all x and y shift values. If we spatially limit the sinusoid in Figure 3.9.3a in both directions along and perpendicular to its wave direction (as shown in Figure 3.9.3b) and convolve this with the image in Figure 3.9.2a, then the result will match the waveforms at their location. Furthermore, we can also use rotations of this mask, shown in Figures 3.9.3c–e, to match other orientations (in 45 degree steps) in Figure 3.9.2a. Thus, by producing a space-limited sinusoid of a certain wavelength, and producing different orientations of this, we can convolve it with a query image to locate different orientations of the same wavelength. These spatially limited sinusoidal waveforms are called *wavelets*, and their use at different chosen wavelengths and across a number of orientations is the essence of wavelet filtering.

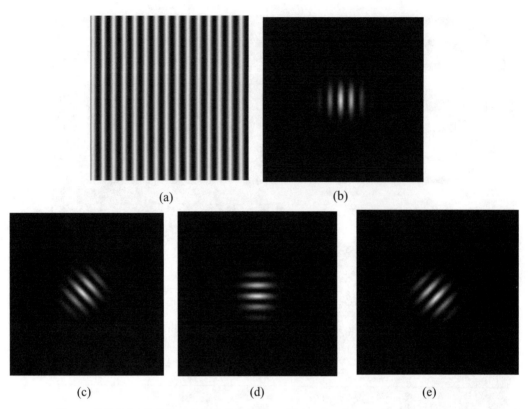

(a) (b)

(c) (d) (e)

Figure 3.9.3. (a) Full image containing sinusoidal wave of a single wavelength. (b) Localized sinusoid, spatially limited by multiplication with Gaussian window of variance ratio 7 to 5 along the x and y axes respectively. (c–e) 45, 90, and 135 degree rotations of spatially limited sinusoid in (b).

A particular type of wavelet that is often used in image analysis is the Gabor wavelet, $G(x, y)$. It is the product of a 2-D, elliptically-shaped Gaussian (for spatial limitation) with a plane wave (for wavelength matching). The real and imaginary Gabor wavelet functions are:

$$\textbf{Real}: \quad G(x, y) = \frac{1}{A} e^{-\pi \left[\frac{(x-x_0)^2}{\sigma_x^2} + \frac{(y-y_0)^2}{\sigma_y^2} \right]} (\cos(2\pi f_0 x))$$

$$\textbf{Imaginary}: \quad G(x, y) = \frac{1}{A} e^{-\pi \left[\frac{(x-x_0)^2}{\sigma_x^2} + \frac{(y-y_0)^2}{\sigma_y^2} \right]} (\sin(2\pi f_0 x))$$

$$where \; A = (2\pi \sigma_x \sigma_y)$$

The exponent term is an elliptical Gaussian, which tapers the spread of the function along the waveform direction by the standard deviation σ_x, and tapers the spread of the function perpendicular to the waveform direction by the standard deviation, σ_y. In both cases, the spread is narrower for smaller σ and wider for larger σ. The cosine and sine terms describe a plane wave with frequency f_0 in the x direction. The real term is symmetric (because a cosine is symmetric about $x = 0$, i.e., $\cos(x) = \cos(-x)$), and the imaginary term is asymmetric (because the sinusoid is asymmetric about $x = 0$,

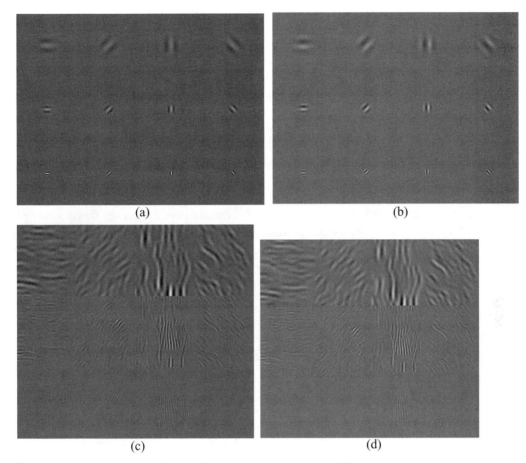

Figure 3.9.4. Gabor wavelet filter masks in (a) real (symmetric) and (b) imaginary (asymmetric). Results of filtering Figure 3.9.2d: (c) symmetric results by using masks in (3a), and (d) asymmetric results by using masks in (3b).

i.e., $\sin(x) = -\sin(-x)$). This Gabor function can be rotated in the spatial plane to produce different orientations of the same shape.

An image analysis task that often arises is to locate a pattern or texture in an image. For this task, Gabor wavelets are produced at several frequencies and orientations, and these are convolved with the image to locate the desired signal. Figure 3.9.4a shows 12 real (symmetric) masks and 3.9.3b shows 12 imaginary (asymmetric) masks. Figure 3.9.4c and 3.9.3d show the result of convolving each of the masks of 3.9.3a and 3.9.3b respectively against Figure 3.9.2d. One can see the stronger intensity regions are at the locations where the stripe patterns match in both frequency and orientation to each appropriately oriented mask. In this case – and in many cases for natural images – there is not much difference between symmetric and asymmetric filtering results, so only the real (symmetric) result can be used.

Besides texture analysis, another major use of wavelets is image compression. These are utilized especially for images consisting of a large amount of pattern or texture information, such as fingerprint images. A wavelet transform can be used to achieve extremely efficient compression, we do not consider compression here.

3.9.1 COMPARING TO OTHER ANALYSIS METHODS

There are similarities and overlaps between Gabor wavelet analysis and other analysis methods described in this text. Multiresolution analysis (Section 3.7) produces images at multiple resolutions or sampling rates, either via low-pass or band-pass filtering. Multiresolution analysis does not localize orientation and is a global method, treating all areas of the image equally. In contrast, Gabor wavelet analysis produces images for a single spatial frequency, but over many orientations, and it does maintain location. Template matching (Section 3.8), finds the presence and location of object shapes in the image. These are usually particular objects and their shape, size, and orientation must be known before processing (or a range of these can be chosen for matching). In contrast to template matching, Gabor wavelet analysis matches a waveform versus an object and usually does so over a range of orientations. Frequency domain analysis (Chapter 7) transforms the image to the frequency domain, where all frequencies are present, but location is not preserved. In constrast to frequency domain analysis, Gabor wavelet analysis examines one frequency over many chosen orientations and preserves location.

Program

gabor

```
     USAGE:   gabor inimg outimg [-s N_SCALES] [-r N_ORIENT]
              [-f1 FREQ_LO] [-f2 FREQ_HI] [-m MASK_SIZE]
              [-d MASK_FILENAME] [-L]

              gabor performs gabor wavelet filtering on input image
              resulting in a gray-scale output image containing
              multiple sub-images. Each sub-image contains a
              different scale and orientation filter result. These
              are arranged in the output image from largest scale on
              top row to smaller scales (each half scales) on each
              lower row, and orientations by columns from 0 degrees
              at left and proceeding counter-clockwise by the
              orientation increment to the right.
  ARGUMENTS:          inimg: input image filename (TIF)
                     outimg: output image filename (TIF)
    OPTIONS:      -s N_SCALES: number of scales [N_SCALES>=2, dflt=3]
                  -o N_ORIENT: number of orientations
                               [N_ORIENT>=2, dflt=4]
                  -f1 FREQ_LO: low frequency of filter
                               [FREQ_LO>=-1717986918, dflt=-0.00]
                  -f2 FREQ_HI: high frequency of filter
                               [FREQ_HI<=0, dflt=-0.00]
               -m MASK_HSIZE: size of half mask [MASK_HSIZE>=2,
                               dflt=10]
          -d MASK_FILENAME: filename to display Gabor masks
                          -L: print Software License for this module
```

References and Further Reading

Tai Sing Lee. Image representation using 2D Gabor wavelets. IEEE Trans. PAMI, Vol. 18, no. 10, Oct. 1996, pp. 1–13.

Figures 3.9.2b and 3.9.2c are copyright Dover Publications and printed by permission from, P. Brodatz, Textures: A Photographic Album for Artists and Designers, Dover Publications, New York, 1999. By Phil Brodatz.

3.10 Binarization

Typical Application(s) – conversion of a gray-scale image into a black-and-white image.

Key Words – thresholding, local, global, contextual.

Related Topics – image intensity histogram statistics (Section 2.1), histogram transformations (Section 2.2), edge detection (Section 3.4).

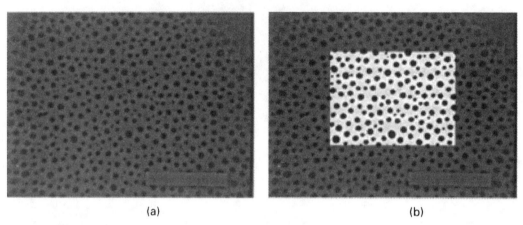

(a)   (b)

Figure 3.10.1. Pictorial Example. Illustration of binarization: (a) original image (previously flat fielded; See Section 3.3), (b) image following binarization, applied to a preselected area of interest.

A gray-scale image often contains only two levels of significant information, namely the foreground level constituting objects of interest and the background level against which the foreground is discriminated. For instance, the text on this page is the foreground against the white background of the page. Even when the information in an image is not inherently binary, binarization is often an indispensable step in extracting pertinent information (see Fig. 3.10.1). A simple example is that of an aerial picture of white sheep in a dark green field. A gray-scale image of this scene will contain a wide range of gray values representing the variation in grass coloring, the sheep's shadows, etc. However, if we are just interested in counting sheep, this information is of no value; instead, an effective approach is to binarize the image and count the number of white blobs it contains. This is precisely the strategy used to perform cell counts in biological images, to estimate areas of particular land usage in satellite images, and to locate areas of interest in radiological images.

Notwithstanding its apparent simplicity, binarization is a very difficult problem. Even in cases in which images contain inherently binary information, the corresponding intensity histograms (Section 2.1) will generally contain many more than simply two values. Figure 3.10.2 shows an image of a page of text and its histogram. While there are two principal peaks of the foreground and the background intensities in the histogram, there are many other intensities present. This is due to nonuniformities in regions of otherwise uniform intensity caused by such effects as nonuniform lighting, shading, and noise. In this example, binarization can be accomplished by the choice of an intensity between the two histogram peaks that is the threshold between all background intensities below and all foreground intensities above.

Thresholding is a synonymous term for binarization, emphasizing that the most important task in performing binarization is the choice of an appropriate threshold level. Also in Fig. 3.10.2, we show examples of the results of poorly chosen thresholds. In Fig. 3.10.2(e), the threshold is chosen at too low an intensity, resulting in larger blobs than desired. In Fig. 3.10.2(f), the threshold is chosen too high, resulting in breaks in what is desired to be complete regions.

A number of conditions can make binarization difficult. Poor image contrast makes it difficult to resolve foreground from background; corresponding histogram peaks will tend to overlap. Worse are spatial nonuniformities in background intensity so that, for example, the image appears light at one side and dark at the other. In this case, it is difficult or impossible to choose a single threshold that works well for the entire image. Another condition is one in which foreground and background are not obvious, perhaps because of multiple levels of different objects. This case requires user interaction for specifying the desired object and its distinguishing intensity features.

The variety of conditions under which binarization is to be performed requires different approaches. There is not a single optimally suited method of binarization for all image types. We describe in this section different approaches for image binarization and indicate the type of image for which each is appropriate. However, it is often best to make the decision by experimentation.

We distinguish two categories of binarization techniques (Fig. 3.10.3). Global techniques use the intensity histogram (Section 2.1) to identify a threshold between foreground and background intensity values. We determine this single threshold by treating each pixel independently of its neighborhood, or without context, and it is applied to all pixels. Locally adaptive techniques examine relationships between intensities of neighboring pixels to adapt the threshold according to the prevailing intensity statistics for different image regions.

Adaptive techniques are applied in an attempt to counter the effects of nonuniformities in background intensity (see also Section 3.3, flat fielding). Images that are captured under controlled lighting conditions (such as those provided by a document scanner over a piece of paper) will have more uniform backgrounds than those imaged by a single, fixed light source or under ambient light. Larger images tend to be less uniform than smaller images.

For images with uniform backgrounds, the optimal global techniques – in particular the methods described below – will ensure better binarization than the locally adaptive techniques because, for global techniques, threshold selection is based on a larger data set of pixels. Also, for adaptive techniques, a size parameter matching the size of uniform background regions must be chosen, but it may not match for different images or even for different locations in the same image.

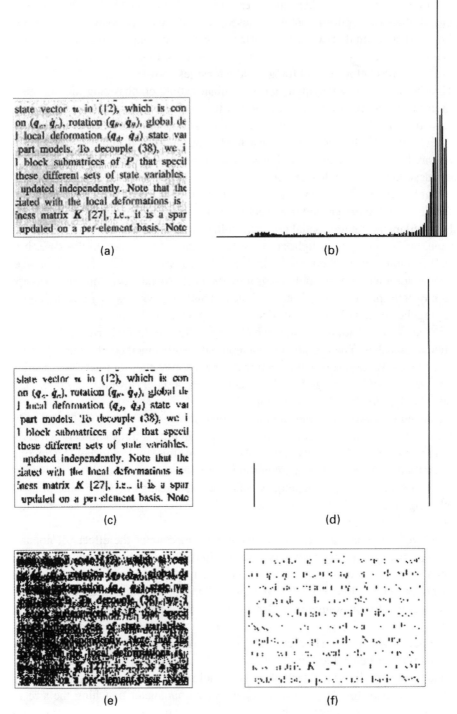

Figure 3.10.2. Text image examples of global binarization and associated histograms: (a) original, (b) histogram of (a), (c) binarized result of (a), (d) histogram of (c), (e) result of binarizing (a) at too low a threshold, (f) result of binarizing (a) at too high a threshold.

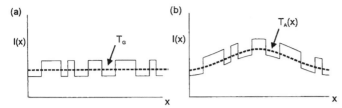

Figure 3.10.3. Illustration of global versus adaptive thresholding technique. Both figures show an intensity cross section through an image that gives a waveform whose high plateaus should be binarized to value ON and whose low plateaus should be binarized to OFF. The dashed line indicates the value of the intensity threshold. (a) Uniform background: global thresholding will binarize as intended; the threshold is fixed for all image locations. (b) Nonuniform background: adaptive thresholding is required; the threshold adapts to the background variation.

3.10.1 GLOBAL TECHNIQUES

Fixed Thresholding

The most straightforward global, noncontextual technique is just to binarize with respect to a specific selected threshold value. This threshold may correspond to half the maximum intensity by default or may be chosen by experimentation. This approach will be satisfactory when foreground and background intensities are clearly distinct and uniform throughout the image. Furthermore, if a number of images are to be processed, this strategy requires that all images in the set have the same intensity distribution characteristics, that is, the same contrast, brightness, and relative foreground and background intensities. This fixed threshold procedure is quite adequate for the many applications involving controlled lighting, fixed backgrounds, and similar foreground objects, for example, sets of images extracted from a video tape recorded under a fixed set of conditions.

A more robust extension to fixed thresholding is to set the threshold to a predetermined fraction between the minimum and the maximum intensity values in the image to account for dark or light intensity shifts. As a further extension, rather than relying on maximum and minimum intensities, it is often preferable to rely on intensity values corresponding to the bottom and the top 1 or 5 percentile levels to reduce sensitivity to outlying intensities. If the approximate relative areas of foreground and background regions are known, the choice of threshold should reflect this proportion. The intensity histogram serves to locate minimum and maximum intensities as well as proportions of above- and below-threshold pixels.

Histogram shape

Histogram shape (Section 2.1) can be helpful in locating the threshold. For foreground and background regions of similar total areas and well-separated intensities, the histogram will exhibit two peaks. The minimum, or valley, between the peaks is the obvious choice for the threshold that best separates the peaks. Given that automatic peak and valley detection are prone to locating local maxima and minima rather than those desired, the search can be aided by additional constraints, requiring, for example, that each peak fall within given intensity bounds or exceed a certain intensity level, etc. Histogram smoothing (Section 2.2) may also be helpful in eliminating extraneous spikes, but this requires some knowledge of the amount of noise and the closeness of peaks so that noise peaks are reduced without merging region peaks.

Optimal Thresholding

Histogram shape is not a reliable indicator for threshold selection when peaks are not clearly resolved. To address this rather common occurrence, several techniques have been designed to determine classes of pixel intensities – foreground and background – by more formal pattern recognition techniques that optimize some measure of separation. For these techniques, a criterion function is devised that yields some measure of separation between classes. The criterion function is calculated for each intensity value and that which maximizes this function is chosen as the threshold.

We describe below advantages and disadvantages of several of these methods. No single method works best for all image characteristics, and it is a matter of continuing research to identify *a priori* a best method for a particular image. In practice, try the different techniques on images that are representative for a particular application and observe the results.

Discriminant Method   In this method, also called Otsu's method, the thresholding problem is formulated as a discriminant analysis. This is a formal pattern recognition procedure in which a particular criterion function is used as a measure of statistical separation between classes. Statistics are calculated for the two classes of intensity values (foreground and background) that are separated by an intensity threshold. The criterion function used here is σ_{Bi}^2/σ_T^2 for every intensity, $i = 0, \ldots, I - 1$, where σ_{Bi}^2 is the between-class variance and σ_T^2 is the total variance. The intensity that maximizes this function is said to be the optimal threshold. This is a classic method (older than the others described here) and it has well-deserved popularity.

Entropy   Entropy, a concept widely used in information theory, measures the amount of information needed to represent a class of data. Entropy also can serve as a measure of separation. For example, a class with up to just two levels requires only 1 bit, up to four levels needs 2 bits, up to eight levels requires 3 bits, etc. This method entails separating the image data into two classes, above and below an intensity threshold, and measuring the entropies of each class. This separation is done for each intensity level, and that level for which the sum of entropies of the two classes (the criterion function in this case) is maximum is the optimal threshold.

Moment Preservation   Another approach to threshold selection is through moment preservation. The objective is to choose a threshold such that the resulting thresholded image best preserves mathematical moments of the original gray-scale image. Although this approach has not received much attention in the literature, we have had very good results with it and present an implementation here.

Moments (see also Section 4.5) are first calculated for the original gray-scale image. Next, they are calculated for images resulting from every possible threshold. The threshold value at which the original and the thresholded images have closest moments is said to be the optimal threshold. The first k moments of the gray-scale image are evaluated directly from the intensity histogram in the form

$$m_k = \sum_{i=0}^{I-1} p_i i^k,$$

where p_i is the probability of gray level i between 0 and $I - 1$, k is the order of the moment, and m_0 is defined to be 1. For binarization, a system of four linear equations ($k = 4$) is required.

Minimum Error Thresholding   This method assumes that the histogram is composed of two normally distributed (i.e., Gaussian) classes of pixel intensities. Two normal distribution curves are determined by an iterative process to fit the two classes of pixels in the histogram and minimize a specified classification error. On each iteration, a prospective threshold value is tested by calculation of the means and the variances from the histogram for the two classes separated by this threshold. The criterion function is minimized to find the best fit between the statistical model and the histogram. One principal advantage of this very effective and popular method is that it yields a robust separation of classes even when one class size is much smaller than the other such that it cannot be visibly resolved from the histogram. A disadvantage is that it is susceptible to error for a noisy image whose histogram contains many spurious peaks that are due to noise. In addition, if the assumption of normally distributed classes is erroneous, this will introduce bias or error in the threshold selection.

Connectivity-Preserving Thresholding   The connectivity-preserving thresholding method is designed to combine aspects of global and locally adaptive thresholding. The method is classified as global since all pixels are thresholded with respect to a single threshold. However, unlike the other global methods that use global statistical measures of pixel intensities retaining no information on pixel locality, this method uses a local measure, connectivity. That is, it attempts to maintain local connected regions of foreground pixels. The method is equivalent to binarizing an image at every intensity value, and choosing a threshold within a range of thresholds where the number of connected regions is relatively constant. This ensures that a threshold is located at an intensity value that least breaks up regions.

The method was designed for document and graphically created images containing text, engineering drawings, and graphics or cartoon pictures with flat valued regions. The objective for these types of images is to create resultant images in which characters and graphics remain connected. This results in a more pleasing image to view, but more importantly image analysis produces optimally connected results for subsequent character or graphics recognition. This method will not perform as well as some of the other methods for natural images in which regions change intensity more gradually between foreground and background.

An additional feature of this method is that it can perform multi-thresholding. If an image contains not only background and foreground, but background and multiple foreground levels, then it will produce multiple resultant images, one for each foreground level. An example of an image containing multiple levels is a document image that has been converted from color to gray-scale, where the color original has multiple colors of text or graphics. As the number of levels increase and as noise increases, the resultant thresholded images are less reliable. However, for a single- or two-foreground level image with textual or graphical components, this method can produce much better images than the other methods.

3.10.2 LOCALLY ADAPTIVE TECHNIQUES

Region Averaging

One approach to locally adaptive thresholding is to subtract a globally nonuniform background from the original image and to perform thresholding on the uniform result. We may implement this by first calculating a running region average, by comparing each pixel value to its local average, and setting that value to either the foreground, if it is much above the average, or to the background, if it is much below. In cases in which the difference between the pixel value and its average is small, the pixel is set to the same value as the previous pixel result. The region size in which the local average is to be calculated is adjusted to reflect the size of expected foreground features: this size should be large enough to enclose a feature completely, but not so large as to average across background nonuniformity. For instance, for a foreground object of diameter d, a region of diameter $2d$ or $3d$ may be appropriate to evaluate the local average.

This choice of proper size for the window over which to evaluate local averages may be problematic when images contain foreground features of variable size. For instance, for a document image with large headline text, but small text in the body, the window must be set with respect to the headline text size: otherwise, the borders of the headline text will be assigned the foreground value while the interior will be assigned the background value. However, averaging over this large size may not be effective in reducing background nonuniformity in the body of smaller text. Another difficulty lies in the determination of how much of a deviation from the average to tolerate before a different threshold is selected. A low level for this deviation will result in much foreground and background noise; a high level will leave some true features undetected. Selection of this parameter must reflect the level of expected noise, and it usually relies on experimentation. Implementation requires one pass through the image in which a $k \times k$ average is computed and local thresholding is performed.

Thresholding on Subimages

Another approach is to use the global thresholding method of Subsection 3.10.1 on subimages throughout the full image. Depending on the degree of background nonuniformity, the image of size $N \times M$ is partitioned into $N/n \times M/m$ subimages of size $n \times m$. Optimal thresholds are determined within each subimage. Any subimage for which the measure of class separation is small is said to contain only one class; no threshold is calculated for these. Instead, the threshold is taken as the average of the thresholds in the neighboring subimages. Finally, the subimage thresholds are interpolated among subimages for all pixels and each pixel value is binarized with respect to the threshold at the pixel.

References and Further Reading

A very extensive literature is dedicated to binarization. Useful reviews are available in

Ø. D. Trier and A. K. Jain, "Goal-directed evaluation of binarization methods," IEEE Trans. Pattern Anal Mach. Intell. **17**(12), Dec. 1995 1191–1201 (Dec. 1995); J. S. Weszka, "A survey of threshold selection techniques," Comput. Vis. Graph. Image Process. **7**, 259–265 (1978); P. K. Sahoo, S. Soltani, A. K. C. Wong, and Y. C. Chen, "A survey of thresholding techniques," Comput. Vis. Graph. Image Process. **41**, 233–260 (1988).

Methods of optimal thresholding are discussed in

J. Kittler and J. Illingworth, "Minimum error thresholding," Pattern Recogn. **19**(1), **41–47** (1986); N. Otsu, "A threshold selection method from gray-level histograms," IEEE Trans. Syst. Man, Cybern. **SMC-9**, 62–66 (1979); T. Pun, "Entropic thresholding: a new approach," Comput. Vis. Graph. Image Process. **16**, 210–239 (1981); J. N. Kapur, P. K. Sahoo, and A. K. C. Wong, "A new method for gray-level picture thresholding using the entropy of the histogram," Comput. Vis. Graph. Image Process. **29**, 273–285 (1985); W-H. Tsai, "Moment-preserving thresholding: a new approach," Comput. Vis. Graph. Image Process. **29**, 377–393 (1985).

L. O'Gorman. Binarization and multi-thresholding of document images using connectivity. Computer Vision, Graphics and Image Processing Journal: Graphical Models and Image Processing, Vol. 56, no. 6, Nov. 1994, pp. 494–506.

Mehmet Sezgin, Blent Sankur, "Survey over image thresholding techniques and quantitative performance evaluation," Journal of Electronic Imaging, Vol. 13, Issue 1, Jan. 2004, pp. 146–68.

Programs

binarize

```
           applies a user-selected threshold to binarize a gray-
           scale input image, setting pixel intensities BELOW
           (darker than) the user-chosen threshold to 0 and those
           ABOVE (lighter than) the threshold to 255 resulting in
           a binary output image.
    USAGE: binarize inimg outimg [-t THRESHOLD] [-i]
           [-a x1 y1 x2 y2] [-L]
ARGUMENTS:  inimg: input image filename (TIF)
           outimg: output image filename (TIF)
  OPTIONS:
 -t THRESHOLD: gray value between 0 and 255 (or other maximum
           intensity value); intensities above THRESHOLD are
           set to ON, all others to OFF; default = 128.
        -i: INVERT: intensities ABOVE (lighter) threshold set
           to 0 and those BELOW (darker) threshold set to
           255
-a x1 y1 x2 y2: upper left, lower right coords for Area of Interest
        -L: print Software License for this module
```

threshm

performs binarization with respect to automatically determined intensity threshold; the input gray-level image is converted to a binary image; threshold determination is made by the moment-preservation method.

USAGE: threshm inimg outimg [-i] [-a x1 y1 x2 y2] [-L]

ARGUMENTS: inimg: input image filename (TIF)

outimg: output image filename (TIF)

OPTIONS: -i: INVERT: intensities ABOVE (lighter) threshold set to 0 and those BELOW (darker) threshold set to 255

-a x1 y1 x2 y2: upper left, lower right coords for Area of Interest

-L: print Software License for this module

thresho

performs binarization with respect to automatically determined intensity threshold; the input gray-level image is converted to a binary image; threshold determination is made by Otsu's moment preservation method.

USAGE: thresho inimg outimg [-L]

ARGUMENTS:

inimg: input image filename (TIF)

outimg: output image filename (TIF)

OPTIONS:

-L: print Software License for this module

threshe

performs binarization with respect to automatically determined intensity threshold; the input gray-level image is converted to a binary image; threshold determination is made by the maximum entropy method.

USAGE: threshe inimg outimg [-L]

ARGUMENTS:

inimg: input image filename (TIF)

outimg: output image filename (TIF)

OPTIONS:

-L: print Software License for this module

threshk

performs binarization with respect to automatically determined intensity threshold; the input gray-level image is converted to a binary image; threshold determination is made by Kittler's minimum error method.

USAGE: threshk inimg outimg [-L]

ARGUMENTS:
     inimg: input image filename (TIF)
    outimg: output image filename (TIF)
  OPTIONS:
        -L: print Software License for this module

threshc
  usage: threshc in.tif out.tif [-f MAX_FLAT_DEV <15>%]
         [-n MIN_NUM_RUNS <100>] [-c MINCONTRAST <10>]
         [-i INCREMENT <1> (must be power of 2)]
         [-h <hist_flag=0>] [-L]
         THRESHC program accepts gray-level image, and produces
         binary output images for each threshold level. Thresholds
         are found maximizes connectivity of image objects at each
         thresholded level.
         NOTE that output filenames have added indication of
         threshold level, as in outL0.tif, outL1.tif, etc., for
         specified output out.tif.
  FLAGS:
      -f MAX_FLAT_DEV max. deviation of intensity within a level
         This is expressed as a percentage of maximum number of
         runs.
         This is set small for threshold images whose levels have
         only small intensity deviations, or larger to give
         tolerance to individual levels with higher deviation due
         to noise.
      -n MIN_NUM_RUNS min. number of runs in a thresh range to keep
         it (approximately sidelength plus sidelength).
         This regulates the ability to disregard threshold levels
         if they are only due to small region sizes.
      -c MINCONTRAST min. intensity diff. between threshold levels.
         This is the worst-case (min.) spacing between intensity
         levels, expressed as a percentage of the occupied
         intensity bins.
      -i INCREMENT is subsampling interval for connected runs calc.
         By increasing this increment, processing is sped up.
         (This must be power of 2, otherwise there is no speedup.)
      -h flag to display histograms on output images.
         help - to show this help message
      -L: print Software License for this module

4 Binary Image Analysis

In this chapter we introduce a variety of techniques for the processing and the analysis of binary images. Pixels in binary images are restricted to two intensity values, generically referred to as OFF and ON. Binary analysis is performed when the information of interest is fully contained in the binarized version of the image and subtleties such as shading and texture are of no concern. For example, in object shape analysis, only the object outline is of interest. In some cases, as for text scanned by a fax machine, images may be initially acquired in binary format; in other cases, they may be generated by binarization of gray-scale originals, as discussed in Section 3.9.

Section Overview

Section 4.1 describes morphological and cellular processing. These operations are used to modify the shape of binary regions. This may include removing noise, smoothing boundaries, joining disconnected regions, and separating regions that are erroneously connected.

Section 4.2 describes binary noise removal. Noise in binary images most commonly consists of small, extraneous pixel regions. The kFill method that is presented here is designed to reduce this noise. Although morphological operations can do similar noise reduction, kFill was designed specifically and exclusively for this task (in contrast to the more general morphological operations). It reduces noise while minimally affecting the rest of the image.

Section 4.3 describes methods for the detection and the segmentation of regions, defined as connected sets of ON pixels on an OFF background (or vice versa). The objective of the methods presented in this section is to find these regions and to store them in a concise representation that facilitates image analysis. Region detection is a required step before such operations as shape detection, convex hull determination, and moment calculation (topics of the following sections).

Section 4.4 describes the analysis of region shape in terms of a variety of characteristic features including geometric descriptors such as perimeter-to-area ratio, boundary curvature, and moments. Moment analysis is a common method of shape analysis that provides a description of a region in terms of quantities such as area, centroid location, orientation, and related quantities. Several methods of evaluating moments of low order are described. The determination of a region's shape facilitates its recognition or differentiation of one region from another. Statistical computation based on all region shapes can also provide a useful global description of an image.

Section 4.5 describes shape analysis in terms of Fourier descriptors that are derived from a spectral analysis of region shape. A one-dimensional version of Fourier transformation, a topic further discussed in Chap. 7, is invoked to obtain a systematic description of region or boundary shape that is complementary to the moment analysis discussed in Section 4.4. The choice of specific descriptors from Sections 4.4 or 4.5 in a specific situation will depend on the types of shapes to be analyzed and the type of desired information. In practice, this often entails comparing shape and moment descriptions and choosing empirically.

Section 4.6 describes a method to construct the convex hull of a region. The convex hull is a geometric object that encompasses a region without any indentations (overhangs). The convex hull simplifies the representation of complex region boundaries, and it delineates the area of influence of a region whose contour has indentations.

Section 4.7 describes thinning of elongated shapes. Thinning replaces a single elongated region by a single-pixel line that traces the region's medial line and thus produces a skeleton of the original shape. Thinning preserves and produces a description of a region's connectivity, as expressed in branching, for example. Thinning is most appropriate for elongated regions or regions that are best described by lines. Fatter regions (blobs) are better described by the shape and moment techniques of Sections 4.4 and 4.5.

Section 4.8 describes line thickness. This is a descriptor of a region that augments that of thinning. The additional information is that of the thickness of lines and elongated portions of a region.

Section 4.9 describes global features that are of specific interest to the analysis of binary images. Whereas the previous sections deal primarily with descriptions of single regions, this section deals with descriptions of an entire image of binary regions. Such descriptions include, for instance, the density of regions in an image and their spatial distribution. When individual regions are not of interest and when there is a characteristic global signature of an image, global features are a good choice.

Section 4.10 describes the Hough transform. This is a global technique used to determine the existence and the location of individual regions whose combined shapes comprise geometrical entities. For instance, the existence of regions in a line or along the circumference of a circle can be determined by use of the Hough transform. The geometrical entities are limited primarily to lines or (much less often) circles, because of practical and computational considerations.

(a)   (b)

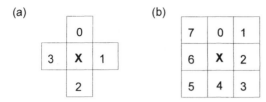

Figure 4.0.1. Neighborhoods and connectivity: (a) A four-connected neighborhood consisting of a center pixel X and its four closest neighbors, (b) an eight-connected neighborhood consisting of a center pixel X and its four closest neighbors plus its next four closest (diagonal) neighbors.

Glossary of Basic Terminology

OFF (black), ON (white) – These are the generic designations of pixel intensities in a binary image. In the figures, OFF-valued pixels are always white. ON-valued pixels may be represented either as black regions or individual pixels, designated by an X. In other literature, OFF and ON are often called 0 and 1, respectively.

Neighbors – One pixel is said to be a neighbor of another if it is its closest horizontal, vertical, or diagonal pixel. Each pixel has eight neighbors, four horizontal and vertical and four diagonal. A pixel and its neighborhood is shown in Fig. 4.0.1.

Connectivity – A pair of pixels is considered to be connected (or to possess connectivity one to the other) if each is ON-valued and they are neighbors. In the literature, two types of connectivity are defined. Four-connected pixels include only the four horizontal and diagonal neighbors, and eight-connected pixels include the remaining four diagonal neighbors. In this text we use connectivity to mean eight-connected, since this yields the more efficient representation for image analysis.

Region or blob – A region or blob is a singly connected group of ON-valued pixels.

Boundary – The boundary of a region consists of all pixels of a region that border one or more background (OFF-valued) pixels.

4.1 Morphological and Cellular Processing

Typical Application(s) – modification of region shapes in binary images.

Key Words – mathematical morphology, cellular logic; erosion, dilation; shrink, expand; region growing; structuring element, binary filtering.

Related Topics – template matching (Section 3.8); binary noise reduction (Section 4.2), thinning (Section 4.7), linewidth determination (Section 4.8).

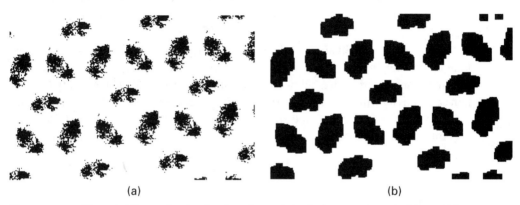

(a)   (b)

Figure 4.1.1. Pictorial Example. Application of morphological processing to isolate and identify regions in a binary image: (a) original image, (b) result after four iterations of opening, one iteration of erosion, and four iterations of dilation with a square, 3×3 structuring element.

In the context of binary image processing, morphological and cellular processing refer to a class of spatial filtering operations that are applied to change the shape of a binary-valued region. That is, morphological operations replace the binary value at a single-pixel location by a value computed from the pixel values within a neighborhood of chosen shape and size around that location. As with convolution, this operation is applied to all pixels in the image. We describe two families of operations yielding similar results but differing somewhat in their methods: morphological processing, also called mathematical morphology (see Fig. 4.1.1), and cellular processing, also called cellular logic.

The most basic morphological and cellular operations are erosion and dilation. Erosion is the reduction in size of ON regions, and this is most readily accomplished by iterative peeling of single-pixel layers from the outer boundary of all ON regions. Dilation is the opposite process, entailing iterative addition of single-pixel layers to the boundary of each ON-valued region to increase region size. Both operations are usually applied iteratively to erode or dilate by many layers. An example for the application of erosion is the removal of a layer, one or two pixels wide, of noisy boundary pixels in a binary image. The analogous dilation process can be applied to fill holes or to join disconnected lines with gaps up to two pixels in width.

Most applications of morphological and cellular processing consist of a sequence of erosion, dilation, and Boolean operations, with a sequence and number of such steps designed for a specific purpose. For the erosion example above, two steps of erosion to erase a noisy boundary layer would usually be followed by two steps of dilation to generate a noise-reduced image containing ON regions of the same width as the original. Combinations of these operations can be designed for such tasks as noise reduction, thinning, contour determination, and feature detection.

Other sections of this chapter describe methods that are specifically designed for particular tasks, although these tasks could also be accomplished with morphological or cellular processing. In most cases, the specific methods are more efficient or effective for their particular task than the more general cellular or morphological equivalent method.

Since erosion and dilation are the basic operations of both morphological and cellular processing, in many cases we can achieve very similar results by using either. One difference in our description below is that cellular processing offers the ability to maintain connectivity within the image. Another difference is that morphological processing offers the option of choosing the size and the shape of features to affect. In practice, it is often a matter of having two similar tools for which the better is determined by experimentation.

4.1.1 MORPHOLOGICAL PROCESSING

Morphological operations are commonly used for noise reduction and feature detection, with the objective that noise be reduced as much as possible without eliminating essential features. We present a narrow and practical exposition of morphological processing particularly geared for binary image analysis by using a simpler notation than that commonly used.

We can think of the structuring element as a filter mask $S(i,j)$ of size $k_1 \times k_2$ in which the coefficients take binary values. These values are chosen so that the mask has a desired shape: The most common shapes for the ON region are a square, a disk, and a line at some chosen orientation. Unlike most filter masks, a structuring element need not be symmetric, and its origin must be explicitly labeled (as we explain below). We apply a morphological operation to a pixel by first placing the designated origin of the structuring element S at that pixel; S defines a neighborhood $N_S(i_0, j_0)$ of the pixel (i_0, j_0).

The two basic morphological operations of erosion and dilation are most conveniently defined in Boolean terms. Figure 4.1.2 shows the results of erosion and dilation.

(a)   (b)   (c)   (d)   (e)

Figure 4.1.2. Morphological erosion and dilation: (a) original image, (b) result of one iteration of erosion, (c) result of one iteration of dilation, (d) difference between (a) and (b) showing pixels removed by erosion, (e) difference between (a) and (c) showing pixels added by dilation.

Erosion

Erosion is used to remove ON pixels at boundaries of regions and to increase the size of holes (OFF-valued regions within ON regions). When a structuring element is over a region, an ON pixel at the origin is set to OFF if the structuring element does not completely overlap ON-valued pixels. This causes a recession or erosion of the boundaries. Set a Boolean ON value for (i_0, j_0) [i.e., $I(i_0, j_0) = $ ON] to OFF if, for all pixels $(i, j) \in N_S(i_0, j_0)$, application of an AND operation between $I(i_0, j_0)$ and the element $S(i - i_0, j - j_0)$ of S, superimposed on pixel (i, j), yields a Boolean OFF value.

More formally, this may be expressed as

$$I^E(i_0, j_0) = \begin{cases} \text{ON} & \text{if } I(i_0, j_0) = \text{ON and } I(i, j) \bigwedge S(i - i_0, j - j_0) = \text{ON,} \\ & \quad \text{for all } (i, j) \in N_S(i_0, j_0). \\ \text{OFF} & \text{otherwise} \end{cases}$$

The size of ON-valued pixel regions can only decrease. In the morphological literature, erosion is denoted by $I \ominus S$.

Dilation

Dilation is used to add pixels at region boundaries or to fill in holes. When a structuring element is over a region, an OFF pixel at the origin is set to ON if any of the structuring element overlaps ON pixels of the image. This causes an enlargement or dilation of the boundaries. Set a Boolean OFF value for (i_0, j_0), [i.e., $I(i_0, j_0) = $ OFF] to ON if, for any of the pixels $(i, j) \in N_S(i_0, j_0)$, application of an OR operation between $I(i, j)$ and the element $S(i - i_0, j - j_0)$ of S, superimposed on pixel (i, j), yields a Boolean ON value.

More formally, this may be expressed as

$$I^D(i_0, j_0) = \begin{cases} \text{ON} & \text{if } I(i_0, j_0) = \text{ON or } I(i, j) \bigwedge S(i - i_0, j - j_0) = \text{ON,} \\ & \quad \text{for any } (i, j) \in N_S(i_0, j_0). \\ \text{OFF} & \text{otherwise} \end{cases}$$

The size of ON-valued pixel regions can only increase. In the literature, dilation is denoted by $I \oplus S$.

Morphological Opening and Closing

Two common morphological transformations consist of combinations of erosion and dilation. The opening operation involves the application of erosion, followed by dilation. The effect of using a square- or disk-shaped structuring element for opening is to smooth boundaries, to break narrow isthmuses, and to eliminate small noise regions. The erosion operation reduces these features (and associated noise), and the subsequent dilation operation restores regions to their original size, now lacking the above-mentioned features.

The companion operation to opening is closing, and this involves the application of dilation, followed by erosion. The effect of using a square- or disk-shaped structuring element for closing is to smooth boundaries, to join narrow breaks, and to fill small holes caused by noise.

The difference between opening and closing is in the initial iteration, erosion, or dilation. The choice of operation depends on the image and the objective. For example, opening is used when the image has many small noise regions. It is not used for narrow regions where there is the chance that the initial erosion operation might disconnect regions. Closing is used when a region has become disconnected and the desire is to restore connectivity. It is not used when different regions are located closely such that the first iteration of dilation might connect them.

An iterative sequence of erosion and dilation, or dilation and erosion, operations will smooth boundaries and reduce noise. Application of the same number of dilation and erosion steps, performed alternately, as for opening and closing, ensures minimal net change in region size. Figure 4.1.1 illustrates the combination of iterations of opening, erosion, and dilation.

Other Morphological Considerations and Operations

The choice of the size of the structuring element S and that of the number of iterations represents a trade-off between the feature size and the number of required iterations. The larger the size, the fewer the number of iterations, but the more computation will be required on a single iteration. The most common sizes range between 3×3 and 5×5. The size of S should be no larger than the features it will affect if side effects are to be avoided.

The determination of whether dilation should precede erosion or vice versa depends on the type of features to be retained and the type of noise to be reduced. Erosion preceding dilation will tend to eliminate regions of ON-valued noise, but will also eliminate long or thin features. Dilation preceding erosion will reduce OFF-valued noise (holes), but will tend to smooth sharp features such as corners. We usually determine a compromise between noise reduction and feature retention by preliminary testing on representative images and visually observing how much noise is reduced and what features are retained.

Additional morphological transformations may be constructed in the form of combinations of the basic operations. We mention them here for the sake of completeness, but provide more specific methods later in this chapter to accomplish the same tasks. These are examples, we believe, in which the specific methods are more efficient than morphological operations.

The hit-or-miss transformation serves to detect features in the image that match the shape of the structuring element. The skeleton transformation reduces a region to its minimum number of connected ON-valued pixels in such a way that the original image may be regenerated when disks of known radii are centered on the skeleton pixels. The thinning transformation (Section 4.7) reduces a region shape to lines that are symmetrically contained within the shape. The boundary transformation (Section 4.3) finds those pixels of a region that are simultaneously adjacent to internal (ON) and external (OFF) pixels. The convex hull transformation (Section 4.6) produces a region representing the convex hull of the original region. The pruning transformation reduces short branches extruding from a region.

4.1.2   CELLULAR PROCESSING

In morphological processing, new pixel values are assigned on the basis of Boolean operations applied to a designated neighborhood of each pixel. In contrast, cellular processing relies on arithmetic evaluations of pixel values over a designated neighborhood and subsequent comparison with preselected thresholds. For each pixel, the result depends on two calculations. First, the number of neighborhood ON-valued pixels is calculated:

$$\phi(i, j, \eta) = \sum_{(n,m)\in\eta} I(n, m),$$

where η describes the neighborhood pixels around the center pixel. The second calculation is the number of eight-connected groups of pixels in the neighborhood $\chi(i, j, \eta)$. These results are compared with user-set thresholds T_ϕ and T_χ, respectively, and the center pixel is set to ON or OFF, depending on the outcome.

As for morphological processing, the most basic operations in cellular processing are erosion and dilation.

Cellular Erosion

Erosion is used to remove ON pixels at boundaries of regions and to increase the size of holes (OFF-valued regions within ON regions). Erosion entails setting an ON pixel to OFF if the number of neighborhood ON pixels falls short of the preselected threshold and if the following connectivity criterion is met:

$$I^{E_C}(i_0, j_0) = \begin{cases} \text{OFF} & \text{if } I(i_0, j_0) = \text{ON and } \phi(i, j, \eta) < T_\phi \text{ and } \chi(i, j, \eta) \text{ op } T_\chi \\ \text{ON} & \text{otherwise} \end{cases},$$

where op stands for an operand ($>, \leq, =$).

The most frequent choice of a threshold T_ϕ for erosion is half of the total number of neighborhood pixels, four. This tends to remove rough spurs and corners from region boundaries. A higher value of T_ϕ leads to faster and coarser erosion. A lower value of T_ϕ leads to slower and finer erosion; a value of 1 or 0 has no effect. In choosing the operand op and the threshold T_χ for connectivity, we must first determine whether to tolerate a change in connectivity in the output image. If connectivity is of no concern, we can neglect it by setting $\chi(i, j, \eta) \leq T_\chi \equiv 8$, a condition that is always true. If connectivity does matter, operand and threshold are chosen as follows.

If the center pixel is to be set to zero only if connectivity remains unchanged, the connectivity condition should be $\chi(i, j, \eta) = 1$. If, on the other hand, the center pixel should be set to zero only if connectivity does change, the criterion should be $\chi(i, j, \eta) > 1$. Other values of T_χ are useful in special cases to aid in the recognition of particular patterns: For instance, a trijunction can be recognized by a connectivity value of three when the mask is positioned at its center. Figure 4.1.3 shows the results of cellular erosion, both with and without retention of connectivity. When connectivity retention is set, the result in Fig. 4.1.3(b) maintains the same shape as the original;

(a)   (b)   (c)

Figure 4.1.3. Cellular erosion: (a) original image, (b) result of erosion with seven iterations, factor threshold of 2, and connectivity retention; (c) result of erosion with seven iterations, factor threshold of 2, and without connectivity retention.

however, when it is not set, the result is breaks in the regions. (To erode down to a single-pixel-width line with perfect retention of connectivity, use a thinning method designed for this purpose; see Section 4.7.)

Cellular Dilation

Dilation is used to add pixels at region boundaries or to fill in holes. Dilation entails setting an OFF pixel to ON if the number of neighborhood ON pixels equals or exceeds the preselected threshold $\phi(i, j, \eta)$ and if the following connectivity criterion is met:

$$I^{D_C}(i_0, j_0) = \begin{cases} ON & \text{if } I(i_0, j_0) = OFF \text{ and } \phi(i, j, \eta) \geq T_\phi \text{ and } \chi(i, j, \eta) \text{ op } T_\chi \\ OFF & \text{otherwise} \end{cases}.$$

The most frequent choice of a threshold T_ϕ for dilation is to set it equal to half of the total number of neighborhood pixels, four. This choice ensures that indentations at the region boundaries are to be filled in. A lower value of T_ϕ leads to faster and coarser dilation. A higher value of T_ϕ, up to eight, leads to slower and finer dilation; a value of eight yields no dilation.

Other Cellular Operations

Cellular processing usually involves several iterations of erosion and dilation operations, perhaps combined with Boolean image operations. Many of these combinations are similar to those for morphological processing. Thus, to perform smoothing, erosion and dilation iterations are alternated. The larger the number of iterations, the more pronounced the smoothing. An iterative sequence that uses erosion preceding dilation will tend to eliminate regions of ON-valued noise, but will also eliminate long or thin features. If dilation precedes erosion, this will eliminate OFF-valued holes, but will tend to smooth sharp features such as corners.

A number of specialized cellular processing operations have been introduced in the literature, although the same caveat applies to their use as that mentioned for morphological processing. That is, better algorithms usually exist that are specialized to the particular task. We will describe some of these in subsequent sections. One task is to find the number of connected regions (blobs) in an image containing many such regions or to locate approximate region centroids, for which a sequence of erosion operations may be used: T_ϕ should be set to 7 and the connectivity criterion should be $\chi(i, j, \eta) = 1$. This ensures that erosion is always performed for boundary cores until only isolated cores with no ON-valued neighbors remain. The number of these remaining cores represents the number of original regions, and their locations give rough estimates of region centroid positions. A faster and more effective method to determine the number of regions and a good measure of their centroids is by contour-based region detection (see Section 4.3).

References and Further Reading

Mathematical fundamentals of morphological transformations are discussed in J. Serra, *Image Analysis and Mathematical Morphology* (Academic, London, 1982); applications to image analysis are considered in R. M. Haralick, S. R. Sternberg, and X. Zhuang, "Image analysis using mathematical morphology," IEEE Trans. Pattern Anal. Mach. Intell. **9**, 532–550 (1987).

The application of cellular logic to image analysis is described in K. Preston Jr., M. J. B. Duff, S. Levialdi, P. E. Norgren, and J.-I. Toriwaki, "Basics of cellular logic with some applications in medical image processing," Proc. IEEE **67** (5), 826–856 (May 1979).

Programs

`cellog`

```
            performs cellular logic of input binary image.
  USAGE:    cl inimg outimg [-f FAC_THRESH] [-c CONNECTIVITY]
                           [-i ITERATIONS] [-o OPERATION] [-L]
ARGUMENTS:          inimg: input image (TIF)
                  outimg: output image (TIF)
 OPTIONS: -f FAC_THRESH: factor number threshold, above which
                         operation is performed (default = 5,
                         range= 0-7); the smaller the threshold
                         value, the larger the effect
        -c CONNECTIVITY: flag to maintain connectivity; default
                         is not set. NOTE: this does not guarantee
                         retention of, only maintains it better
                         than not. If absolute retention is
                         required, use KFILL or THIN program,
                         depending on purpose
        -i ITERATIONS: number of iterations to perform
                         (default = 1)
         -o OPERATION: dilation (1), erosion (2), closing(3),
                         or opening(4) closing operation is
                         alternating dilation--erosion; opening
                         operation is alternating erosion
                         --dilation; for opening or closing, the
                         number of iterations is the sum of the
                         comprising dilation and erosion
                         iterations.
                    -L: print Software License for this module
```

morph

performs morphological processing on binary image.

USAGE: morph inimg outimg [-i ITERATIONS] [-o OPERATION] [-s Size] [-L]

ARGUMENTS:

imimg: input image (TIF)

outimg: output image (TIF)

OPTIONS: -i ITERATIONS: number of iterations to perform (default = 1)

-o OPERATION: dilation(1), erosion(2), closing(3), or opening(4); closing operation is alternating dilation--erosion; opening operation is alternating erosion--dilation; for opening or closing, the number of iterations is the sum of the comprising dilation and erosion iterations.

-s Size: size of structuring element--must be odd (≥ 3) (default = 3);

-L: print Software License for this module

4.2 Binary Noise Removal

Typical Application(s) – reduction of noise in binary images.

Key Words – salt and pepper, *k* Fill.

Related Topics – noise reduction (Section 3.2).

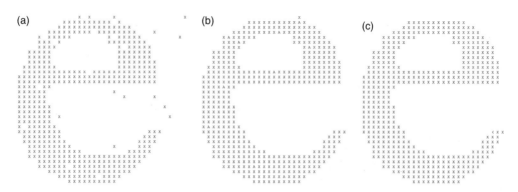

Figure 4.2.1. Pictorial Example. Removal of salt-and-pepper noise: (a) original image (original ON-valued pixels have the symbol X); (b) an intermediate result after several iterations: added pixels are marked by the symbol A, (c) final result: small regions are eliminated, small holes are filled, and noisy indentations and bumps on the region contours are smoothed.

Most image analysis methods benefit greatly from the reduction of noise in the preprocessing step. Noise reduction can mean a substantial improvement in the reliability and the robustness of feature detection and object recognition and also can serve to reduce the memory requirement for image storage.

A common manifestation of noise in binary images takes the form of isolated ON pixels, or pixel regions, in a background of OFF pixels, or vice versa. We use the descriptive name salt-and-pepper noise, also sometimes referred to as speckle noise, shot noise, or simply dirt. The process of removing this type of noise is called filling: Each isolated salt-and-pepper "island" is filled in by the value of the surrounding "sea."

4.2.1 STRATEGY: THE *k* FILL FILTER

The most straightforward and commonly used method for binary noise reduction is just the elimination of a single pixel of noise. A 3×3 mask is applied to each pixel, and if the center pixel value is different from all the neighboring pixels, the center value is set to that of the neighbors.

While this simple technique eliminates single-pixel noise features, larger noise features will not be eliminated, particularly when these noise features occur at the boundary of a region in the form of "peninsulas" or "inlets."

A more general filter, called k Fill, is designed to reduce isolated noise and noise on contours up to a selected limit in size. The k of kFill refers to a size adjustment parameter. Other kFill parameters can be set to control rounding of the filtered features. Many synthetic shapes display 90° corners: To preserve these, rounding must be minimized,

139

and the default parameters of kFill are chosen accordingly to retain corners of $90°$ or greater. Greater noise reduction can be achieved at the risk of rounding these corners, and kFill parameters may be set accordingly.

4.2.2 IMPLEMENTATION

Filling operations are performed within a $k \times k$ window that is applied at each image pixel in raster-scan order. This window consists of an interior $(k-2) \times (k-2)$ region, the core, and the exterior $4(k-1)$ pixels on perimeter, referred to as the neighborhood. The filling operation entails setting all values of the core to ON or OFF, depending on pixel values in the neighborhood.

Noise reduction is performed iteratively. Each iteration consists of two subiterations, one performing ON fills and the other OFF fills. When no filling occurs on two consecutive subiterations, the process stops automatically. Iterations of noise reduction are shown in Fig. 4.2.1.

The decision to fill with ON (OFF) requires that all core pixels to be OFF (ON) and it furthermore depends on three variables, n, c, and r. These are determined from the neighborhood pixels. For a fill value equal to ON (OFF), n equals the number of ON (OFF) pixels in the neighborhood, c denotes the number of connected groups of ON pixels in the neighborhood, and r represents the number of corner pixels that are ON (OFF). The conditions on n and r derive from the window size k.

In the default implementation, filling occurs when the following conditions are met:

$$(c = 1) \text{ AND } [(n > 3k - 4) \text{ OR } (n = 3k - 4) \text{ AND } r = 2].$$

The significance of these conditions is as follows:

- $n > 3k - 4$: This controls the degree of smoothing: A reduction of the threshold for n leads to enhanced smoothing.
- $n = 3k - 4$ AND $r = 2$: This ensures that corners of $\leq 90°$ are not rounded. In the absence of this condition, greater noise reduction would occur, but corners may be rounded.
- $c = 1$: This ensures that filling does not change connectivity (i.e., does not join two regions together or separate two parts of the same connected region). If connectivity were allowed to change (i.e., no constraint on c), greater smoothing would occur, but without the assurance that the number of distinct regions would remain constant.

In general, these conditions are conservative, erring on the side of retaining image features at the expense of foregoing greater noise reduction. A procedure for establishing conditions in a particular application is to start with the conditions as stated here and to experiment with relaxing them until a middle ground is found with good noise reduction and little unwanted alteration of the image.

Our implementation performs run-length encoding of the input raster image on a first pass to yield more efficient performance in subsequent iterations. Since each iteration is still costly and since much less noise is removed on later iterations, a practical choice in applying the method is to set a maximum number of iterations (determined by experimentation on a particular image) or to stop the iterations when noise is reduced by a chosen percentage.

The *k*Fill filter has been used extensively to reduce noise in document images. The *n* variable is set relative to the text size to remove as much noise as possible while still retaining dots, periods, and serifs. By performing *k*Fill in the preprocessing step, we may achieve a significant data reduction, resulting in an improvement in subsequent tasks such as transmission and compression of the image or in analyzing the features (in the manner performed for optical character recognition). For document images, data size is reduced by an amount of 5%–20% (this is very dependent on the original number of regions, application, and quality of the data).

Program

kfill

```
           removes spatial noise from binary input image
     USAGE: kfill inimg outimg [-f FDIFF] [-c] [-n NITER] [-r R]
                           [-k K][-l] [-d] [-L]
ARGUMENTS: inimg: input image (TIF)
           outimg: output image with values in range [0, 255] (TIF)
   OPTIONS: -f FDIFF: amount of increase or decrease (negative) of
                      factor no. from default
               -c: when set, retain connectivity (default is
                   otherwise)
               -e: when set, do NOT retain end points (default is
                   otherwise)
             -r R: percentage of max noise to leave remaining on
                   any iteration before stopping (default is 0)
         -n NITER: maximum number of iterations
                   (default max = 20)
            -k K: window size for kxk mask
                  (k >= 3, default = 3)
              -l: first iteration fill value opposite from
                  default (default is 0 value)
              -d: to display results of each iteration (< 40x40
                  image)
              -L: print Software License for this module
```

4.3 Region Detection

Typical Application(s) – segmentation of images by delineation of regions.

Key Words – region detection, segmentation; contour representation, cumulative angular bend, curvature point; region filling or coloring, connected component labeling, contour detection.

Related Topics – object shape analysis (Section 4.4), thinning (Section 4.7).

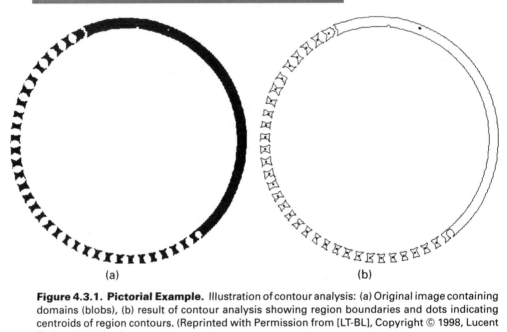

(a) (b)

Figure 4.3.1. Pictorial Example. Illustration of contour analysis: (a) Original image containing domains (blobs), (b) result of contour analysis showing region boundaries and dots indicating centroids of region contours. (Reprinted with Permission from [LT-BL], Copyright © 1998, Lucent Technologies Inc.)

Although humans view a binary image as comprising regions that convey characters, lines, or portions of a picture, the image as presented to a computer is no more than an array of individual ON and OFF pixels. To go from these individual pixels to connected binary regions, or blobs, we must identify groups that consist of mutually adjacent or connected ON pixels. This process is referred to as region detection, binary segmentation, or simply region finding.

A given region may or may not correspond to the object of interest in its entirety. For instance, the letter i is composed of two regions, and the process of recognizing this letter is first to find each region, and then to associate them as a dot over a vertical line, which together represent the letter.

There are two basic strategies to region detection. One is contour based: A region is identified by determination of its boundary pixels. The other is region based: Each image pixel is examined in the context of its neighbors and a region is grown by addition of new pixels if they are connected. Both strategies have relative advantages and disadvantages, but the choice is dictated primarily by the convenience of either representation for subsequent analysis.

4.3.1 CONTOUR-BASED DETECTION

The contour of a region, also referred to as border or perimeter, consists of those of the region's ON pixels that are adjacent to OFF pixels. The binary image in Fig. 4.3.1(a) contains readily identifiable regions; the panel in Fig. 4.3.1(b) contains the contour image of the original. We describe two procedures to determine the contour of regions in an image.

Contour Tracing ("Tracking Bug")

A search is made in raster-scan order for an ON pixel that follows an OFF pixel; we presume that the search is begun from an OFF-valued background. This is guaranteed to lie on a region contour. We label this with an integer value, CONTOUR! = ON! = OFF. We begin contour tracing by centering a 3×3 mask on the CONTOUR pixel: all eight neighbors of the central pixel are examined in clockwise order around the mask. The first ON pixel that is followed by an OFF pixel is chosen to be the next pixel in the contour and accordingly labeled. The mask is now placed on this new contour pixel and the clockwise examination repeated to identify the next contour pixel. By this iterative process, the contour is traced around a region. The contour is fully traced when examination of the 3×3 mask encounters the starting pixel of the contour. Following completion of this trace, the raster search is continued from its previous location to locate the next OFF/ON pixel pair. In this manner, all region contours will have been traced by the end of the raster search.

The outcome of this contour detection procedure is a clockwise sequence of contour pixel coordinates for each region and a counterclockwise sequence for each hole. A hole is an OFF region enclosed in an ON region. Nesting refers to the occurrence of regions within regions, as in the example of recursive nesting of regions and holes is a bull's-eye pattern. The general rule for this is that a contour encloses an ON region in counterclockwise order and an OFF region in clockwise order. If only the outermost region contour is desired, then the contour detection process can be modified as follows. When performing the raster-scan search within the boundary of a contour, no other contour is started until that outer contour region is exited. This is recognized by a CONTOUR-pixel followed by an OFF-pixel.

Closely related to the contour representation of a region is the representation in terms of the backbone or skeleton, obtained by way of a thinning operation (Section 4.7). Both contour and skeleton representations facilitate more efficient storage of objects. While either representation may be chosen for any binary object, contour detection is more suitable for compact (rounded) shapes, while thinning is preferable for elongated shapes. There is no sharp distinction between the two types of shapes; however, viewing object shapes as filled regions or as lines drawn at arbitrary thickness, we might say that contour detection is most useful for the analysis of regions, while thinning focuses on lines. We return to this issue from the other perspective, that of thinning, in Section 4.7.

Boundary Detection and Representation by Local Curvature Points

More efficient than storing the contour as a sequence of pixel coordinates for all contour points is a representation based on the directions between neighboring contour pixels. A procedure of this type that combines detection and efficient representation of region contours is based on the identification of special contour points marking changes in the local contour directions [Zahn, 1969]. Only these local curvature points are retained in the contour representation; all others are ignored. For smooth shapes, this representation has the advantage of being very sparse.

Local curvature points are detected by application of suitable Boolean masks applied to each boundary point. Boundary points are first detected from the image, as previously described. Boolean operations, listed in Table 4.3.1, are applied to the resulting pattern of CONTOUR pixels to extract the directions of the incoming and outgoing segments of the contour to which a given boundary point belongs. Each of these directions is represented according to a widely used chain code (further discussed in Section 5.1) by an integer between 0 and 7 (Fig. 4.3.2). If incoming and outgoing directions differ, the inspected boundary point is termed a curvature point (the more general concept of curvature and the detection of curvature maxima in a given line are topics that we discuss in more detail in Section 5.4.).

Each curvature point is entered into a list in which it is represented by four integers: its (x, y) pixel coordinates and the direction values for incoming and outgoing contour

Table 4.3.1.   Boolean Determination of Local Edge Direction with Six-Element Masks [Lunscher and Beddoes, 1987]

	HORIZONTAL MASKS: $D1$ $D2$ $D3$ $D4$ $D5$ $D6$		
NEIGHBORHOOD CONFIGURATION	IN-EDGE DIRECTION	NEIGHBORHOOD CONFIGURATION	OUT-EDGE DIRECTION
$D5 \wedge \bar{D}1 \wedge \bar{D}2 \wedge \bar{D}4$	1	$D3 \wedge D5 \wedge \bar{D}5$	1
$D2 \wedge D6 \wedge \bar{D}5$	3	$D2 \wedge \bar{D}1 \wedge \bar{D}4 \wedge \bar{D}5$	3
$D2 \wedge D3 \wedge \bar{D}5 \wedge \bar{D}6$	4	$D1 \wedge D2 \wedge \bar{D}4 \wedge \bar{D}5$	4
$D2 \wedge \bar{D}3 \wedge \bar{D}5 \wedge \bar{D}6$	5	$D2 \wedge D4 \wedge \bar{D}5$	5
$D1 \wedge D5 \wedge \bar{D}2$	7	$D5 \wedge \bar{D}2 \wedge \bar{D}3 \wedge \bar{D}6$	7
$D1 \wedge D4 \wedge \bar{D}1\bar{D}2$	0	$D5 \wedge D6 \wedge \bar{D}2 \wedge \bar{D}3$	0
	VERTICAL MASKS: $D2$ $D3$ $D5$ $D6$ $D7$ $D8$		
$D6 \wedge D7 \wedge \bar{D}5$	1	$D6 \wedge \bar{D}3 \wedge \bar{D}2 \wedge \bar{D}5$	1
$D6 \wedge D8 \wedge \bar{D}5 \wedge \bar{D}7$	2	$D3 \wedge D6 \wedge \bar{D}2 \wedge \bar{D}5$	2
$D6 \wedge D8 \wedge \bar{D}5 \wedge \bar{D}7$	3	$D6 \wedge D2 \wedge \bar{D}5$	3
$D3 \wedge D5 \wedge \bar{D}6$	5	$D5 \wedge \bar{D}6 \wedge \bar{D}8 \wedge \bar{D}7$	5
$D2 \wedge D5 \wedge \bar{D}3 \wedge \bar{D}6$	6	$D5 \wedge D7 \wedge \bar{D}6 \wedge \bar{D}8$	6
$D5 \wedge \bar{D}3 \wedge \bar{D}6\bar{D}2$	7	$D8 \wedge D5 \wedge \bar{D}6$	7

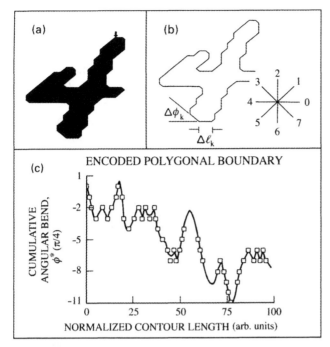

Figure 4.3.2. Boundary encoding in terms of tangential representation $\phi^*(l)$, as discussed in the text. The binary image in (a) is described by the polygonal contour $\{(\Delta\phi_k, \Delta l_k), 1 \leq k \leq N\}$, displayed in (b). The arrow in (a) marks the (arbitrarily chosen) origin of arc length $l = 0$. The cumulative angular bend function $\phi^*(l_k)$ for the object in (a) is shown in (c), along with a cubic spline fit (solid curve) to guide the eye. (Reprinted with Permission from [Seul et al., 91], Copyright © 1991 American Physical Society.)

segments. Following completion of the raster scan, the collected list of curvature points, generally not all part of the same domain boundary, must be linked and sorted. The result is a linked list of curvature points for each of the distinct boundary curves in the scanned image.

The data structure representing each linked list contains the coordinates of a starting point, generally the first curvature point of the given boundary encountered in the raster scan, and the direction of the boundary segment leaving this point, as well as two values for each curvature point or vertex: the length Δl_k of the linear segment connecting it to the previous vertex and the change in direction $\Delta\phi_k$ between incoming and outgoing edges, measured in units of $\pi/4$ (Fig. 4.3.2). This (integer) representation permits a very efficient compression of the information contained in the original image into one-dimensional objects, namely, a representation of the form $\{(\Delta l_k, \Delta\phi_k), 1 \leq k \leq N\}$ for each contour. This stored representation suffices to reconstruct the vertex coordinates $\{(x_k, y_k), 1 \leq k \leq N\}$ and the lines joining these vertices of the original contour.

Tangential and Radial Contour Representations

On the basis of a region contour, a number of region features can be determined, and this is discussed in detail in Section 4.4. To extract region shape features that are unbiased as to a given contour length or to position, orientation, or size of a region, it is common practice to base shape analysis on a representation of the region boundary that is periodic in the contour length (L) and invariant under translation, rotation, and dilation. We now introduce two representations that are suitable for this purpose.

Referring to Fig. 4.3.2, we define $\phi(l)$ as a function measuring the cumulative angular bend along a boundary curve; l denotes arc length along this curve, $0 \leq l \leq L$. The function $\phi(l)$ measures the difference between the angular direction $\theta(l)$ of the contour tangent at point l and the angular direction $\theta(l = 0) \equiv \theta_0$ at an (arbitrarily chosen) starting point: $\phi(l) \equiv \theta(l) - \theta_0$. This is illustrated in Table 4.3.2 for the image of the numeral 4 shown in Fig. 4.3.2. For a clockwise-oriented contour, $\phi(L) = -2\pi$; for the opposite orientation, $\phi(L) = 2\pi$.

We also define an azimuthal angle t that corresponds to arc length: As a closed contour is traced from the starting point, t varies between 0 and 2π. For a circular object, the difference in tangent angles at two points along the contour equals the azimuthal angle, $t \in [0, 2\pi]$: $\theta(t) - \theta_0 = -t$. This feature of the circle may be used to define a corrected cumulative angular bend function:

$$\phi^*(t) \equiv \phi\left\{\frac{L}{2\pi}t\right\} + t; \quad t \in [0, 2\pi]. \tag{4.1}$$

By construction, $\phi^*_{\text{circle}} = 0$. That is, ϕ^* represents a measure of the deviation of a contour configuration from that of a circle. This tangential contour representation is based entirely on angular measurements along the contour. This has the advantage of not requiring the choice of a reference point for the measurement of radial distances from that point to a given position on the contour. Such a choice is easy when symmetric distortions of a circular object are to be analyzed, but it can be difficult in other cases when the shape of a distorted object is to be analyzed in such a way that no bias in the form of a net translation is introduced. This is further discussed in connection with the evaluation of Fourier descriptors in Section 4.4.

A disadvantage of the tangent representation is its sensitivity to degradations of region contours such as those generated in the process of global thresholding and image binarization (Section 3.9): This pixel noise gives region contours a jagged appearance and thus greatly increases the number of curvature points defining a boundary. Before further analysis, it may therefore become desirable to apply a smoothing procedure to this noisy polygonal boundary. We discuss a variety of available methods in Section 5.3.

Alternatively, we may elect a radial contour representation that is less sensitive to pixel noise. This representation, $\delta\rho^*(l)$, is constructed to measure radial deviations of a given contour from that of a circular reference state. Importantly, the radial representation requires the choice of a reference point; we define it here with respect to the centroid, $\mathbf{v}_c \equiv (x_c, y_c)$ of the region of interest.

Referring to Fig. 4.3.3, we define for a closed polygonal contour $\{(x_k, y_k);$ $1 \leq k \leq N\}$, the representation $\delta\rho^*(l_k)$ in terms of the distance between its centroid and the vertex $\mathbf{v}_k \equiv \mathbf{v}(l_k) = (x_k, y_k)$, corrected by the quantity ρ_0, which denotes the radius of a reference circle whose area is identical to that of the singly connected region bounded by the polygonal contour $\{(x_k, y_k)\}$. That is, $\rho_0 \equiv \sqrt{m_{00}/\pi}$, where m_{00}

Table 4.3.2. Tangential Encoding of the Numeral 4 (see Fig. 4.3.2)

INDEX (K)	OUT-EDGE DIRECTION	BEND [†](CW:−, CCW:+)		EDGE LENGTH	
		$\Delta\Phi_k\,(\pi/4)$	$\Phi_k\,(\pi/4)$	Δl_k	$l_k \equiv \sum_{i=1}^{i=k}\Delta l_i$
1	7	−1	−1	2	2
2	6	−1	−1	$\sqrt{2}$	$2+\sqrt{2}$
3	5	−1	−3	4	$6+\sqrt{2}$
4	6	+1	−2	$4\sqrt{2}$	$6+5\sqrt{2}$
5	5	−1	−3	4	$10+5\sqrt{2}$
6	6	+1	−2	$2\sqrt{2}$	$10+7\sqrt{2}$
7	7	+1	−1	4	$14+7\sqrt{2}$
8	0	+1	0	$\sqrt{2}$	$14+8\sqrt{2}$
9	7	−1	−1	4	$18+8\sqrt{2}$
10	5	−2	−3	$\sqrt{2}$	$18+9\sqrt{2}$
11	4	−1	−4	$3\sqrt{2}$	$18+12\sqrt{2}$
12	5	+1	−3	4	$22+12\sqrt{2}$
13	6	+1	−2	$3\sqrt{2}$	$22+25\sqrt{2}$
14	5	−1	−3	2	$24+15\sqrt{2}$
15	6	+1	−2	$2\sqrt{2}$	$24+17\sqrt{2}$
16	5	−1	−3	4	$28+17\sqrt{2}$
17	6	+1	−2	$2\sqrt{2}$	$28+19\sqrt{2}$
18	5	−1	−3	2	$30+19\sqrt{2}$
19	4	−1	−4	$\sqrt{2}$	$30+20\sqrt{2}$
20	3	−1	−5	4	$34+20\sqrt{2}$
21	2	−1	−6	$3\sqrt{2}$	$34+23\sqrt{2}$
22	1	−1	−7	2	$36+23\sqrt{2}$
23	2	+1	−6	$2\sqrt{2}$	$36+25\sqrt{2}$
24	1	−1	−7	2	$38+25\sqrt{2}$
25	2	+1	−6	$2\sqrt{2}$	$38+27\sqrt{2}$
26	3	+1	−5	2	$40+27\sqrt{2}$
27	4	+1	−4	$\sqrt{2}$	$40+28\sqrt{2}$
28	3	−1	−5	12	$52+28\sqrt{2}$
29	2	−1	−6	$\sqrt{2}$	$52+29\sqrt{2}$
30	1	−1	−7	2	$54+29\sqrt{2}$
31	0	−1	−8	$9\sqrt{2}$	$54+38\sqrt{2}$
32	1	+1	−7	2	$56+38\sqrt{2}$
33	0	−1	−8	$2\sqrt{2}$	$56+40\sqrt{2}$
34	7	−1	−9	2	$58+40\sqrt{2}$
35	6	−1	−10	$\sqrt{2}$	$58+41\sqrt{2}$
36	5	−1	−11	2	$60+41\sqrt{2}$
37	7	+2	−9	$4\sqrt{2}$	$60+45\sqrt{2}$
38	0	+1	−8	$\sqrt{2}$	$60+46\sqrt{2}$
39	1	+1	−7	4	$64+46\sqrt{2}$
40	2	+1	−6	$3\sqrt{2}$	$64+49\sqrt{2}$
41	1	−1	−7	4	$68+49\sqrt{2}$
42	2	+1	−6	$2\sqrt{2}$	$68+51\sqrt{2}$
43	1	−1	−7	2	$70+51\sqrt{2}$
44	2	+1	−6	$2\sqrt{2}$	$70+53\sqrt{2}$
45	1	−1	−7	2	$72+53\sqrt{2}$
$46 \equiv 0$	0	−1	−8	$5\sqrt{2}$	$72+58\sqrt{2}$

[†] CW, clockwise; CCW, counterclockwise.

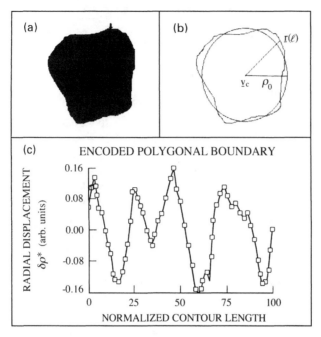

Figure 4.3.3. Boundary encoding in terms of radial representation $\delta\rho^*(l)$, as discussed in the text. The binary image in (a) is described by the polygonal contour $\mathbf{v} = \mathbf{v}(l)$, shown in (b). The arrow in (a) marks the arbitrarily chosen origin of arc length, $l = 0$. When the centroid $\mathbf{v}_c$ is selected as an origin, the function $\delta\rho^*(\mathbf{v}_k)$ is constructed from $|\mathbf{v}(l_k) - \mathbf{v}_c|$; ρ_0 denotes the radius of a reference circle whose area is identical to that enclosed by the polygonal contour. The radial displacement function $\delta\rho^*(l_k)$ describes the radial distortion of the object in (a) relative to a circle and is shown in (c) along with a cubic spline to guide the eye (solid curve). (Reprinted with Permission from [Seul et al., 91], Copyright © 1991 American Physical Society.)

is the zero-order moment of $\{(x_k, y_k)\}$ giving the enclosed area (see also Section 4.4). By construction, $\delta\rho^*_{\text{circle}} = 0$.

The radial contour representation thus has the form:

$$\delta\rho^*(l) \equiv |\mathbf{v}(l) - \mathbf{v}_c| - \rho_0. \tag{4.2}$$

4.3.2   REGION-BASED DETECTION

In comparison with contour-based methods, the principal benefit of region-based methods of detection and analysis is their relative insensitivity to shape degradation and noise. For example, pixel noise often introduces spurious features that seriously affect the fidelity of tangential or radial representations of a region contour and can eliminate some of the advantages of a sparse representation. Region-based methods of detection and analysis, relying on the entire set of interior region pixels, offer an alternative approach.

Here we present two region-based methods. The first, connected component labeling or filling, is initiated in the interior of a region and relies on pixel connectivity to identify and mark every region pixel; the second, disk scanning, is based on a raster scan of the image and provides a simple and rapid procedure for the common task of locating centroids of circular regions.

Component Labeling

For purposes of displaying or analyzing a region, it may be desirable to identify explicitly each pixel in that region, rather than just the contour. This is referred to as connected component labeling, region filling, or region coloring. As these names imply, pixels within each region of an image are assigned a specific label or color. As a result, different regions in the image are readily discriminated. Should it be necessary, for example for purposes of shape or moment analysis (Sections 4.4 and 4.5), to obtain coordinates for all pixels in each region, this is a suitable data structure.

The method involves a two-pass process. First, the image is examined in raster order. When a pixel is found to be ON, neighboring pixels to the left (1) and above (3) are examined. Four situations can occur: None of these neighbors is ON, and the current pixel is set to a new label; one of the neighbors is ON, and the current pixel is given the same label; more than one neighbor is ON and labeled equivalently, so the current pixel is given the same label; finally, two or more neighbors are ON, but labeled differently, so the current pixel is set to one of the labels, and all labels are put in an equivalence class to indicate that these regions are in fact the same region on the basis of being connected. When this first pass is completed, the number of connected components is given by the number of equivalence classes, plus the number of labels not in an equivalence class. All labels in each equivalence class are therefore merged to a single label. In the second pass, each image pixel is visited again and labels are reassigned to match the final label to which all have been merged.

Centroids of Convex Regions

In many instances, it may be of interest to analyze the spatial statistics of collections of objects such as cells or colloidal spheres, a topic we discuss in detail in Chap. 6. The principal objective in this case is to detect the presence of all objects and to assign each a precise location (see also Section 3.6). In many cases of practical interest, objects have convex shapes, and it is straightforward to determine their centroid location by the following simple variant of region filling.

Consider an image containing a set of objects corresponding to ON regions on an OFF background. A raster scan of the image is performed. When a contour point is encountered, the scan is interrupted and a line-by-line filling routine is initiated. A particularly simple variant of region filling is adequate in the common situation in which regions of interest are singly connected (no holes) and convex: In this case, a horizontal line will intercept the region contour at most twice (once in tangent points). This information forms the basis for a very simple line-by-line filling algorithm that proceeds to fill the encountered region by setting its pixels to OFF. As with the more general region filling methods, this procedure has the advantage of being largely unaffected by contour noise in making a robust estimate of the region centroid.

Fast Peak Detection for Circular Regions

In some commonly encountered instances – analyzing a collection of cells or colloidal spheres in a microscope image, say – the objects, while spatially extended, are known to be of a given (average) size and to exhibit a circular shape or, more generally,

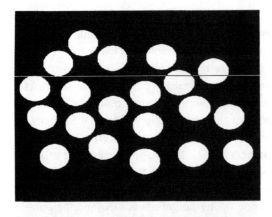

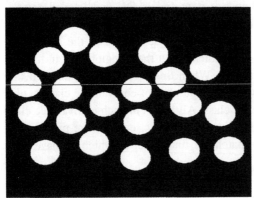

Figure 4.3.4. Illustration of fast peak detection (disk-scanning) algorithm. A scan line is stepped through the image, intersecting several circular domains; two successive positions of the scan line are shown. From the set of intersection points of the scan line with the contour of each disk, a disk centroid is estimated.

an azimuthally symmetric intensity distribution. While this information may be taken into account in optimizing the design of template matching strategies (Section 3.8), a far more efficient strategy, described here, takes explicit advantage of this knowledge in constructing estimates of a circular region's centroid location from a small number of contour points.

Thus, for perfect circular object of known radius, defined by an uncorrupted contour, the centroid location may be computed from just three contour points. If such a contour is available, for example as a result of advanced edge detection methods (Section 3.5) or other preceding processing steps, the centroid may be identified from just two (scan) lines intersecting the object. In general, a larger sample of contour points will be required to compensate for uncertainty in the definition of noisy contours or for uncertainty in the definition of object radii. The statistical character of this procedure implies that the fidelity of the estimate increases with the size of the sample of contour points. In comparison with template matching methods, very substantial speed enhancements result from the fact that the number of samples scales linearly with object radius, rather than with object area as in template matching. The present method of fast peak detection combines aspects of contour-based detection and region-based detection.

The underlying simple strategy of the algorithm is illustrated in Fig. 4.3.4. For a given collection of (near-) circular objects, or more generally, objects of azimuthally symmetric intensity distribution, a set of raster scans is executed. To ensure that each object is intersected at least n times and thus to generate at least $2n$ contour points from which to estimate the center location, the horizontal scan interval Δi is simply chosen such that $n\Delta i \leq 2\langle R \rangle$, where $\langle R \rangle$ denotes the mean radius of the objects in the image. Hence, to ensure that six contour points will be available to estimate each region's center location, Δi must be chosen according to the condition $2\langle R \rangle/4 < \Delta i \leq 2\langle R \rangle/3$.

Two lists of data structures are maintained during a single pass of rastering the image with scan interval Δi: In the course of each horizontal scan, region contour points are detected by application of an appropriate one-dimensional edge detection filter of suitably chosen size to match the prevailing boundary width (Section 3.5).

Systematic improvements may be obtained if the desire to increase precision in localizing the region center outweighs the computational cost incurred in the execution of additional raster scans, in the application of more extended edge detection filters to locate contour pixels, or in more extensive fitting of circular arcs to the set of contour points associated with a given object. Finally, the set of region center coordinates may be subjected to additional refinement by subsequent region matching.

References and Further Reading

[Lunscher and Beddoes, 1987] W. H. H. J. Lunscher and M. P. Beddoes, "Fast binary-image boundary extraction," Comput. Vis. Graph. Image Process. **38**, 229–257 (1987).

[Seul et al., 1991] M. Seul, M. J. Sammon, and L. R. Monar, "Imaging of fluctuating domain shapes: methods of image analysis and their implementation in a personal computing environment," Rev. Sci. Instrum. **62**, 784–792 (1991).

[Zahn, 1969] C. T. Zahn, "A formal description for two-dimensional patterns," in *Proceedings of the International Joint Conference on Artificial Intelligence*, D. E. Walker and L. M. Norton, eds., Washington, D. C. (1969), pp. 621–628; C. T. Zahn and R. Z. Roskies, "Fourier descriptors for plane closed curves," IEEE Trans. Comput. **C-21**, 269–281 (1972).

Programs

contour

```
            identifies contours, or boundaries, of regions in a
            binary image, and determines features of the regions.
     USAGE: contour inimg outimg [-d DISPLAY][-m MAX_CONT_LENGTH]
            [-L]
 ARGUMENTS: inimg: input image (TIF)
            outimg: output image (TIF)
   OPTIONS: -d DISPLAY: display just centroids (1) or both contours
                        and centroids (2); default displays just
                        contours
     -m MAX_CONT_LENGTH: maximum contour length in pixel connections
                        (default = 10000)
                    -L: print Software License for this module
```

xcp

> detects and encodes contours, or boundaries, of regions in a binary image, using an algorithm by C. T. Zahn [1969].

USAGE: xcp inimg [-L]

ARGUMENTS: inimg: input image (TIF format)

OPTIONS:
> -L: print Software License for this module

---

xah

XAH (Area Histogram)

> constructs area moments and area histogram, of image containing domains ("blobs"); the output image is a point pattern representing domain centroids; also evaluates histogram of domain areas

USAGE: xah inimg outimg [-a x1 y1 x2 y2] [-L]

ARGUMENTS: inimg: input image filename (TIF)

outimg: output image filename (TIF)

OPTIONS: -a x1 y1 x2 y2: upper left (x1, y1), lower right (x2, y2)
> coordinates of area to scan (int)
>
> -L: print Software License for this module

---

xcc

> determines, in an efficient manner, estimates of center positions for circular objects by taking advantage of knowledge about the circular object shape and object size, the latter in the form of known radius, R, or mean and variance of the distribution of radii;

USAGE: xcc inimg outimg [-t file] [-w file] [-s rad]
> [-n lines] [-b] [-z pix] [-i imgtype]
> [-f len] [-L]

ARGUMENTS: inimg: input image filename (TIF)

outimg: ouput image filename (TIF)

OPTIONS: -t file: read test input from file fn (.tpl)
> -s, -n options disabled
>
> -w file: write disk parameters to file fn.dsk
>
> -s rad: supply estimated (mean) disk radius (in pix)
>
> -n lines: supply minimum number of scan lines to sample each disk (must exceed 2)
>
> -b: do not display disk boundary
>
> -z pix: zero a border strip of width pix (default: 1)
>
> -i str: specify image type, B(BINARY) (default) or G (GRAY)
>
> -f len: specify edge filter length, (odd)l (default: 5)
>
> -L: print Software License for this module

---

xrg (Region Growing)

> evaluates the area of each region in a binary image by pixel filling using an interior point as the starting point.

USAGE: xrg inimg point_file [-L]

ARGUMENTS: inimg: input image filename (TIF)

> point_file: List of (x,y) coordinates of one interior point for each region

OPTIONS: -L: print Software License for this module

4.4 Shape Analysis: Geometrical Features and Moments

Typical Application(s) – contour analysis of objects or domains and evaluation of simple geometrical features as well as moments.

Key Words – polygonal representation, tangential and radial contour representation, global shape descriptor, curvature point, curvature energy; moments of inertia, moment invariants, recursive evaluation.

Related Topics – edge detection (Section 3.4); spectral shape analysis (Section 4.5), convex hull (Section 4.6); polygonalization (Section 5.3), critical point detection (Section 5.4).

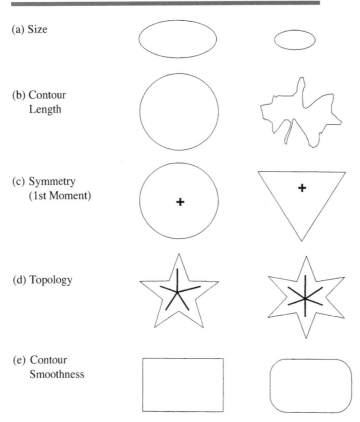

(a) Size

(b) Contour
   Length

(c) Symmetry
   (1st Moment)

(d) Topology

(e) Contour
   Smoothness

Figure 4.4.1. Pictorial Example. Illustration of object discrimination on the basis of simple shape descriptors.

Within the areas of pattern recognition and computer vision, a variety of powerful algorithms have been developed in recent years to recognize and match image objects by attributes and features of their shapes. In the specialized field of optical character recognition, there exists a highly sophisticated level of feature extraction to handle the variety of shapes that describe alphabets and other symbols. However, in contrast to the complexities involved in analyzing printed characters, in applications in physics, chemistry, and biology, one most often encounters a smaller set of simple shapes and patterns for which circles and lines serve as the reference shapes.

In this section, we describe shape descriptors that are constructed from either the interior pixels or the boundary of a binary region (Section 4.3). These include area, perimeter, and compactness as well as contour curvature. The spectral analysis of region boundary configurations in terms of Fourier descriptors represents an advanced mode of shape analysis, which is described in Sections 4.5.

4.4.1 GLOBAL SHAPE DESCRIPTORS

There are many global shape descriptors, and we cannot hope to offer a comprehensive list here. When a region shape exhibits very detailed features by which it must be distinguished from other regions, as in the case of optical character recognition, special descriptors and other tools of analysis are generally required. In other cases, objects of known shape are to be detected in the presence of obscuring elements, a task for which template matching (Section 3.8) may be a more appropriate alternative to shape analysis. In this subsection, we present a sampling of shape descriptors that are commonly encountered and prove particularly useful in practice when the application calls for the comparison with simple reference shapes such as circles and lines.

The objective in selecting or designing a shape descriptor is that it appropriately describe the inherent features of the shapes of interest and that it serve as a good discriminator to differentiate among all the shapes involved. Figure 4.4.1 shows a number of images containing pairs of shapes. Associated with each image are the values of a shape descriptor that discriminates well between the two shapes.

Geometrical Descriptors: Perimeter, Area, Compactness

One of the simplest ways to describe shapes is by shape metrics. For instance, the area measurements (number of ON-valued pixels) of a region helps to distinguish between differently sized objects. The most obvious example is the discrimination between smaller regions of noise and larger objects of interest. This serves the same purpose as the noise reduction procedures of Section 4.2, which use more general methods.

Instead of evaluating the area of a region by adding the number of interior pixels, we may also determine the area from a polygonal representation of the contour. If the contour pixels are in a sequence, $C \equiv \{(x_k, y_k); 1 \leq k \leq N\}$, then the area is determined by the expression $A = 1/2 \sum_{k=1}^{N} (y_k x_{k-1} - x_k y_{k-1})$. As discussed in Section 4.6, this is also the zero-order moment of the region.

Another measurement of region size is the perimeter. We measure perimeter as the sum of distances between boundary pixels, where 1 is the vertical or horizontal distance between adjacent pixels and $\sqrt{2}$ is the distance between pixels on diagonals.

A larger perimeter of a region does not necessarily imply a larger interior area. A measure that combines area with perimeter is compactness. Given a perimeter length L and area A, compactness is defined as $C = L^2/4\pi A$. The most compact shape is that of the circle, with $C = 1$. Larger values indicate less compact shapes.

Boundary Curvature

The total bending energy E_c, stored in a contour of given configuration, has been proposed as a robust global shape descriptor that may be preferable to the commonly employed compactness descriptor C defined above. E_c is defined in terms of a sum over values of local curvature. This is a global descriptor using boundary curvature; in Section 5.4, we discuss features of lines based on *local* curvature.

For present purposes, it is useful to define the curvature as a simple geometrical property of a curve. For any smooth curve, three consecutive vertices, $\mathbf{v}_{k-1}$, $\mathbf{v}_k$ and $\mathbf{v}_{k+1}$, define the so-called osculating circle of radius $R_c(\mathbf{v}_k)$ at $\mathbf{v}_k$. The curvature at $\mathbf{v}_k$ is simply defined as the inverse of the radius of curvature, $R_c(\mathbf{v}_k)$. For a polygonal contour, defined in terms of its vertices, $\{\mathbf{v}_k; 1 \leq k \leq N\}$, E_c is readily evaluated in the form

$$E_c = \sum_{k=1}^{N} \Delta s_k / R_c(v_k)^2 \tag{4.3}$$

where $\Delta s_k = 1/2(d(\mathbf{v}_k, \mathbf{v}_{k-1}) + d(\mathbf{v}_k, \mathbf{v}_{k+1}))$, and $\mathbf{v}_0 = \mathbf{v}_N$, $\mathbf{v}_{N+1} = \mathbf{v}_1$.

An estimate of E_c may also be obtained directly from the tangential contour representation $\{(\Delta l_k, \Delta \phi_k); 1 \leq k \leq N\}$, introduced in Section 4.3. For any smooth curve, the unit tangent τ at l_0 is given by $d\mathbf{v}/dl|_{l_0}$. Given that $d\tau/dl = [(d\tau/d\phi)\cdot(d\phi/dl)] = [\mathbf{n}\cdot(d\phi/dl)]$, where $\mathbf{n}$ denotes the unit normal, we have $|d\tau/dl|^2 = |1/R_c(l)|^2 = |d\phi/dl|^2$, so that E_c may be written in the form

$$E_c = \sum_{k=1}^{N} \left| \frac{\Delta \phi_k}{\Delta l_k} \right|^2 \Delta s_k \tag{4.4}$$

where Δs_k, as above, refers to the length of the chord connecting vertices $\mathbf{v}_{k-1}$ and $\mathbf{v}_{k+1}$. If not based on a normalized representation of the boundary contour, E_c may be normalized to region size by dividing by the boundary length. The evaluation of E_c may also be based on a spectral representation of the contour, to be discussed below.

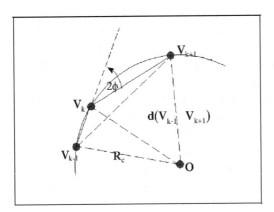

Figure 4.4.2. The osculating circle of radius R_c, defined in terms of three vertices, $\mathbf{v}_{k-1}$, $\mathbf{v}_k$, and $\mathbf{v}_{k+1}$, serves to evaluate the local curvature at $\mathbf{v}_k$.

Topological Features

Shapes can also be characterized by topological features. These are defined as features that do not change with elastic deformations of the object. Topological invariance precludes connecting or separating different regions. For binary regions, topological features include the number of holes in a region and its number of indentations and protrusions. These are determined from the results of region detection (Section 4.3).

A more precise expression than holes is subregions within a region. This is because regions can occur recursively. That is, regions can contain subregions that can contain subregions, etc. A simple example to illustrate the discriminating power of topology is evident in our alphanumeric symbols. Symbols O and 4 have a single subregion, and B and 8 have two subregions. A bull's-eye pattern has recursive subregions.

Another example of the use of topology for differentiating objects is for chromosomes. X chromosomes have a different topology than Y chromosomes by definition of their shapes' having four and three branches, respectively. Measurement of this feature, however, is better performed by thinning (Section 4.7) rather than the region and contour shape measurements described in this section.

4.4.2   MOMENT ANALYSIS

The evaluation of moments represents a systematic method of shape analysis. The most commonly used region attributes are calculated from the three moments of lowest order. Thus the area is given by the zeroth-order moment and represents the total number of region-interior pixels. The centroid, determined from the first-order moments, provides a measure of object location. The orientation of a (noncircular) region is determined from the principal axes that are determined from the second-order moments.

Knowledge of these low-order moments permits the evaluation of central moments, normalized central moments, and moment invariants. These quantities convey shape attributes that are independent of object position, size, and orientation, and are thus useful for object recognition and matching when position, size, and orientation are not pertinent to the object identity.

Moment analysis may be based on region-interior pixels, and a region growing or filling procedure to compile all of the region's interior pixels (Section 4.3) is then a prerequisite step. Alternatively, moment analysis may be based on the region contour, and this requires the detection of the contour (Section 4.3).

Region-Based Moments

For the binary images of interest in this chapter, region-interior pixels are assigned the value 1 (ON), and the definition of the moments m_{pq} of a binary region is

$$m_{pq} = \sum_R x^p y^q. \tag{4.5}$$

where the sum runs over all region-interior pixels. The definition of moments suggests an iterative evaluation, and a corresponding algorithm has been described.

Table 4.4.1. Moments and Vertex Coordinates

By Symmetry, an Expression for $m_{01}(m_{02})$ Follows from that for $m_{10}(m_{20})$ When x and y are Exchanged; all Expressions Assume Closed Boundaries, with $(x_0, y_0) \equiv (x_N, y_N)$. [(Seul et al. 91)]

$$m_{00} = \frac{1}{2} \sum_{k=1}^{N} y_k x_{k-1} - x_k y_{k-1},$$

$$m_{10} = \frac{1}{2} \sum_{k=1}^{N} \left\{ \frac{1}{2}(x_k + x_{k-1})(y_k x_{k-1} - x_k y_{k-1}) - \frac{1}{6}(y_k - y_{k-1})\left(x_k^2 + x_k x_{k-1} + x_{k-1}^2\right) \right\},$$

$$m_{11} = \frac{1}{3} \sum_{k=1}^{N} \frac{1}{4}(y_{k-1} - x_k y_{k-1})(2x_k y_k + x_{k-1} y_k + x_k y_{k-1} + 2x_{k-1} y_{k-1}),$$

$$m_{20} = \frac{1}{3} \sum_{k=1}^{N} \left\{ \frac{1}{2}(y_k x_{k-1} - x_k y_{k-1})\left(x_k^2 + x_k x_{k-1} + x_{k-1}^2\right) \right.$$
$$\left. - \frac{1}{4}(y_k - y_{k-1})(x_k^3 + x_k^2 x_{k-1} + x_k x_{k-1}^2 + x_{k-1}^3) \right\}.$$

More generally, the definition of moments m_{ij} refers to pixel locations x, y and pixel values $f(x, y)$:

$$m_{pq} = \int_{-\infty}^{\infty} \int_{-\infty}^{\infty} x^p y^q f(x, y)\, dx\, dy. \tag{4.6}$$

Contour-Based Moments

For binary, simply connected (but not necessarily convex) regions, one readily derives explicit expressions for moments up to quadratic order in terms of the vertex coordinates that define the bounding region contour. Thus, if a polygonal representation of a region contour is available, then area, centroid, and principal axes' orientation may be readily computed from the expressions given in Table 4.4.1.

Moment-Derived Shape Descriptors

The lowest-order moment m_{00} simply represents the sum of pixels interior to the region and thus gives a measure of the area. This is useful as an object descriptor if an object of interest is particularly larger or smaller than other objects in the image. However, area should not be used indiscriminately because a given object may occupy a smaller or larger portion of an image, depending on the scale of the picture, distance of the object from the observer, and perspective.

The first-order moments in x and y, normalized by the area, yield the x and y centroids: $x_c = m_{10}/m_{00}$ and $y_c = m_{01}/m_{00}$, respectively. These determine the region's average location.

The central moments μ_{pq} represent descriptors of a region that are normalized with respect to location. They are defined in terms of the centroid location:

$$\mu_{pq} \equiv \sum_{R} (x - x_c)^p (y - y_c)^q. \tag{4.7}$$

Commonly, the central moments are also normalized with respect to the zeroth moment to yield the normalized central moment:

$$\eta_{pq} = \mu_{pq}/\mu_{00}^{\gamma}, \qquad \gamma = (p+q)/2 + 1. \tag{4.8}$$

The most commonly used normalized central moment is η_{11}, the first central moment in x and y. This provides a measure of the deviation from a circular region shape. That is, a value close to 0 describes a region that is close to circular, and one with a larger value is increasingly noncircular.

Principal major and minor axes are defined to be those axes that pass through the centroid, about which the moment of inertia of the region is, respectively, maximal or minimal. Their directions are given by the expression

$$\tan \theta = \frac{1}{2} \left(\frac{\mu_{02} - \mu_{20}}{\mu_{11}} \right) \pm \frac{1}{2\mu_{11}} \sqrt{\mu_{02}^2 - 2\mu_{02}\mu_{20} + \mu_{20}^2 + 4\mu_{11}^2}. \tag{4.9}$$

Figure 4.4.2 shows regions with centroids and principal axes superimposed. A list of equations for central moments and normalized central moments up to third order is given in Table 4.4.2.

The evaluation of the direction of the major principal axis provides an independent way to determine the orientation of a nearly circular object. It is thus a suitable parameter by which to monitor, for example, orientational motion of distorted contours for objects whose shape changes in time.

The normalized and the central normalized moments introduced above were normalized with respect to scale (area) and translation (location). Normalization with respect to orientation is provided by a family of moment invariants. Table 4.4.2 lists the first four moment invariants as calculated from the normalized central moments.

A common empirical procedure uses these moment invariants for object recognition and preliminary classification. That is, the list of possible objects of interest in a given

Table 4.4.2. Central Moments and Moment Invariants

Central Moments

$$\mu_{10} = \mu_{01} = 0,$$
$$\mu_{11} = m_{11} - m_{10}m_{01}/m_{00},$$
$$\mu_{20} = m_{20} - m_{10}^2/m_{00},$$
$$\mu_{02} = m_{02} - m_{01}^2/m_{00},$$
$$\mu_{30} = m_{30} - 3x_c m_{20} + 2m_{10}x_c^2,$$
$$\mu_{30} = m_{03} - 3y_c m_{02} + 2m_{01}y_c^2,$$
$$\mu_{12} = m_{12} - 2y_c m_{11} - x_c m_{02} + 2m_{10}y_c^2,$$
$$\mu_{21} = m_{21} - 2x_c m_{11} - y_c m_{20} + 2m_{01}x_c^2.$$

Moment Invariants

$$\phi_1 = \eta_{20} + \eta_{02},$$
$$\phi_2 = (\eta_{20} - \eta_{02})^2 + 4\eta_{11}^2,$$
$$\phi_3 = (\eta_{30} - 3\eta_{12})^2 + (3\eta_{21} - \eta_{03})^2,$$
$$\phi_4 = (\eta_{03} + \eta_{12})^2 + (\eta_{21} + \eta_{03})^2.$$

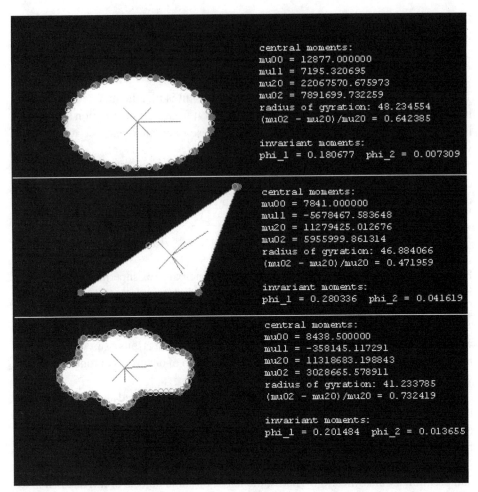

Figure 4.4.3. Three binary regions of differing shapes with superimposed centroids and principal axes. Centroids, computed from first-order moments, are marked by an X symbol; principal axes, computed from second-order moments, are indicated by lines. Also listed for each shape are the first four central moments, μ_{00}, μ_{11}, μ_{20}, and μ_{02}; the radius of gyration; a measure for shape eccentricity $(\mu_{02} - \mu_{20})/\mu_{20}$; and invariant moments ϕ_1 and ϕ_2.

image is established based on region segmentation, and moment invariants are calculated for each. On examination, only those invariants that effectively differentiate the objects of interest from other objects are retained. Figure 4.4.3 shows a set of geometric shapes and the values of their moment invariants. A similar picture with these objects scaled, translated, or rotated would result in similar values (with small differences due to discretization error).

References and Further Reading

[Seul et al., 1991] M. Seul, M. J. Sammon, and L. R. Monar, "Imaging of fluctuating domain shapes: methods of image analysis and their implementation in a personal computing environment," Rev. Sci. Instrum. **62**, 784–792 (1991).

Curvature energy is introduced as a global shape descriptor in I. T. Young, J. E. Walker, and J. E. Bowie, "An analysis technique for biological shape," Inf. Control **25**, 357 (1974).

The use of moments for image analysis is discussed in many reference texts, including: R. C. Gonzalez and R. P. Woods, *Digital Image Processing*, 2nd ed. (Addison-Wesley, Reading, MA, 1992); and A. Rosenfeld and A. C. Kak, *Digital Picture Processing* (Academic, Orlando, FL, 1982), Vol. II, Chap. 12. Additional background on moments may be found in R. N. Bracewell, *The Fourier Transform and Its Applications* (McGraw-Hill, New York, 1978).

Programs

xpm

> extracts global features and performs a moment analysis on planar boundaries

     USAGE: xpm inimg outimg [-L]

ARGUMENTS: inimg: input image filename (TIF)
           outimg: output image filename (TIF)

   OPTION: -L: print Software License for this module

xfm

> iteratively evaluates moments of convex shape up to 3rd order

     USAGE: xfm inimg [-d] [-w file] [-t] [-L]

ARGUMENTS: inimg: input image (TIF)

  OPTIONS: -d: default mode: evaluate moments of binary region
         -w file: write file (.mdt) to disk
             -t: generate test filter array
             -L: print Software License for this module

4.5 Advanced Shape Analysis: Fourier Descriptors

Typical Application(s) – quantitative analysis of object shape and shape distortions.

Key Words – shape analysis, spectral shape analysis, Fourier descriptors.

Related Topics – region growing (Section 4.3), shape descriptors (Section 4.4).

Fourier descriptors up to 10th order are:

0.271	0.330
0.358	-0.203
0.246	-0.963
-0.449	-0.126
-0.050	-0.580
0.264	0.419
0.421	-0.022
-0.125	-0.102
-0.144	-0.180
0.031	-0.074

Polar descriptors up to 10-th order are:

0.427	0.884
0.412	5.766
0.994	4.962
0.466	3.416
0.582	4.626
0.495	1.009
0.421	6.231
0.162	3.825
0.231	4.038
0.080	5.111

Figure 4.5.1. Pictorial Example. Binary shape in the form of the numeral 4 (see Fig. 4.3.2), with superimposed curvature points and marked centroid (symbol "X"). Listed in the inset text are Fourier shape descriptors up to tenth order, computed by direct summation from the tangential boundary representation (Section 4.3) as follows: Fourier coefficients a_n (top, left column) and b_n (top, right column), and coefficients in the more common polar form, namely magnitude (bottom, left column) and phase (bottom, right column).

In many application in physics and chemistry, a quantitative evaluation of shape distortions can yield detailed information regarding the interactions that determine a particular shape. For example, the analysis of shapes of red blood cells has been invoked to determine elastic moduli of the cells.

Fourier descriptors represent a useful and common set of shape descriptors that are determined by a spectral analysis of contour deviations from a suitable reference state such as a circle. Contour representations that lend themselves to this type of analysis were introduced in Section 4.3.

This section makes use of Fourier methods that are discussed in Chap. 7 and specifically the one-dimensional Fourier transform that is reviewed in Section A.1. The Fourier transform provides a tool to obtain a spatial frequency representation of a particular contour representation. Fourier descriptors are derived from the Fourier transform and describe what are essentially the curvature features of the region contour including the number of protrusions, the sharpness of turns and corners, as well as the number of smaller excursions (see Fig. 4.5.1).

4.5.1 SPECTRAL ANALYSIS

For a polygonal contour that is represented by a sequence of vertices, $P \equiv \{(x_k, y_k); 0 \leq k \leq N - 1\}$, we first represent each coordinate pair as a complex number, $u_k = x_k + iy_k$, to convert the sequence of two-dimensional vertex coordinates into a one-dimensional sequence of complex numbers. Next, we determine the one-dimensional Fourier transform of this sequence according to the definition

$$a(u) = 1/N \sum_{k=0}^{N-1} u_k \exp(-i2\pi uk/N). \tag{4.10}$$

The resulting sequence of complex coordinates $a(u)$ yields Fourier descriptors in the form of the power spectrum $|a(u)|^2$ (see also Section A.1). Lower-order terms describe the slowly varying, or low-frequency, distortions, and hence characterize the global shape of the boundary. Higher terms describe increasingly finer detail.

Fourier descriptors have been defined on the basis of both the tangential (ϕ^*; see Fig. 4.5.2) and the radial contour representations ($\delta\rho^*$; see Fig. 4.5.3) of Section 4.3.

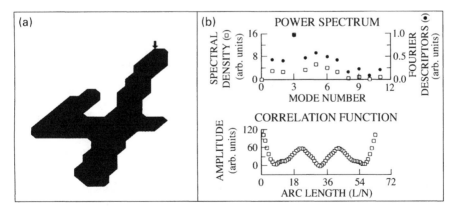

Figure 4.5.2. Binary shape in the form of the numeral 4 with the power spectrum and the correlation function derived from a tangential representation of the domain contour (see Fig. 4.3.2); both the spectral densities, computed by application of the FFT, and the descriptors computed by direct summation (Fig. 4.5.1, inset) are depicted. (Reprinted with Permission from [Seul et al., 91], Copyright © 1991 American Physical Society.)

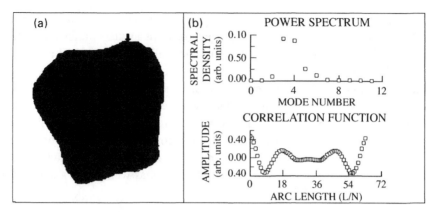

Figure 4.5.3. Binary shape, with the power spectrum and the correlation function derived from the radial representation of the domain contour (see Fig. 4.3.3). (Reprinted with Permission from [Seul et al., 91], Copyright © 1991 American Physical Society.)

For simple shapes, higher sensitivity to local features of the boundary can be expected from the tangential description. This reflects the relation of $\phi^*(l_k)$ to local curvature and implies the relative predominance of higher modes in the distribution of spectral weight in the power spectrum of ϕ^*. By the same token, the tangential representation also is more susceptible to random pixel noise: Presmoothing may thus be advisable (Sections 4.1 and 4.2; see also Section 5.3).

4.5.2   PRACTICAL CONSIDERATIONS

Evaluation of Fourier Descriptors from Tangential Representation   Many applications encountered in practice involve simple shapes, and the number N of vertices in the tangential representation of these shapes may be sufficiently small to consider the direct evaluation of the Fourier coefficients for $\phi^*(l_k)$ from the tangential representation of Section 4.3. In contrast to fast Fourier transform (FFT) methods (see Section 7.1), this has the advantage of being applicable for any N. For the numeral 4 with the representation listed in Table 4.3.2, the requisite summations yield the spectral densities listed in Fig. 4.5.1 and displayed in Fig. 4.5.2.

For large data sets, the most efficient computation of the power spectrum of $\phi^*(l_k)$ and, in any event, of $\delta\rho^*(l_k)$, relies on the (one-dimensional) FFT (Section A.1). In general, this will require padding, a concept further discussed in Section 7.1, or resampling of the original or a smoothed contour to generate 2^n contour points. In the examples of Fig. 4.5.2, $2^n = 64$.

Choice of Origin for Radial Contour Representation   As indicated in Section 4.3, the definition of the radial representation requires the choice of an origin; in Eq. (4.2) the centroid serves this function. However, the requirement that $\delta\rho^*$ be invariant under translation requires careful inspection of this choice: In the Fourier domain, translation invariance imposes the condition $|a_1|^2 = 0$ for the spectral density of the first mode. The proper origin for the definition of $\delta\rho^*$ is thus the point $\mathbf{v}_0 = (x_0, y_0)$ that guarantees compliance with that condition. In general, $\mathbf{v}_0$ coincides with $\mathbf{v}_c$ only for a circle. In practice, $\mathbf{v}_0 \simeq \mathbf{v}_c$ for nearly circular shapes. In the absence of an analytic solution, a trial-and-error procedure is the only choice to locate $\mathbf{v}_0$ in the vicinity of $\mathbf{v}_c$ so as to minimize $|a_1|^2$. This may be costly, but unavoidable for large deviations from a circular shape.

Application of the inverse FFT to the power spectrum yields the autocorrelation function $\langle f(l_k) f(0) \rangle$, where f denotes ϕ^* or $\delta\rho^*$. The autocorrelation function has long been known to constitute a very efficient device (Section A.1) for the identification of periodic components in a noisy environment.

References and Further Reading

[Seul et al., 1991] M. Seul, M. J. Sammon, and L. R. Monar, "Imaging of fluctuating domain shapes: methods of image analysis and their implementation in a personal computing environment," Rev. Sci. Instrum. **62**, 784–792 (1991).

[Zahn and Roskies, 1972] C. T. Zahn and R. T. Roskies, "Fourier Descriptors for Plane Closed Curves," IEEE Trans. Computers **C-21**, 269–281 (1969).

The relative merits of several spectral representations in optical character recognition are discussed in some detail in E. Persoon and K.-S. Fu, "Shape discrimination using Fourier descriptors," IEEE Trans. Syst. Man Cybern. **SMC-7**, 170–179 (1977) and in T. Pavlidis, *Structural Pattern Recognition* (Springer-Verlag, Berlin, 1977), Chap. 7.

The application of spectral shape analysis to investigate elastic properties of biological membranes is described in M. B. Schneider, J. T. Jenkins, and W. W. Webb, "Thermal fluctuations of large cylindrical vesicles," Biophys. J. **45**, 891–899 (1984); and in H. P. Duwe, J. Kaes, and E. Sackmann, "Bending elastic moduli of lipid bilayers: modulation by solutes," J. Phys. **51**, 945–962 (1990).

Program

xbdy

```
          detects and encodes planar boundaries in a given image,
          implementing an algorithm by C. T. Zahn [Zahn, 1969]

    USAGE: xbdy inimg outimg [-L]
ARGUMENTS: inimg: input image filename (TIF format)
           outimg: output image filename (TIF format)
  OPTIONS:    -L: print Sotfware License for this module
```

4.6 Convex Hull of Polygons

Typical Application(s) – delineation of polygonal region; association of a shape with a group of points forming the vertices of a polygon.

Key Words – extreme points, convex hull, shape.

Related Topics – shape analysis (Section 4.4).

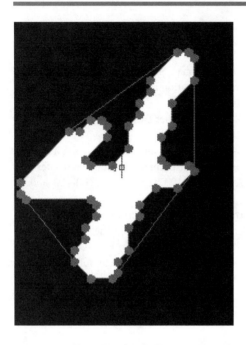

Figure 4.6.1. Pictorial Example. Illustration of the convex hull for a region shape delineating the numeral 4. The polygon joining a subset of vertices is the convex hull. Note that this convex hull contains the original polygon and all vertex angles are convex with respect to the interior of the polygon.

The concept of the convex hull of a region (in the two-dimensional plane) is a very intuitive one: for a polygon **P**, the convex hull is the smallest convex polygon containing **P**. Thus, if we regard **P** as having pegs at its vertices $\{v_k\}$, then a rubber band stretched to encompass these pegs assumes the shape of the convex hull. This concept finds application in diverse contexts such as pattern recognition, linear programming, and statistics, as discussed in the literature.

In the context of image analysis, the convex hull is used for mainly two reasons. One is to represent a complex polygonal shape (i.e., one with convex and concave sections of contour, as in Fig. 4.6.1) by the simplest polygon that still completely encompasses the original. This simpler shape often suffices to perform matching or recognition. By the same token, the relative complexity of the original shape might adversely affect the results of matching or recognition because of extraneous detail. The second reason is that the convex hull polygon delineates the area of influence of a region. If another region (or its convex hull) overlaps this convex hull, then it is said to encroach on that first region's area of influence.

In practice, we can construct the convex hull by starting from the entire set of original boundary points identified in a region detection step (Section 4.3) or from a reduced set of boundary points obtained in a polygonal fit of the original points (Section 5.3).

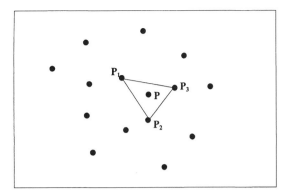

Figure 4.6.2. Illustration of the definition of an external, or extreme, vertex point. Point *P* is enclosed by a triangle (P_1, P_2, P_3) of points in the set and is therefore an internal point.

4.6.1 CONCEPT: EXTREME POINTS

The desired construction requires the identification of hull vertices that form a subset of the given set of polygon vertices. Although there are faster methods for particular cases, as discussed in the references, most practical applications are well served by the procedures we discuss below. Since they do not appeal to special constraints concerning the arrangement of vertices, they acquire more general validity.

An appropriate criterion to facilitate identification of hull vertices is furnished by the distinction between internal and external points, or extreme points, in the given set of vertices, as illustrated in Fig. 4.6.2. A point $E \in \mathbf{P}$ is said to be an extreme point of $\mathbf{P}$ if no triangle containing E can be constructed that uses the other points in the set of polygon vertices. The convex hull polygon of $\mathbf{P}$ consists of the set of extreme points of $\mathbf{P}$, which are sorted in angular order about any interior point of $\mathbf{P}$. Angular sorting is described below.

4.6.2 HULL CONSTRUCTION

At least two general algorithms for the construction of the convex hull polygon are based on the essential insight that a substantial gain in efficiency results if the two steps, namely the identification of extreme points and their sorting in angular order with respect to an interior point O, may be combined. This is possible if the sorting step is performed first.

Angular Sorting of Polygon Vertices

The efficient implementation of both algorithms depends on an efficient method to sort a set of points in angular order with respect to a point O internal to the set. This internal point is often chosen to be the set's centroid, or the centroid of a subset formed by any three noncollinear points of the original set. First, all polygon coordinates are transformed so as to make O the new origin by subtracting from each the location of O. Next, angular sorting is accomplished by pairwise comparison of the azimuthal angles subtended by the two chords drawn from the origin to any two vertices. In angular

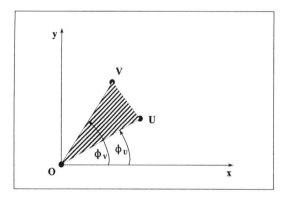

Figure 4.6.3. Angular sorting of points with respect to given reference point. To determine the relative magnitude of the two azimuthal angles Φ_U and Φ_V, it suffices to calculate the signed area (see text) of the triangle (O, U, V): U [or the chord (O, U)] subtends a smaller angle Φ_U with the abscissa if the triangle has a positive signed area.

sorting, a vertex U precedes a vertex V if the azimuth defined by (O, U) is smaller than that defined by (O, V) (see Fig. 4.6.3).

For any two points, P_1, P_2, in question, we may perform this comparison without explicit numerical evaluation of the respective azimuthal angles by simply considering the signed area of the triangle spanned by $O \equiv (x_0, y_0)$, $P_1 \equiv (x_1, y_1)$ and $P_2 \equiv (x_2, y_2)$. P_2 subtends a strictly smaller azimuthal angle with the real axis than does P_1 if and only if the triangle $T \equiv \{O, P_2, P_1\}$ has strictly a positive-signed area. We obtain the signed area A_Δ of a triangle spanned by any three (noncollinear) points O, P_1, and P_2 by evaluating the determinant $\Delta \equiv (x_0 y_1 - x_2 y_1) + (y_0 x_2 - x_0 y_2) + (x_1 y_2 - x_1 y_0) = 2A_\Delta$.

Graham Scan

For the construction of the convex hull, we implement the Graham scan as follows.

The essential first step of the hull construction algorithm proposed by Graham involves the sorting of the given point set by azimuthal angle and by distance with respect to an internal point O serving as the origin. Distance serves as the secondary sorting category, that is, a distance comparison is required only for points whose azimuthal angles coincide; a pair of such points lies on the same straight line that also contains the origin, rendering the requisite comparison trivial.

Following the arrangement of the sorted points into a cyclic, doubly linked list, corresponding to the sketch in Fig. 4.6.4, a single scan of the ordered points is performed to eliminate all internal points: An internal point is one that is not on the convex hull and is thus internal to some triangle spanned by points in the set, as indicated above. Given the ordered sequence of vertices obtained from the sorting procedure just discussed, the requisite test is simple. From Fig. 4.6.4, it is apparent that if a point V in the ordered set is not on the convex hull, then it is internal to the triangle $T \equiv \{O, V, W\}$, where V and W are consecutive hull vertices. Assuming a counterclockwise scan, this condition in turn implies that the sequence of points UVW defines a right-hand turn because the angle subtended by the line segments UV and VW exceeds π. This violates an immediate consequence of convexity, which requires that a counterclockwise traversal of a convex polygon contain only left-hand turns. Whether the point W lies to the right of the line segment UV, implying a right-hand turn, may be ascertained by examination of the signed area of the triangle UVW.

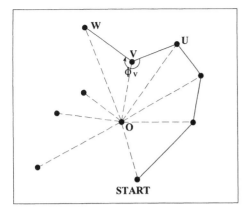

Figure 4.6.4. Sketch indicating Graham scan: Vertex V is eliminated on examination of the triple (U, V, W) of consecutive vertices because the triple defines an internal angle Φ_V that exceeds π.

The Graham scan thus proceeds from the angular sorting step to the scan of the sorted point set, effectively wrapping the hull polygon around the set of extreme points. In this scan, starting at a START point known to be a hull vertex, such as the rightmost vertex with the smallest ordinate, consecutive triples of points are examined: If $P_1 P_2 P_3$ involves a right-hand turn, vertex P_2 is eliminated and the triple $P_0 P_1 P_3$ is inspected; if $P_1 P_2 P_3$ indicates a left-hand turn, the scan is advanced and $P_2 P_3 P_4$ is examined. The scan terminates after completion of the traversal, that is, when START is encountered, leaving the hull vertices in sorted order in the doubly linked list set up initially. The Graham scan permits construction of the convex hull in $O(N \log N)$ time, with linear storage requirement.

References and Further Reading

This section follows the discussion in F. P. Preparata and M. I. Shamos, *Computational Geometry*, 2nd ed. (Springer-Verlag, New York, 1985). This reference contains an in-depth discussion of the convex hull, its uses in computational geometry, and alternative construction algorithms. Another useful reference is J. O'Rourke, *Computational Geometry in C* (Cambridge U. Press, New York, 1993).

Program

xph (Polygon Hull)

```
        constructs convex hull of polygonal shape
   USAGE: xph outimg [-r file] [-L]
ARGUMENTS:  outimg: output image filename (TIF)
  OPTIONS: -r file: input data file containing delta_phik and
                    delta_lk; if filename not specified, will use
                    internal data for test
              -L: print Software License for this module
```

4.7 Thinning

Typical Application(s) – thinning (skeletonization) of elongated regions, lines, and contours.

Key Words – skeleton, medial axis transform.

Related Topics – polygonalization (Section 5.3), line fitting (Section 5.5).

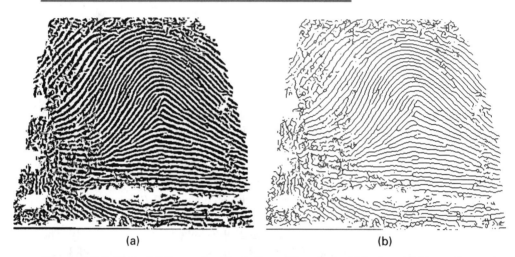

(a)   (b)

Figure 4.7.1. Pictorial Example. Illustration of the effect of thinning: (a) binary fingerprint image containing readily identifiable lines, (b) result of thinning, in which all original lines now have maximum one-pixel thickness; this image required five iterations.

Thinning is an image processing operation in which binary-valued image regions are reduced to skeletons of the regions. These skeletons approximate center lines with respect to the original region boundaries. In the ideal case, this set of points represents the medial lines of the original boundaries, defined to contain all points that are equidistant from two points on the original boundary (see Fig. 4.7.1).

Although the thinning operation can be applied to binary images containing regions of any shape, it is most suitable for elongated, as opposed to convex, or bloblike shapes. Although this is merely a qualitative distinction, it is usually obvious when thinning is appropriate. For instance, chromosomes are classified as X and Y on the basis of their characteristic elongated shapes and patterns of line branching. Images of chromosomes can be thinned to retain this pertinent information. In contrast, liver cells usually assume a bloblike shape: Here, area and contour configuration are the important descriptors (Sections 4.3–4.6); thinning would not provide any useful information.

4.7.1 OBJECTIVES

Five objectives must be met in performing a thinning operation:

1. Connected image regions must thin to connected line structures. To generate a reliable topological representation, it is essential that the thinning operation

preserve the connectivity of the original. This guarantees the result of exactly one thinned, connected line structure for each disjoint region in the original image.

2. The thinned result should be minimally eight connected. This requirement stipulates that the line segments obtained by thinning will always contain the minimal number of pixels that maintain eight-connectedness.

3. Approximate end-line locations should be maintained. The end locations of lines are shortened to some degree because of thinning. This shortening should be minimized.

4. The thinning result should approximate the medial lines. Because of the discrete nature of a digital image and because of noise, thinned structures will not be located exactly in the center of the original lines. The objective is to be as close as possible.

5. Extraneous "spurs" introduced by thinning should be minimized. Extraneous features are undesirable.

4.7.2 IMPLEMENTATION

A common thinning approach is to peel the region boundaries, iteratively one layer at a time, until the regions have been reduced to thin lines. Figure 4.7.2 shows the sequence of thinning iterations for the image of a symbol. In each iteration, every image pixel is inspected in raster-scan order, and single pixels that are not required for preserving connectivity or maintaining end lines are erased (set to OFF). The decision whether to erase a specific pixel is based on the values of its neighbors within a 3×3 window. For each window, the following criteria must be satisfied for an ON-valued center pixel to be erased:

1. The connectivity, defined as the number of chains of connected ON pixels in the neighborhood, is equal to 1. This condition ensures that the connectivity of a structure is not altered. If the condition is true, then the neighborhood contains a single, unclosed chain of connected ON pixels, and erasure of the core will not destroy connectivity within any ON chains in the neighborhood.

2. The maximum length of a chain of four-connected ON pixels in the neighborhood is greater than 1. This condition maintains the end locations of the end lines.

3. The maximum length of a chain of four-connected OFF pixels in the neighborhood is also greater than 1. This condition may be viewed as the inverse of the previous condition; it prevents the inward erosion of OFF regions into ON regions.

Erased pixels are set to the value ERASED on the image being thinned, where this value is different from that of ON or OFF. This temporary pixel designation enables intermediate output to be stored in the same memory as the original image; future 3×3 mask operations treat an ERASED value as if it were an ON value. (Otherwise, many layers could be eroded unevenly on a single iteration.) If the thinning test fails, the

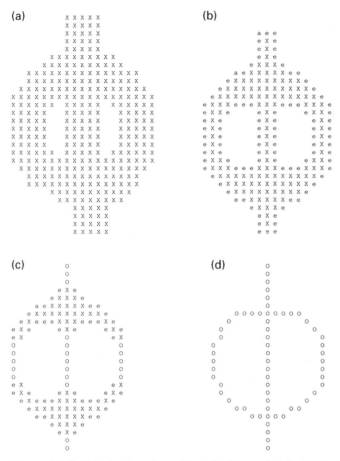

Figure 4.7.2. Thinning iterations: (a) original image (original ON-valued pixels have the symbol X); (b), (c) intermediate results after several iterations (eliminated pixels are marked by the symbol e, anchor pixels by a, and pixels that have reached the final thinned state by O; (d) final thinned result (skeleton).

examined pixels retain their original values. This processing is applied to the entire image. At the end of an iteration, the ERASED values are set to OFF values. Iterations then continue until no pixels are erased on an iteration, whereupon the process is stopped and thinning is complete.

Because of its iterative nature, thinning can be a time-consuming procedure. In the most straightforward implementation, each pixel is examined on each iteration, so the number of operations is of the order of $M^2 n_I$ where M is the number of pixels and n_I is the number of iterations, which is in turn proportional to the maximum width of regions.

One method of reducing the computation is the following. During the first iteration runs, ON-valued pixels are marked by their first location (run-length coded); for subsequent iterations, OFF values are not repeatedly checked. This substantially increases the speed of subsequent iterations, especially for sparse images of few ON regions.

References and Further Reading

Thinning is reviewed in L. Lam, S.-W. Lee, and C. Y. Suen, "Thinning methodologies – A comprehensive survey," IEEE Trans. Pattern Recog. Mach. Intelligence, 56–75 (1992).

Useful thinning (skeletonization) algorithms are discussed in C. J. Hilditch, "Linear skeletons from square cupboards," Mach. Intelligence **4**, 403–420 (1969); and in L. O'Gorman, " $k \times k$ thinning," Comput. Vis. Graph. Image Process. **51**, 195–215 (1990).

Program

`thin`

```
           performs iterative thinning of binary objects in input
           image to produce skeleton image with values OFF (0) and
           ON (255)
    USAGE: thin inimg outimg [-k K] [-n MAXITER] [-d DISPLAY] [-L]
ARGUMENTS: inimg: input image filename (TIF)
           outimg: output image filename (TIF)
  OPTIONS:  -k K: window size for kxk mask (k >= 3, default = 3)
           -n NITER: maximum number of iterations (default max = 20)
               -d: display results of each iteration (< 40x40 image)
               -L: print Software License for this module
```

4.8 Linewidth Determination

Typical Application(s) – thinning of an image with simultaneous retention of linewidth information.

Key Words – augmented thinning, line image reconstruction.

Related Topics – thinning (Section 4.7).

(a)

(b)

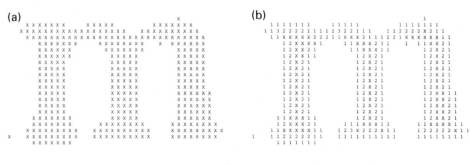

(c)

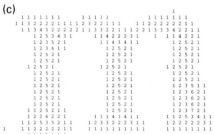

Figure 4.8.1. Pictorial Example. Linewidth determination: (a) original image; (b) intermediate result after several iterations (pixel values show distances to the closest outer boundary); (c) final result (pixel values show the narrowest widths of the original lines at these locations).

Section 4.7 introduced thinning as a procedure to extract the skeleton of an image composed of elongated objects. This analytical tool is particularly well suited when the predominant concern is with topological information, reflecting the connectivity of the original pattern: The geometrical proportions of connected regions are usually of no interest in that context. There are, however, applications in which both topological and geometrical image features must be considered. An example is an image of handwriting. If the meaning of the letters is the primary concern, analysis of the skeleton suffices; however, if there is an additional requirement to determine the width of the pen nib, a measure of the typical width of lines in the original line image must be made available.

The solution presented here is an augmented thinning algorithm with the additional provision to retain linewidth information. This is stored in the pixel values along the thinned lines. An example of thinning with line width retention is shown in Fig. 4.8.1.

Strategy

The approach we describe below assumes knowledge of the method described in Section 4.7. A simple modification of the thinning algorithm suffices. The crucial step concerns the action to be taken whenever a pixel or core has been marked for elimination. Whereas in the thinning algorithm, the value of each of the corresponding pixels is set to ERASED, we now set pixel values within the core to the sum of the distance to the closest erased pixel plus the distance value of that closest pixel. This ensures that the value of each erased pixel equals the distance to the closest region boundary.

At the last stage, the final thin line pixels are set not to a distance value, but to a width value. The width for a pixel in the middle of a line is just the sum of distance values of two neighbors on either side of the line, plus one, to account for its own thickness. Note that the resultant width is approximate. This is in large part because, as a matter of practicality, we store not the actual Euclidean distance in each pixel (image byte) but simply the closest integer.

As described in Section 4.7 for $k \times k$ thinning, values of k equal to 4 or 5 may improve the speed. However, just as the thinning results are less accurate, so are the width results.

Implementation

Pertinent details of the width-augmented $k \times k$ thinning algorithm are as follows. While in the simple thinning algorithm, it is sufficient to keep track of OFF, ON, and ERASED values, we must now associate distance values with the ERASED designation. That is, an erased pixel must store, in addition to its erased status, its distance to the closest boundary. Furthermore, we also use a permanent-ON (P-ON) designation to mark pixels that have already been tested and can never be erased. This field for P-ON designation can be used to store the final width value in the course of thinning, thereby eliminating the necessity of a post-iteration step to calculate width values from distance values.

We represent parameter values by taking advantage of the available range of 256 possible gray values of an image byte. We split the 256 values for ON and OFF values plus the ranges of values needed for ERASED and P-ON, such that ERASED can range over 51 values and P-ON over 102 values. (There are two other pixel designations that we do not mention because they are used only for intermediate thinning steps. These require the rest of the range of 256 values.) Therefore, this method can accommodate thicknesses up to 102 pixels. This should not be a restriction. In practice, if thicknesses are large (say, greater than 10 or 15 pixels), then it is more appropriate to use the shape analysis methods of Section 4.4.

Line Image Reconstruction

Retention of width information facilitates reconstruction of the original image from the skeleton – albeit not exact reconstruction. The method is straightforward. For each pixel in the width image that contains a nonzero width value, a circular area of neighborhood pixels, whose diameter matches the width value, is set to ON around the pixel.

References and Further Reading

The algorithm discussed here (and in Section 4.7) is discussed in L. O'Gorman, "$k \times k$ thinning," Comput. Vis. Graph. Image Process. **51**, 195–215 (1990).

Program

thinw

performs iterative thinning of binary objects in input image to produce skeleton image with values OFF (0) and ON (255); skeleton pixel values are widths of the original image lines.

USAGE: thinw inimg outimg <-k MASK_SIZE> <-d>[-L]

ARGUMENTS: inimg: input image filename (TIF)

outimg: output image filename (TIF)

OPTIONS: -k MASK_SIZE: window size for kxk mask (k >= 3, default = 3)

-d: display results of each iteration (< 40x40 image); this only displays for images <= 40x40 to fit on screen.

NOTE: image output values are the following:
-- thinned lines -- width value plus offset of 152
-- OFF -- 0
-- ON -- 255

-L: print Software License for this module

4.9 Global Features and Image Profiles

Typical Application(s) – global analysis of collections of multiple objects.

Key Words – statistical features, image moments, image projection profiles, intensity signatures.

Related Topics – global image features (Section 2.1); multiresolution analysis (Section 3.7); shape analysis (Section 4.4), moments (Section 4.4); 2D Fourier transform (Section 7.1).

(a) Number of
ON pixels

(b) 1st Moment

(c) 2nd Moment

(d) Uniformity

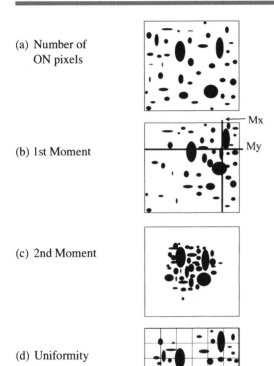

Figure 4.9.1. Pictorial Example. Illustration of global shape features providing a basis to differentiate among images: (a) number of ON pixels, (b) first moment indicating that image objects predominate in upper right-hand corner, (c) small second moment [relative to images of (a) and (b)] indicating a cluster of image objects, (d) grid used to detect nonuniform object density.

Rather than analyzing features of individual objects or regions, as in the previous sections, it is sometimes of interest to evaluate features of an entire set of regions within an image. For example, we may wish to determine the density of regions or the particular pattern formed by regions in the area (see also Chap. 6). In this section, we examine some of these global features.

The global analysis introduced here is related in scope to that of Fourier domain processing, described in Chap. 7. For example, a two-dimensional Fourier transform produces frequency characteristics of an entire image from which we can derive such information as the density of objects and any regularity in their arrangement. This information is inherently global and relates to an entire group or set of objects. The

177

features we describe here are global in the same sense, but manifest themselves in the spatial domain.

Most readily accessible are those global features that simply reflect an averaging operation over the set of individual region features. The image is strictly global, in the same sense as the intensity histogram (Section 2.1). This profile is a transformation of the image from two-dimensional space to a one-dimensional histogram to describe its density along a chosen axis.

4.9.1 STATISTICAL FEATURES

Many descriptive features of an image can be obtained by summing and averaging or by other statistical operations performed over the set of individual features. The most direct features relate to quantities of individual pixels, such as the number of ON pixels and the ratio of ON pixels to OFF pixels. These can be used to determine how densely the image is filled, whether the density of ON pixels is close to that expected, to ascertain whether the image contains anything of interest, or simply to ascertain if there are any ON pixels at all.

The region detection methods of Section 4.3 can be used to yield a number of global features such as the total number of blobs, the density of blob areas relative to OFF pixels, the average location of blobs, the range of blob sizes, the average and the standard deviation of blob sizes, the total number of holes, the average number of holes per blob, the ratio of hole area versus filled area, etc. A number of global features characterizing an image are shown in Fig. 4.9.1.

In many instances, the features of interest are known and calculations can then be limited to these. In other cases, the pertinent features must be determined. An example is the task of making a distinction between diseased and healthy tissue on the basis of the most prominent visual features. To determine which are the most suitable characteristics, the first step is to evaluate a great many features of prelabeled diseased and healthy tissue images. This is followed by a statistical analysis, ranking features on the basis of their reliability in discriminating between the two tissue types. This training step will help to narrow down the set of critical features, and subsequent diagnosis is based on an inspection of only these significant features.

When regions are not uniformly distributed over the image, it may be of interest to determine the degree of this nonuniformity. An approach to this problem is afforded by a moment calculation (see Section 4.4) on the locations of blobs or lines in the image. The first moment gives the average location, and if this lies close to the center of the image, the spatial distribution of blobs is symmetric with respect to this center; this is the analog of the moment of inertia. The second moment of the differences of locations from the average location gives a measure of variability in the spatial distribution.

An alternative approach to establishing an estimate of the variability in the spatial distribution of features is to partition the image by a grid overlay and to determine features visually within each grid partition (see Fig. 4.9.2). From a comparison of grid populations, a measure of variability can be determined. The number of grid fields may vary from two to many more, depending on the expected size of blobs and their expected spatial configuration.

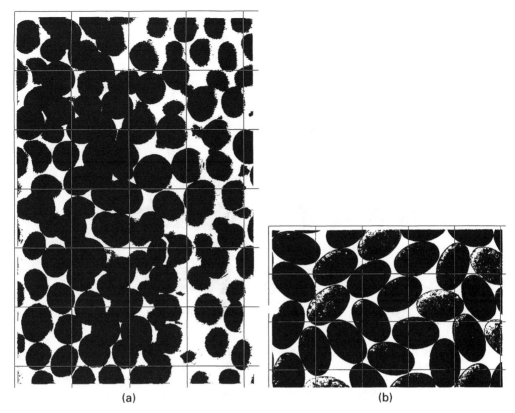

Figure 4.9.2. Grid overlay on image for manual analysis: (a) grid used for counting objects in all or a sampling of grids, (b) grid used to measure object sizes. (This product/publication includes images from Corel Stock Photo Library, which are protected by the copyright laws of the U.S., Canada, and elsewhere. Used under license.)

4.9.2 IMAGE PROFILES

Another approach to probing the distribution of blobs in an image is to determine the image projection profiles (also referred to as the image signature). A profile results from the summation of ON pixels taken along some axis across the entire image. While a profile may be evaluated along an axis of any orientation, the most common profiles are horizontal and vertical. The horizontal profile results from determining the sum of ON pixels in each column of the image; the vertical profile results from determining the sum of ON pixels in each row of the image. The profile types, horizontal and vertical, refer to their axis orientations, not the directions of summation.

One use of the profile is to identify the presence of an oriented texture or pattern. In Figure 4.9.3(a), the image has a discernible pattern. This pattern is made more discernible by inspection of the profile in Fig. 4.9.3(b) in which peaks are at regular intervals corresponding to the rows of blobs. The period of these rows can be measured by the distance between peaks in the profile. In Fig. 4.9.3(c), we have rotated the image (Section 2.4) such that the rows are exactly horizontal. For this case, the profile in Fig. 4.9.3(d) has even sharper peaks corresponding to the same rows as those in Fig. 4.9.3(b), but there are secondary peaks corresponding to the rows of fewer blobs. These secondary peaks were lost in Fig. 4.9.3(b) because of the image skew angle.

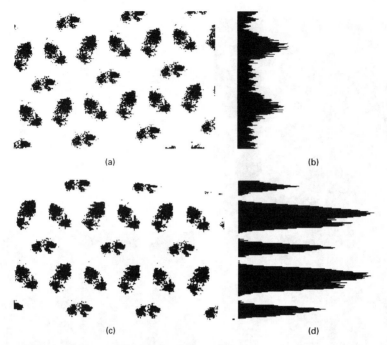

Figure 4.9.3. Vertical image profiles: (a) original image containing pattern of blobs, (b) vertical image profile with peaks corresponding to rows, (c) image obtained by a 15° rotation of the original to make rows of blobs horizontal, (d) vertical image profile with sharper peaks than in (b) corresponding to primary and secondary rows.

Another use of the profile is to determine the region of highest density, which in turn may be the chosen region of interest. As an example, high bin heights at the left of a horizontal profile indicate that most of the regions are in the left of the image. High bin heights at the upper bins of a vertical profile indicate that most of the regions are in the top of the image. Combining this information, we know from the profiles that most regions are in the top left of the image. This may be the end purpose of this analysis, or we can use it to concentrate further processing in the top left area of the image.

A projection profile may also be used to extract features for an individual image region if that region is the only one in the image. For example, a circular disk will yield a bell-shaped profile, both vertically and horizontally, while a vertically elongated region will yield a narrow, sharp peak in the horizontal profile and a wider, lower peak in the vertical profile; projection profiles may be used in this way in lieu of using region moments (Section 4.4).

References and Further Reading

Aspects of image profiling and its applications are discussed in J. M. Chambers and T. J. Hastie, *Statistical Models in S* (Wadsworth and Brooks/Cole Advanced Books and Software, Pacific Grove, CA, 1992); and in R. P. Kruger, J. R. Towne, D. L. Hall, S. J. Dwyer, and G. S. Ludwick, "Automatic radiographic diagnosis via feature extraction and classification of cardiac size and shape descriptors," IEEE Trans. Biomed. Eng. **BME-19** (1972).

Programs

globalfeats

determines some global features of input image and prints
these results to stdout; global features determined are
-- image size;
-- total number of pixels;
-- number of ON-valued pixels;
-- percentage of ON-valued pixels to total pixels;
-- 1st moment (average) of x,y locations of ON pixels;
-- 2nd moment of x,y locations of ON pixels;
-- number of edge pixels;
-- percentage of edge pixels to total pixels.

NOTE: calculations are performed on all pixels EXCEPT the
1-pixel-wide borders of pixels at the sides of the
image.

USAGE: globalfeats imimg [-L]
ARGUMENTS: imimg: binary input image (TIF)

OPTIONS: -L: print Software License for this module

profile

determines the horizontal or vertical profile of a
binary image, that is the summation of pixel values
along the y axis or the x axis respectively;
default is horizontal.

USAGE: profile inimg outimg [-h || -v] [-s SIZE] [-L]
ARGUMENTS: inimg: input image filename (TIF format)

outimg: output image filename (TIF format)

OPTIONS: [-h || -v]: horizontal or vertical profile, respectively.
-s SIZE: height of horizontal profile or width of
vertical profile (default = 200)
-L: print Software License for this module

imggrid

creates test image of grid with chosen spacing.

USAGE: imggrid inimg outimg [-s SPACING] [-i] [-L]
ARGUMENTS: inimg: input image filename (TIF)

outimg: output image filename (TIF)
OPTIONS: -s SPACING: spacing of grid lines; default=20 pixels.
-i INVERT_FLAG: inverts grid high/low for low/high image.
-L: print Software License for this module

4.10 Hough Transform

Typical Application(s) – detection of lines (to a lesser degree of other shapes such as circles) in noisy images.

Key Words – Hough transform, line fitting.

Related Topics – template matching (Section 3.8); shape analysis (Section 4.4); line fitting (Section 5.5); two-dimensional Fourier transform (Section 7.1).

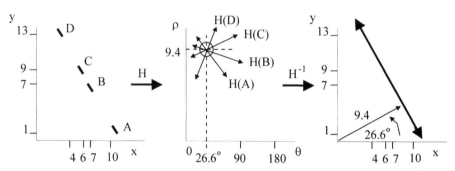

Figure 4.10.1. Pictorial Example. Illustration of the Hough transform: The plot on the left shows a broken line in the spatial domain. The middle plot shows a peak in the Hough domain at the (ρ, θ) coordinates of the original line. The plot on the right shows a reconstruction of the line – without the breaks – from the Hough representation.

Sometimes we want to identify geometric shapes within a binary image containing disconnected points. A cluster of such points may assume the shape of a line, a circle, or a more complex geometric shape. Although the shape is often readily recognized by eye, the task of shape recognition by computer is more difficult in this situation than it is in the case of identifying a solid region or connected line: Here, the task requires the explicit selection of those points that make up the object.

A popular procedure for grouping points into geometric shapes is the Hough transform (see Fig. 4.10.1). This method is useful for identifying clusters of points defining shapes that can be expressed parametrically. For instance, the parameters of a line are its slope and its intersect; for a circle, the parameters are its center coordinates and its radius. In theory, the Hough transform can be used to detect arbitrarily complex geometric shapes, but practical considerations of computation time and memory usage limit this method predominantly to the task of finding lines and, much less frequently, of finding circles.

4.10.1 LINE DETECTION

We first describe the approach to the problem of detecting a line by application of the Hough transform, then generalize to other parametric shapes.

The central idea of the Hough transform is to represent a line made up of many pixels by a single peak in parametric space – the Hough domain, also referred to as

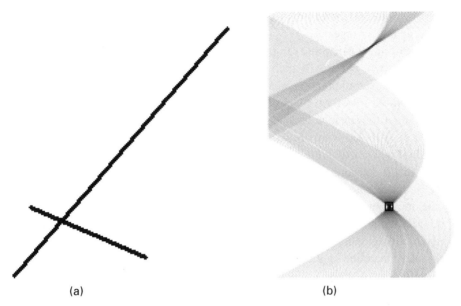

(a)   (b)

Figure 4.10.2. Example of Hough transform: (a) Image with two lines, one longer than the other; (b) Hough space corresponding to (a) with two peaks. The dominant peak, marked by the superimposed box, corresponds to the longer line in (a).

the Hough space, Hough transform plane, or accumulator array. This single peak has coordinate values in Hough space of two parameters necessary to describe the line, such as slope and intersect. Similarly, for an image of multiple lines, its Hough transform will have multiple peaks corresponding to the line parameters of each of the lines.

One question remains to be addressed, namely, how does one construct lines from a set of multiple, disconnected points without *a priori* knowledge of slope or intersect and in the absence of a one-to-one transformation? The solution is to determine the parameters of *all* lines that could possibly intersect each given point and accumulate all these possible parameter results in the Hough space. Out of all these possible lines, only the truly collinear points will share a common line. That is, one of all the possible lines associated with each individual point will have the same parameters for all points on a line and will therefore map to the same location in parameter space. Because each of the collinear points will map to the same location in the Hough space, this will result in a multiple number of points – the number of ON pixels along the line – accumulating in a particular location in the transform plane. This is illustrated in Fig. 4.10.2.

Procedure

Implementation of the Hough transform for line detection requires attention to the following practical details.

Since the slope of a given line can be infinite, we parameterize the line trigonometrically: $x \sin\theta + y \cos\theta = r$, where (x, y) denotes coordinates in the spatial domain and (r, θ) denotes the transform domain parameters. To permit construction of only a finite number of lines for each point, parameter space is discretized to the desired accuracy, perhaps 360 increments for θ (for $360°$) and some number of increments for r, depending on its range.

Because of image noise and sampling inaccuracy, a cluster of points that represents a linear pattern will not usually map into a single bin in the parameter space; instead, there will be a cluster of points around a peak. Suitable peak detection procedures must be invoked in the transform plane, and this is usually preceded by application of a smoothing filter to reduce noise and to improve the accuracy of the peak localization (Section 3.4). If we expected only a single line, we restrict peak detection to the maximum peak; if we want all significant lines we set a threshold for the peak accumulation at a value that is determined by the minimum number of pixels in a line.

The Hough transform technique for line detection finds only lines of infinite length, not finite line segments. Put differently, the Hough transform yields two line parameters that describe a line, but not end-line coordinate parameters that would be needed to describe finite segments. End lines can be found by examination of the overlap of the Hough-determined line on the points of the original image.

4.10.2 CIRCLE DETECTION

For detection of a circle, parameterized by $(x - a)^2 + (y - b)^2 = r^2$, the Hough transform space parameters are (a, b, r). The procedure is the same as that for line detection. That is, every point in the spatial domain is mapped to all possible points in the parameter space; a peak in that space indicates the presence of a circle with parameters equal to the coordinates of that location. Compared with the case of line detection, in which the Hough space is two dimensional, circle detection requires examination of a three-dimensional parameter space. This extra dimension is the cause of additional computational and memory expense, and it is for this reason that the Hough transform is far less popular for geometrical objects other than the line.

4.10.3 ENHANCEMENTS

Even line detection can be computationally expensive, and methods that speed up execution of the classic algorithm are worthwhile. These methods usually entail estimating the slope of a line at each point and mapping to only that slope value in Hough space. This requires additional information beyond the coordinates of disconnected points, but this is often available. For instance, often the raw data consist of not only isolated points, but also dashed or broken lines. That is, many points are connected and form smaller line segments of the same slope as the larger line.

In this case, since two or more points are connected, a slope can be estimated and the corresponding group of points can be mapped to a single slope in the Hough space. If an edge map generated from a gray-scale image (Sections 3.4 and 3.5) is to be subjected to a Hough transform analysis, it is computationally advantageous to store the direction of the edge, along with the location of edge points. This edge direction provides an estimate of the slope to find lines (or higher-order curves) representing the edge.

References and Further Reading

The Hough transform is introduced in P. V. C Hough, "Method and means for recognizing complex patterns," U.S. Patent 3,069,654 (1962).

Applications of the Hough transform are described in D. H. Ballard and C. M. Brown, *Computer Vision*, (Prentice-Hall, Englewood Cliffs, NJ, 1982), pp. 123–128; and in E. R. Davies, *Machine Vision: Theory, Algorithms, Practicalities* (Academic, London, 1990).

Program

hough

```
            transforms binary image from spatial domain [(x,y)
            coordinates) to polar coordinate domain ((rho, theta)
            coordinates]; peak in the polar ("Hough") domain
            indicates a dominant line in the spatial domain;

            NOTE: origin (0,0) of Hough transform image is in
                  top-left corner; rho increases along horizontal
                  axis to maximum rho equal to image diagonal
                  length; theta increases downward from 0 radians
                  to PI radians.

     USAGE: hough inimg outimg [-b BORDER] [-d]

 ARGUMENTS: inimg: input image filename (TIF)
            outimg: output image filename (TIF)

   OPTIONS: -b BORDER: remove noise at borders of image by omitting
                  image for this number of rows/cols [default
                  = 2].
            -d: display peak in Hough transform space.
            -L: print Software License for this module
```

5

Analysis of Lines and Line Patterns

This chapter introduces a set of methods for the encoding and analysis of images containing lines. We use the terms line and line structure in this chapter to describe a chain of ON-valued pixels that are connected from one pixel to neighboring pixels, and whose width is a single pixel. A line can be straight or curved, and is simply connected, that is, without branches. A line structure may consist of a single or many line segments that are connected at junctions. One can also say that a line is made up of one line segment with two end points or no end points (loop); and a line structure with one line segment is equivalent to a line.

A line image can be a result of previous processing steps such as binarization (Section 3.9), thinning (Section 4.7), or edge detection (Sections 4.4, 4.5), or it can be an unprocessed image containing features in the form of lines such as those in a diagram or flow chart. As with gray-scale (Chap. 3) and binary (Chap. 4) images, filtering is the first step of processing line images to reduce noise introduced by prior processing steps or contained in the original image.

We describe methods of filtering line images that facilitate subsequent analysis. We also introduce strategies to construct piecewise linear approximations and other procedures to construct an approximate representation of lines. The objective here is to simplify the description of lines by use of parameters such as the end points or the slope and intercept of a straight line for global statistical analysis (such as average line length) and for line identification and matching.

Features such as branch and end points are of special interest. For instance, in the field of statistical physics these can indicate topological defects and provide the basis for morphological pattern analysis. A global statistical analysis facilitates the identification of structural motifs in the pattern morphology. To illustrate the salient concepts, we discuss (in Section 5.7) the analysis of labyrinthine patterns, formed in certain magnetic films and related materials. The examples will also demonstrate the fact that it is especially the investigation of disordered patterns that benefits from the application of direct-space methods of analysis.

Section Overview

Section 5.1 introduces the chain code: this provides a transformation from pixel space – individual pixels of ON and OFF values – to a concise representation of all lines in that image. This representation provides a lossless compression, preserving all topological and morphological information. As a result, the chain code provides a substantial advantage in speed and effectiveness for the analysis of line patterns.

Section 5.2 introduces the representation of line segments, portions of lines delimited by junction features such as branch points and end points and describes methods of noise reduction suitable to this representation. Lines contain characteristic noise of a type that differs from that encountered in gray-scale images, for example in the form of line breaks or short, spurious segments (spurs) emanating from longer lines. These can be removed in an approach that uses the context and information describing the line structure, such as its length or area. This is an analytical step that relies on the recognition of features and thus goes beyond simple pixel processing.

Section 5.3 describes polygonalization of lines. The objective is to fit connected, straight segments to portions of lines so as to simplify their representation. This simplified representation facilitates subsequent analysis.

Section 5.4 is devoted to the important topic of detecting critical points, or landmark points, along lines. These features, including corners and curvature maxima joined by otherwise featureless lines capture pertinent shape information. As discussed in connection with object shape analysis (Section 4.4), this reduction of line objects to a set of connected critical points creates an even more concise description than that generated by polygonalization.

Section 5.5 introduces the simplest approximation to a line feature, a linear fit between end points or junction features. The purpose is to obtain the closest straight-line fit to a given line segment.

Section 5.6 introduces cubic spline fitting, a more general method of approximating curvilinear objects. Splines are used primarily for smoothing curves to reduce noise present in a ragged chain of points and so to bring out the general shape of the underlying curve. While smooth appearance by itself is not an important result of image analysis, the spline fit often yields a parameterization of the curve by which to identify and match lines visually.

Section 5.7 provides methods for the analysis of the morphology and topology of line patterns. In contrast to previous sections in this chapter, the focus of the analysis is not on individual lines but on patterns formed by multiple lines or multiple distinct (but connected) line segments. For example, groups of parallel lines, such as those commonly seen in contour maps, can form shapes that connote information about underlying structure. The topology of lines, manifesting itself in the form of branches, cuts, or line junctions on road maps, reflects line connectivity.

5.1 Chain Coding

Typical Application(s) – efficient representation of line patterns such as contour maps, engineering diagrams, fingerprints, and magnetic domain patterns.

Key Words – directional coding, chain code, primitives chain code (PCC).

Related Topics – region detection (Section 4.3), thinning (Section 4.7).

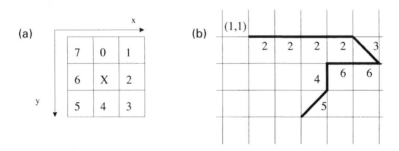

(c) Freeman: (1,1) 2, 2, 2, 2, 3, 6, 6, 4, 5, 1

Figure 5.1.1. Pictorial Example. The Freeman chain code: (a) chain direction codes as emanating from central pixel X, (b) an example line structure with starting code coordinate and Freeman directions, (c) Freeman code for (b): the last integer marks the end of the chain (see text).

When objects are described by their contours (Section 4.3) or skeletons (Section 4.7), they can be represented more efficiently than by listing all ON- and OFF-valued object pixel locations. The most common method to achieve this more efficient representation is chain coding, which retains only the ON-valued pixels marking the object contour or skeleton. These are represented not as absolute pixel coordinates but as a sequence of directions from one ON-valued pixel to its neighbor along lines and curve. In a discrete pixel grid, there are eight possible directions from any pixel to its neighbors.

There are two advantages to chain coding by direction instead of simply storing the raster coordinates of each ON-valued pixel. One is increased storage efficiency. For images exceeding 256 × 256 pixels in size, coordinates would need to be represented as two 16-bit words; in contrast, an eight-direction code can be packed into 3 bits (although they are often just stored one per 8-bit byte to avoid packing and unpacking). A more important advantage arises from the fact that the chain represents the connectedness of the line, and this offers a convenient basis for processing (e.g., smoothing of deviations from an otherwise smooth curve) and analysis (e.g., junction detection and straight-line recognition).

5.1.1 FREEMAN CHAIN CODE

A popular chain code is illustrated in Fig. 5.1.1(a) [Freeman, 1970]. A line connecting a given pixel to its next ON neighbor to the north is assigned the direction (chain) code 0, northeast is 1, east is 2, etc. The beginning ON pixel of the chain is coded by its coordinate location. The end of a chain is encoded by the reverse of the previous direction, a sequence of directions that is otherwise impossible; for example, $(\ldots, 0, 4)$, $(\ldots, 1, 5)$, $(\ldots, 2, 6)$, etc., represent ends of chains in the directions 0, 1, 2, etc.; see Fig. 5.1.1.

Thus Freeman chain coding is performed as follows. Image pixels are searched in raster order until the first ON pixel is located. Its coordinate location is stored. The eight-connected neighbors are examined, the direction of the first ON neighbor is stored, and that pixel is erased (set to OFF). The neighbors of this next pixel are examined in the same way, and the chain is coded until its end is found. Then raster searching for the next chain is resumed at the starting location of the chain just completed. This process is continued to the last pixel of the image.

The Freeman chain code is highly effective for compression of line images. It is also useful for analyzing simple lines and contours. However, since there is no provision for maintaining branching line structures, it is less useful for composite lines joined at junctions. This is because – when applying the Freeman code – one handles branches by following the first path found when testing the neighbors at each junction; remaining branches are encoded as separate lines. That is, there is no distinction in the code between connected branches and separate lines. While this is acceptable for image compression, image analysis of the type described in the following subsections depends on a knowledge of the topology of the complete line structure with all its branches.

5.1.2 PRIMITIVES CHAIN CODE

The primitives chain code (PCC) is an extension of Freeman chain code designed to preserve information on branching and junction topology [O'Gorman, 1992]. PCC introduces codewords in the form of a packed representation of zero to three single-connection direction codes. That is, for chains of one to three pixel connections, there are corresponding PCC codewords; longer chains are represented by a sequence of PCC codes. There are 241 possible configurations of zero to three pixel connections,

Table 5.1.1. PCC Features

FEATURE	SYMBOL	PCC CODEWORD	BRANCHES Incoming	BRANCHES Outgoing
End	E	242	1	0
Line	L	243	1	1
Bifurcation	B	244	1	2
Cross	C	245	1	3
Start	S	246	0	1
Line break	R_L	247	0	2
Bifurcation break	R_B	248	0	3
Cross break	R_C	249	0	4
Stop	—	255	—	—

Table 5.1.2. PCC Codes (†– indicates chain sequences with <3 connections.)

PCC	$d_0 d_1 d_2$	PCC	$d_0 d_1 d_2$	PCC	$d_0 d_1 d_2$	PCC	$d_0 d_1 d_2^{\dagger}$	PCC	$d_0 d_1 d_2$
1	000	51	202	101	431	151	645	201	7–
2	001	52	207	102	432	152	646	202	00–
3	002	53	210	103	433	153	653	203	01–
4	006	54	211	104	434	154	654	204	02–
5	007	55	212	105	435	155	655	205	06–
6	010	56	213	106	442	156	656	206	07–
7	011	57	217	107	443	157	657	207	10–
8	012	58	220	108	444	158	660	208	11–
9	013	59	221	109	445	159	664	209	12–
10	017	60	222	110	446	160	665	210	13–
11	020	61	223	111	453	161	666	211	17–
12	021	62	224	112	454	162	667	212	20–
13	022	63	231	113	455	163	670	213	21–
14	023	64	232	114	456	164	671	214	22–
15	060	65	233	115	457	165	675	215	23–
16	065	66	234	116	464	166	676	216	24–
17	066	37	235	117	465	167	677	217	31–
18	067	38	242	118	466	168	700	218	32–
19	070	69	243	119	467	169	701	219	33–
20	071	70	244	120	531	170	702	220	34–
21	075	71	245	121	532	171	706	221	35–
22	076	72	310	122	533	172	707	222	42–
23	077	73	311	123	534	173	710	223	43–
24	100	74	312	124	535	174	711	224	44–
25	101	75	313	125	542	175	712	225	45–
26	102	76	317	126	543	176	713	226	46–
27	106	77	320	127	544	177	717	227	53–
28	107	78	321	128	545	178	753	228	54–
29	110	79	322	129	546	179	754	229	55–
30	111	80	323	130	553	180	755	230	56–
31	112	81	324	131	554	181	756	231	57–
32	113	82	331	132	555	182	757	232	60–
33	117	83	332	133	556	183	760	233	64–
34	120	84	333	134	557	184	764	234	65–
35	121	85	334	135	560	185	765	235	66–
36	122	86	335	136	564	186	766	236	67–
37	123	87	342	137	565	187	767	237	70–
38	124	88	343	138	566	188	770	238	71–
39	131	89	344	139	567	189	771	239	75–
40	132	90	345	140	570	190	775	240	76–
41	133	91	346	141	571	191	776	241	77–
42	134	92	353	142	575	192	777	242	end
43	135	93	354	143	576	193	—	243	line
44	170	94	355	144	577	194	0–	244	bif
45	171	95	356	145	600	195	1–	245	cross
46	175	96	357	146	601	196	2–	246	start
47	176	97	421	147	606	197	3–	247	line br
48	177	98	422	148	607	198	4–	248	bif br
49	200	99	423	149	643	199	5–	249	cross br
50	201	100	424	150	644	200	6–	255	stop

and there are thus 241 corresponding PCC codewords to represent pixel chains. Unlike the conventional chain code, PCC contains, in addition to those codes required for representing pixel chains, special codes for junction and end-point features (Table 5.1.1).

These line features include start and end points of lines, bifurcation and cross junctions, and breaks. Start or end points of a line just differ in the direction of coding. Bifurcation and cross junctions are characterized by the junction of three and four branches, respectively. A break is a feature that is created artificially by coding. For example, a loop has no beginning or end so it must be broken at some point for the process of coding. A break in a line is called a line break, a break at a trijunction is called a bifurcation break, and a break at a quadjunction is called a cross break. There are eight feature codes, plus a stop code, so there are 250 PCC codewords in total.

Table 5.1.2 contains a complete list of all codes. Pixel chains are represented by codes 1 to 241. Features are assigned codes 242 to 249; code 255 is the stop code. Each PCC codeword is an 8-bit byte. Figure 5.1.2 shows two examples of pixel chains along with the corresponding Freeman code and PCC.

To generate a PCC representation of a line image, a raster-order search is performed until an ON pixel is encountered. The (x, y) image location of the point is stored as well as a PCC code for a start or a break feature. Then PCC chain coding is begun from this point: sets of up to three successive pixels are PCC coded and erased, any features are coded when they are encountered, branches are followed to their ends, loops are followed to their initial coding location, and the process ends when the last pixel of the entire line structure is coded and erased. The suspended raster search for other line structures then continues after the starting position of the previous line structure, and this searching and coding process continues for the entire image.

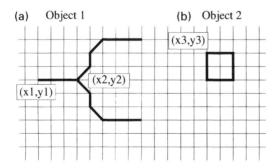

(a)   Object 1   (b)   Object 2

(c) Freeman: $(x1,y1)$ 2,2,2,1,0,1,2,2,2,6; $(x2,y2)$ 3,4,3,2,2,2,6
      PCC: S$(x1,y1)$, 60, B, 25, 60, E, 88, 60, E

(d) Freeman: $(x1,y1)$ 2,2,4,4,6,6,0,0,4
      PCC: B$(x3,x3)$ 62, 118, 194, E, E

Figure 5.1.2. Examples of Freeman and PCC line coding: (a) line object 1, (b) line object 2, (c) Freeman and PCC coding for (a) [note the features: S (start), B (bifurcation), and two E's (end points)], (d) Freeman and PCC coding for (b) [note the features: B (line break) and two E's (end points)].

Decoding of a PCC representation to recover an image is performed in a straightforward manner. A start or break code is first read, followed by the beginning (x, y) coordinates of the line structure. Each subsequent PCC codeword is decoded, and the corresponding chain is written to the image. When a junction or break feature code is read, the current (x, y) coordinate is pushed into a stack – once for a bifurcation feature and twice for a cross feature – and decoding is resumed. Once a line end feature is reached, the beginning location of the next pending branch is popped off the stack and chain decoding is resumed at that new position; or if the stack is empty, further remaining line structures are decoded. The last PCC codeword is the stop code indicating that no line structures remain in the image.

PCC requires $\sim$10%–20% less memory, on average, compared with the Freeman chain code, mainly because of the richer syntax of features. However, the major advantage of PCC from our perspective of image analysis is that features are identified and connectivity is maintained for later image analysis. A disadvantage is that PCC is somewhat more complex to code and decode than the Freeman code.

References and Further Reading

[Freeman, 1970] H. Freeman, "Computer processing of line drawing images," Comput. Surv. **6**(1), 57–98 (1974).

[O'Gorman, 1992] L. O'Gorman, "Primitives chain code," in *Progress in Computer Vision and Image Processing*, A. Rosenfeld and L. G. Shapiro, eds. (Academic, San Diego, 1992), pp. 167–183.

Programs

pcc

produces Primitives Chain Code (PCC) for input line image and writes this output to a file containing PCC code.

NOTE: input image should be a line image.

USAGE: pcc inimg outfile [-L]

ARGUMENTS: inimg: input image filename (TIF)
outfile: output file containing PCC (BINARY)

OPTIONS: -L: print Software License for this module

pccde

decodes Primitives Chain Code (PCC) in input file and generates an output image.

USAGE: pccde infile outimg [-L]

ARGUMENTS: infile: input filename (.pcc) containing PCC
outimg: output image filename (TIF)

OPTIONS: -L: print Software License for this module

pccdump

prints out PCC code and features stored in .pcc file
and writes this output.

USAGE: pccdump infile [-L]

ARGUMENTS: infile: input file (.pcc) containing PCC

OPTIONS: -f FEATURES_FLAG: when set, display only PCC features,
not chains.

-L: print Software License for this module

---

pccfeat

decodes Primitives Chain Code (PCC) in input file and
writes output file with superimposed squares over
features, or lists features and chain codes

USAGE: pccfeat infile outimg [-f or -c or -s] [-L]

ARGUMENTS: infile: input filename (PCC)
outimg: output image (TIF)

OPTIONS: -f FEATURES_FLAG: when set, list only PCC features,
not chains

-c COORDS_FLAG: when set, list only (x,y) coordinates

-s SUMMARY_FLAG: when set, list only total number
of features.

-L: print Software License for this module

5.2 Line Features and Noise Reduction

Typical Application(s) – removal of spurious line features from skeletonized patterns.

Key Words matched line filters, chain code, primitives chain code (PCC), thin line code (TLC).

Related Topics – noise reduction (Section 3.2), binary noise removal (Section 4.2), thinning (Section 4.7), chain code (Section 5.1).

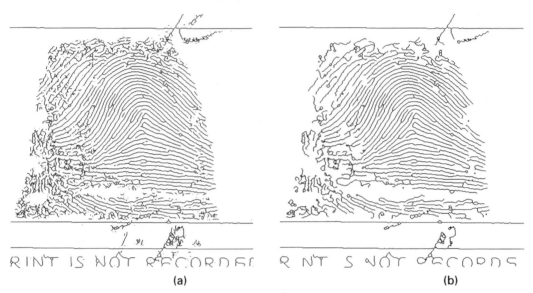

(a)   (b)

Figure 5.2.1. Pictorial Example. Noise reduction applied to a line image: (a) a thinned finger-print image contains many desired ridges plus much undesired noise; (b) many of the "spurs," short lines emanating from longer lines and bridges and short lines joining two parallel lines, are removed.

Noise reduction has been previously discussed for gray-scale (Section 3.2) and binary images (Section 4.2). Line images are derived from gray-scale and binary images, and the question arises as to the necessity for yet another stage of noise reduction. Part of the answer lies in the fact that each processing step, transforming an image from a given representation such as gray-scale or binary to a new representation such as the skeleton, may introduce noise.

5.2.1 LINE NOISE

Noise in line patterns typically has the form of isolated, short lines, spurious lines branching from longer lines "spurs," gaps in lines, and small loops in lines. These can have their origin in noise present in the initial gray or binary image, in discretization error, or in imperfections of the processing technique. A culprit in the latter category is the thinning algorithm (Section 4.7), which may introduce spurs that are due to small extrusions (bumps) on otherwise smooth, elongated regions. As with all noise reduction, the objective here is to eliminate as much of the noise as possible without also removing parts of the signal.

Noise reduction at the stage of a line pattern representation can benefit from the use of a higher level of image representation – lines versus pixels. In the above-mentioned example, once thinning has generated a line representation, bumps on lines in pre-thinned images are transformed into spurs. Now, instead of referring to a nondescript feature (small bump), the transformation yields a more quantitative description that includes spur length and orientation with respect to the line from which it emanates. This is an example of the general principle that irreversible image modifications are to be avoided at early processing stages when a subsequent, higher level of analysis can facilitate more informed, and thus better, judgment.

As always, the first and critical step in performing noise reduction is to distinguish noise and signal: noise in one application may be signal in another. In the present context, noisy lines might be isolated lines, spurs, and loops that are shorter than given thresholds, or lines may have gaps (see Fig. 5.2.1).

One approach to reducing this type of noise in line images is to use matched line filters. This involves convolution of the image with filters matching one-, two-, or three-pixel lines and marking them for elimination, and, in an analogous step, application of filters matching one-, two-, or three-pixel gaps between lines and joining them. This is suitable and effective for small features, but it is impractical for longer line features, given that the number of possible matched filter masks greatly increases with line length.

5.2.2 THIN LINE CODE

We describe a more versatile approach based on a hierarchical representation called the thin line code (TLC) [Jagadish and O'Gorman, 1989]. TLC operates on chains and line features instead of pixels. The lowest level in TLC is the PCC (see Section 5.1); higher levels are line segments, line composites (made up of multiple, joined line segments), and complete line structures (containing all joined lines) at the top level. When this encoding scheme is used, noise reduction can benefit from the use of contextual information to aid the analysis and can so remove features whose size or type would make them difficult to identify by a matched filter. In addition, the hierarchical representation permits more complex operations such as isolation and retention of only long, short, or midlength lines, analogous to low-pass, high-pass, and bandpass filtering, respectively (Chap. 3).

TLC Hierarchy

To introduce the TLC, we refer to the line structure representing the letter H in Fig. 5.2.2. According to the discussion in Section 5.1, six PCC features are identified in this structure: four end points and two bifurcations. These features plus the interfeature PCC codewords make up the lowest level of TLC, level 0.

Level 1 is said to contain line segments, that is, those chains between end-point or junction features. In the example of the letter H, there are five line features: four

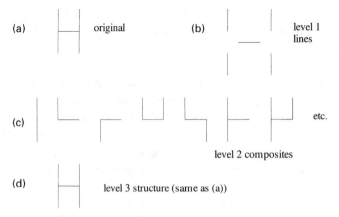

Figure 5.2.2. TLC describing the letter H and portions thereof: (a) original line structure, (b) TLC level 1 line segments, (c) some TLC level 2 line composites, (d) TLC level 3 line structure [complete object, same as (a)].

vertical line segments from end points to bifurcations and one horizontal line segment between bifurcation features.

We may wish to consider subgroupings of elements in this line structure, composed of more than one, but of less than all, connecting line segments: We refer to such subgroupings as composites at TLC level 2. In the example of the letter H, each of the two vertical strokes is a composite that consists of two vertical line segments, and there are many more possible composite configurations, some shown in Fig. 5.2.2. The top TLC level, level 3, contains line structures that are complete objects of connected lines. Figure 5.2.3 summarizes the TLC hierarchy with an example of a line diagram and its components at each level.

Associated with each TLC level are feature descriptors. At the lowest level, there are the PCC features that we have already discussed. Aside from type descriptors, the only additional feature descriptor at this level is the (x, y) location.

Figure 5.2.3. TLC hierarchy, object types, and example.

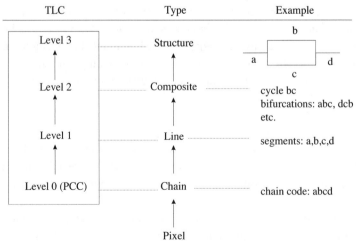

At the line level, there are more descriptors: a line has two end points, each with (x, y) coordinate locations, PCC feature type, and direction; it has a length, expressed in pixels, and an area, defined by the bounding box that completely contains the line. These feature descriptors can be selected, depending on the application, to eliminate noise or to obtain particular objects.

At the composite level, each composite object is specified by its type, which is specified with respect to the features it encompasses, for instance, bifurcation composite (three line segments meeting at a junction), a cross-junction composite (four line segments meeting at a junction), a loop (of multiple lines) composite, etc. A composite has such descriptors as the number of level 1 lines, the total length of pixels, bounding box, etc. Because of the huge number of potential composites, they tend to be defined specifically for each particular application.

The top level line structure has features similar to those on the lower levels. We do not attempt here to make a comprehensive list of all features at a given level of the hierarchy, simply because additional particular features may become desirable in a new application. In general, it is not necessary to evaluate any but those features at each level that are of direct interest in the specific task at hand.

Application of TLC

TLC is used in the same manner as filters are used. First, features of interest are selected and the range of feature dimensions, or of noise dimensions, is set. This is done with knowledge of the characteristics of desired signal or undesired noise. Then, starting with the PCC representation of the image, TLC features and their parameters are identified. These feature parameters are compared with the specified range of feature parameters, and features are retained or removed accordingly. The result is the filtered PCC of the image, which can be decoded to yield the filtered image.

Figure 5.2.1 shows an example of TLC filtering applied to remove noise in a fingerprint image. The TLC features marked for elimination are short isolated lines and spurs. Here, each attribute, such as "small," "short," etc., is quantified by some number that is chosen by the user to yield the true fingerprint features that are used for matching – these are true long line end points and true bifurcations. It is apparent from the result that noise removal facilitates the task of locating these.

Figure 5.2.4 shows an example of TLC filtering at the structure level. The image is first binarized (Section 3.9) and a contour image determined (Section 4.3). This contour image is coded in PCC and subjected to filtering on the basis of structure size. One can extract regions on the basis of the bounding box area that the structure encompasses or the arc length of the structure contour.

In addition to the advantage of feature-based and contextual filtering, TLC also provides a computational advantage over pixel-based processing because there are many fewer TLC structures and features than there are pixels. Take the example of performing document processing on a page of text. Typical images for document processing applications are rendered at a resolution of 300 dots per inch; consequently, an 8.5×11 in. page contains 2550×3300, or 8.4 million pixels. If processing were performed with a variety of matched filters to extract features, computation time would be large. However, if thinning is performed on the pixel image, followed by PCC coding,

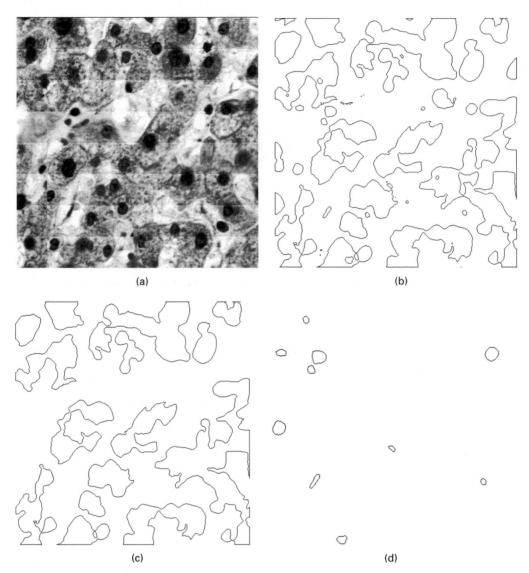

(a)   (b)

(c)   (d)

Figure 5.2.4. The result of line filtering based on structure area and contour length: (a) original liver tissue image; (b) image after low-pass filtering, binarization, and region determination by means of contour lines; (c) result of filtering based on area, retaining all structures whose area exceeds 2000 pixels; (d) result of filtering based on contour arc length, retaining all structures whose length is above 27 and below 90. (Reprinted with Permission from [O'Gorman and Sanderson 85], Copyright © 1986 by IEEE.)

the resulting representation typically requires only ~1% of the memory required to store the original image. TLC encoding leads to further compression: For the example of the text page, there are perhaps 500 to 1000 line structures including characters, dots, periods, etc., and hence only ~0.01% of the original image. Searching for a line type or feature in this representation is of course much faster than doing so on the original image. This is an example for which feature-based image analysis (TLC processing) is preferable to pixel-based image processing (matched filtering).

References and Further Reading

[Jagadish and O'Gorman, 1989] H. V. Jagadish and L. O'Gorman, "An object model for image recognition," IEEE Comput. **22** (12), 33–41 (1989).

[O'Gorman et al., 1985] L. O'Gorman, A. C. Sanderson and K. Preston, Jr., "A system for automated liver tissue damage analysis: methods and results," IEEE Trans. Biomedical Engineering Vol. BME-32(9), pp. 696–706 (Sept. 1985).

Programs

linerid

```
        filters out small lines of two types:
        EE lines - isolated end-line-to-end-line segments;
        FE lines - attached feature-to-end-line segments.
 USAGE: linerid infile outfile [-e MINLINE_EE] [-f MINLINE_FE]
        [-L]
ARGUMENTS:        infile: input filename (PCC)
                 outfile: output filename (PCC)
OPTIONS: -e MINLINE_EE: minimum length of isolated end-end lines;
                        default=30;
         -f MINLINE_FE: minimum length of feature-end, attached
                        lines; default=10;
                    -L: print Software License for this module
```

linefeat

```
        lists line features.
 USAGE: linefeat infile [-s] [-L]
ARGUMENTS:     infile: input filename (PCC)
OPTIONS:          -s: print summary only
                  -L: print Software License for this module
```

linexy

```
        lists (x,y) coordinates for each line segment in all
        image line structures; a line segment connects two PCC
        features, including end points and junctions.
 USAGE: linexy infile [-c] [-s] [-L]
ARGUMENTS: infile: input filename (PCC)
OPTIONS:        -c: display (x,y) coordinates only;
                -s: display summary of features only;
                    default is to display both coordinates and
                    summary.
                -L: print Software License for this module
```

structrid

> filters line structures, that is, objects made up of one or many connected line segments, removing those whose features are outside of the range given.

USAGE: structrid infile outfile [-ll] [-hl] [la] [-ha] [-L]

ARGUMENTS: infile: input filename (PCC)
outfile: output filename (ASCII)

OPTIONS: -ll LOW LENGTH: lowest length structure to retain;
-hl HIGH LENGTH: highest length structure to retain;
-la LOW AREA: lowest bounding box of structure to retain;
-ha HIGH AREA: highest bounding box of structure to retain;

-L: print Software License for this module

---

structfeat

> lists line structures, that is, objects made up of one or many connected line segments.

USAGE: structfeat infile [-s] [-L]

ARGUMENTS: infile: input filename (PCC)
OPTIONS: -s: print summary only;

-L: print Software License for this module

5.3 Polygonalization

Typical Application(s) – smoothing of noisy lines, concise approximation of a curve.

Key Words – straight-line approximation, curve representation.

Related Topics – thinning (Section 4.7), chain coding (Section 5.1), critical point detection (Section 5.4), line fitting (Section 5.5), curve fitting (Section 5.6).

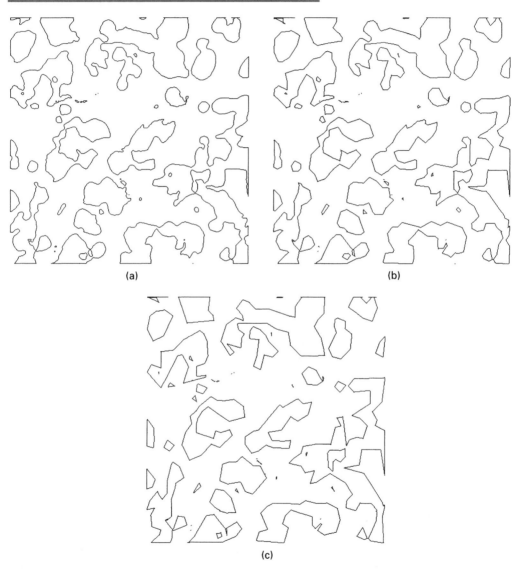

(a)

(b)

(c)

Figure 5.3.1. Pictorial Example. Illustration of polygonalization: (a) contour image of binarized and contour-coded liver tissue image as in previous section, (b) polygonal fit with small error value, (c) polygonal fit with larger error value. (Reprinted with Permission from [O'Gorman et al. 85], Copyright © 1986 by IEEE.)

It is often unnecessary, or even undesirable, to retain all pixel locations along lines and contours in an image. Instead, an approximate representation in terms of straight-line segments may suffice.

5.3.1   POLYGONALIZATION AS CURVE APPROXIMATION

A set of connected linear segments yields an approximation to any (smooth) curve; the closeness of the approximation depends on the degree of curvature and the chosen minimum segment length. This approximation procedure, known as polygonalization, produces a simplified parameterized representation of a line image. It thereby not only considerably reduces storage requirements, but also facilitates subsequent image analysis.

Figure 5.3.1 depicts the result of polygonalizing a contour image of a binarized and contour-coded liver tissue image (the same as shown in Section 5.2). The images illustrate what is intuitively obvious: the closer the approximation, the higher the number and the shorter the length of the line segments produced by the polygonalization.

The user controls the degree of approximation by specifying a limit to the tolerable deviations from the original curve; Fig. 5.3.2 illustrates the effect of selecting different tolerance thresholds. The trade-off is between the simplicity of the approximate representation and its fidelity in reproducing salient shape features.

5.3.2   POLYGONALIZATION METHOD

One general approach to polygonalization is to start from an end-line point and to join a line segment between it and the adjoining point on the line (Fig. 5.3.3). The area between this line and the curve is measured. If the area divided by the line length is smaller than a user-chosen threshold, then the same process is repeated for the next point. This is repeated for increasingly distant points along the curve until the ratio exceeds a preselected threshold. A segment is then constructed between the starting point and the current point, called the break point. The break point becomes the new starting point, and the same process is then repeated (Fig. 5.3.3).

A modification of this strategy improves the closeness of approximation. After a break point is found, the previous break point is adjusted in the following way. A line is joined between the current break point and the break point two previous to it. Then the maximum perpendicular distance is found from that line to a point on the curve. The previous break point is adjusted to that point. This break-point adjustment takes advantage of the fact that the perpendicular distance between the line and the original curve provides a better measure of error than does the area/length measure.

This modified procedure combines two published methods, relying on the perpendicular distance measure, as used by Ramer (1972), and on the area/length ratio, as used by Wall and Danielsson (1984). While the former yields a better approximation, the latter is faster to compute. In either case, there is a trade-off in the choice of the threshold value: A smaller threshold will generally ensure a closer approximation, but it also will produce a greater number of polygonal segments and will consume more computation time.

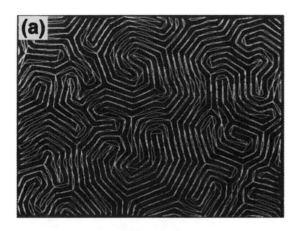

Figure 5.3.2. Polygonal approximation to the skeleton of a labyrinthine magnetic stripe pattern: (a) coarse approximation consisting of 427 segments, (b) closer approximation consisting of 708 segments. (Reprinted with Permission from [Seul et al., 92], Copyright © 1992 Taylor & Francis.)

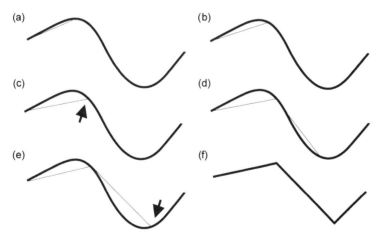

Figure 5.3.3. Progression of polygonalization along a curve: (a) original curve; (b)–(f) each line segment is fixed to the initial point on the curve and is extended along the curve until the area subtended by line segment and curve exceeds a preset threshold. Another segment is then begun.

Implementation

Consider a chain of pixels for which a straight-line approximation is desired. To simplify calculations, we first translate the chain such that the starting point coincides with the origin. The first chord is drawn between the starting point and the third point in the chain. The triangular area between the line and the chain is the first error measure (if the first three points are collinear, the error will be zero). Chords to each subsequent point in the chain are constructed, and the error is accumulated according to $A_i = A_{i-1} + \Delta A_i$, where A_i denotes the current subtended area, A_{i-1} the area subtended at the previous point (x_{i-1}, y_{i-1}), and ΔA_i the error increment incurred in reaching the current point (x_i, y_i); ΔA_i is given in terms of the increments $\Delta x_i = x_i - x_{i-1}$ and $\Delta y_i = y_i - y_{i-1}$ in the form

$$\Delta A_i = x_i \times \Delta y_i - y_i \times \Delta x_i. \tag{5.1}$$

The length of the chord is simply

$$L_i = \sqrt{x_i^2 + y_i^2}. \tag{5.2}$$

The action to be taken is dictated by the comparison of the error reading and the user-set threshold T. If $|A_i|/L_i \leq T$, proceed to the next point in the chain and perform the same comparison. Otherwise, mark the current point as a break point in the polygonal approximation.

Once a new break point has been found, the previous break point is adjusted as follows. Since the area/length error tends to cause the location of the break point to be further along the line than desired, only contour points preceding the middle break point need be examined. Proceeding from the middle break point in the backward direction, we adjust that break point to the previous point and measure the perpendicular distance between the new chord and the curve. When this distance is smaller than the previous, the break point is adjusted to the contour point where the perpendicular distance is greatest.

The error is calculated in an efficient manner as follows. For a straight line subtending an angle θ with the x axis, the perpendicular distance $d_\perp$ to the curve is

$$d_\perp = \Delta y \cos\theta, \quad -\pi/4 < \theta < \pi/4, \tag{5.3}$$

where Δy is just the difference in the y coordinate value between a given contour point and the point on the chord directly above or below. To find the maximum perpendicular distance, it suffices to evaluate Δy for each contour point. Thus the maximum distance is obtained with the above multiplication only once per chord. When chords are nearly vertical, $\cos\theta$ is small and the calculation may be simplified to

$$d_\perp = \Delta x \sin\theta, \quad \theta > \pi/4, \theta \leq -\pi/4, \tag{5.4}$$

where Δx is the difference in x coordinates between a given contour point and the point on the chord directly to the right or the left.

Polygonalization is performed in the same manner for closed and for open contours. For a closed curve, the first and the last points are considered identical. Thus polygonalization for a circle would begin with a line segment, then proceed to four line segments of a square, then to an octagon and so on, until the error threshold is attained.

5.3.3   POLYGONALIZATION VERSUS OTHER METHODS

Vectorization

Polygonalization can be distinguished from another method of approximation, vectorization. When performed on the original image in lieu of thinning, vectorization produces thin, straight-line segments that run entirely within the (nonthinned) linewidths of the original data. In contrast, polygonalization is performed following thinning with the aim to compress data for storage and to facilitate further analysis. Unlike vectorization, polygonalization produces a representation that usually is not contained within the original linewidths.

When results of the two methods are compared, polygonalization generally will yield more concise results than vectorization, given the constraint of the latter to confine lines within original linewidths. Vectorization is often used for compressing engineering drawings because these drawings can be very large and the elimination of pre-processing and simplicity of implementation make this method the preferable one with respect to computational efficiency. However, polygonalization is the more versatile of the two methods because the level of approximation can be adjusted and the preprocessing steps enable noise reduction. For these reasons, we suggest the use of polygonalization for all but specialized applications.

Critical Point Detection and Curve Fitting

Polygonalization also must be distinguished from critical point detection (Section 5.4) and spline fitting (Section 5.6); differences are evident from a comparison of figures in these respective sections with figures of this section. Critical point detection is appropriate for images containing particular geometric shapes composed of straight lines and curves, especially for synthetic pictures or pictures of human-made objects. For these, the objective is to obtain a precise replication of contours and lines that accurately reproduce the distinct (human-made) geometric shapes. Conversely, precise feature detection is less important for natural curvilinear shapes lacking sharp corners and precise shapes. In this case, polygonalization provides a reliable and convenient tool to achieve data reduction and to encode the general shape of a contour for subsequent analysis.

Polygonalization and curve fitting represent methods of approximating general shapes adequately. However, for polygonalization, correspondence between the original curve and the features reproduced by the approximation may be poor or nonexistent. Furthermore, the number and the location of segments will depend on the starting point and on the direction of chain traversal. In contrast, curve fitting best reproduces all features in the original and remains least affected by different orientation and scale of the object.

References and Further Reading

[O'Gorman et al., 1985] L. O'Gorman, A. C. Sanderson and K. Preston, Jr., "A system for automated liver tissue damage analysis: methods and results," IEEE Trans. Biomedical Engineering Vol. BME-32(9), pp. 696–706 (Sept. 1985).

[Ramer, 1972] U. Ramer, "An iterative procedure for the polygonal approximation of plane curves," Comput. Graph. Image Process. **1**, 244–256 (1972).

[Seul et al., 1992] M. Seul, L. R. Monar, and L. O'Gorman (see p. 279) "Pattern Analysis of Magnetic Stripe Domains: Morphology and Topological Defects in the Disordered State," Philos. Mag. **B 66**, 471–506 (1992).

[Wall and Danielsson, 1984] K. Wall and P.-E. Danielsson, "A fast sequential method for polygonal approximation of digitized curves," Comput. Graph. Image Process. **28**, 220–227 (1984).

Program

`fitpolyg`

        performs polygonal line fitting to image lines; straight-line fits are made to portions of each segment to best approximate the chain of points comprising the segment; a segment is a chain of points between features, either end-point features or junction features.

    USAGE: `fitpolyg infile outimg [-t THRESHOLD] [-c] [-L]`

ARGUMENTS: infile: input file name (PCC)
               outimg: output image filename (TIF)

OPTIONS: -t THRESH: threshold on error for polygonal fit (default = 5); the smaller is this threshold, the closer will be the polygonal fit to the original data, but the more straight-line fit segments will be required.

                -c: when set, print out (x,y) coordinates of the polygonal fit endlines

                -L: print Software License for this module

5.4 Critical Point Detection

Typical Application(s) – curve description, shape description, identification of curvature maxima along contours, identification of pertinent curve features.

Key Words – critical points, dominant points, curvature, curvature plot, curvature extrema, corner detection, difference of slopes (DoS), *k* curvature.

Related Topics – shape analysis (Section 4.4), polygonalization (Section 5.3), line fitting (Section 5.5), curve fitting (Section 5.6).

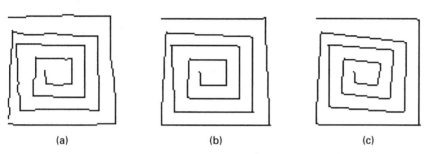

(a)   (b)   (c)

Figure 5.4.1. Pictorial Example. Illustration of the role of critical points in defining features, in this case to determine straight lines between corners of a hand-drawn spiral: (a) original hand-drawn square spiral; (b) the result of critical point detection-corners of the spiral are found, and straight lines are drawn between these corners; (c) the result of performing a polygonal fit – not a critical point fit – to (a). Note that the critical point fit locates the true corner locations better because it is a symmetrical operation with respect to the corner locations. In contrast, since the polygonal fit operation is not a symmetrical operation, the corners are found slightly behind their true corner locations (i.e., behind as the method traverses the line); therefore the final result gives the appearance of the spiral having a right tilt.

The concept of critical points (see Fig. 5.4.1) or dominant points relates to the observation that humans recognize shape in large part by curvature maxima in the contour, that is, points with high curvature or corners. In the classic experiment [Attneave, 1954], illustrated in Fig. 5.4.2, observers were shown a drawing of a sleeping cat in which curvilinear sections between curvature maxima of the outline were replaced by straight

Figure 5.4.2. This line drawing of a sleeping cat is simple to recognize even though the drawing is constructed of only straight lines between curvature maxima (adapted from [Attneave, 1954]). (Reprinted with Permission from [Attneave 54] – Published by American Psychological Association, in the public domain.)

lines. Observers easily identified the animal despite the only crudely reproduced out-line. Accordingly, the identification of critical points addresses the analytical task of recognizing objects or representing object shapes more succinctly.

5.4.1 CRITICAL POINT DETECTION BY CURVATURE ESTIMATION

A popular approach for critical point detection begins with curvature estimation. Curvature (see also Section 4.4) is measured as the angular difference θ between the slopes of two line segments that are fit to portions of the contour around each curvature point (Fig. 5.4.3). The measure so obtained for all points along a line is displayed in a curvature plot, also referred to as the θ plot.

Given that curvature for straight portions of a curve will be low, peaks in this plot indicate corners, with peak height proportional to the corner angle. Note that for a corner, the angle of curvature is defined as the complement of the corner angle ($\theta = 180° -$ corner angle); that is, the sharper the corner, the larger the corresponding value on the θ plot.

Next, a threshold is applied to the curvature plot to eliminate spurious curvature maxima; curvature values above threshold serve as the basis for the determination of critical points. Each corner is parameterized by its location, defined as the intersection

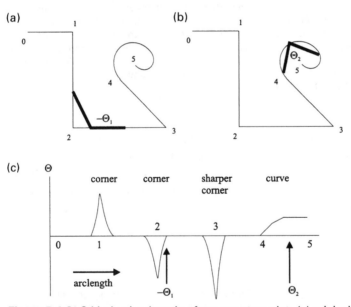

Figure 5.4.3. Critical point detection from curvature plot: (a) original line image with each of the critical points at or between numbers and DoS fit applied just beyond feature 2 (the DoS fit has two segments of arc length W and overlap $-M$ between them (the overlap is set to 0 in this example, $-M = 0$), and the measured angle between the two segments is θ); (b) original line image with DoS fit applied within the curve between 4 and 5; (c) θ plot showing peaks for corners (up or down, depending on direction of curvature) and a plateau for the curve [O'Gorman, 1988].

of the two straight lines bounding the corner; because of corner rounding, this location usually lies outside the curve.

The result of fitting straight lines between critical points is shown in Fig. 5.4.1(b). The critical points are the corner points. We have shown a comparison against polygonalization in Fig. 5.4.1(c). Note how the critical point fit locates the true corner locations better, resulting in a truer fit. This is due to the symmetrical nature of curve fitting, as described below. In contrast, the polygonal fit finds break points at locations slightly after the true corner points. This is due to the nature of the error accumulation method described in Section 5.3, which is not symmetrical.

5.4.2 IMPLEMENTATION: DIFFERENCE OF SLOPES

As indicated above, curvature may be evaluated as the angular difference of slopes (DoS) between two straight-line segments that are fit to portions of the contour before and after each curvature point (Fig. 5.4.3). Given this strategy, the procedure is referred to as the DoS method [O'Gorman, 1988]. Curvature is determined for each point of the data curve; then the curvature plot is usually smoothed to further reduce noise.

Both line segments have arc length W, and the overlap between segments along the original contour is $-M$; that is, segments are spaced apart along the contour if $M > 0$. If L denotes the arc length from the beginning of one segment to the end of the other, one has the simple relation $L = 2W + M$.

We must set two parameters. First, there is M, the overlap length: This is chosen as the maximum arc length of a corner, as we explain next. Although an ideal corner does not have any associated curve – it is a single point – digitization errors and noise will usually introduce some degree of rounding. If M is chosen to be the largest length of a curve that is to be associated with a true corner, then measured curvature will correspond to the true corner angle.

The second parameter to be set is L, the length of the line segment to be fit along the original contour in the process of determining the curvature. This is called the region of support because it relates to the extent of contour that affects the curvature measurement at any single point.

A trade-off governs the choice of L: It should be as long as possible to smooth out effects that are due to noise, but not so long as to average out features. Hence we should choose L as the minimum arc length between critical points. This choice requires the specification of what constitutes noise and what constitutes a true feature, for example the minimum corner angle or curve length. Problems may arise if noise or feature characteristics are not the same for different images subjected to analysis: This situation will require adjustments in parameters. Worse yet, if feature characteristics vary over a large range in a single image, for example coexisting gentle and sharp curves, the choice of a single set of parameters can be problematic. In such cases, it is very difficult to determine all critical points accurately by the methods described here.

One final consideration concerns the problem of borders and line intersections. No feature can be estimated within $L/2$ of beginnings or ends of lines because this L-length region of support is needed for each curvature estimation.

References and Further Reading

[Attneave, 1954] F. Attneave, "Some informational aspects of visual perception," Psychol. Rev., **61**, 183–193 (1954).

[O'Gorman, 1988] L. O'Gorman, "Curvilinear feature detection from curvature estimation," in *Proceedings of the Ninth International Conference on Pattern Recognition* (IEEE Computer Society Press, Los Alamtos, CA, 1988), pp. 1116–1119.

Several approaches to the detection of critical points along contours on the basis of estimating local curvature are discussed by A. Rosenfeld and E. Johnston, "Angle detection on digital curves," IEEE Trans. Comput. **C-22**, 875–878 (1973); A. Rosenfeld and J. S. Weszka, "An improved method of angle detection on digital curves," "IEEE Trans. Comput. **C-24**, 940–941 (1975); H. Freeman and L. Davis, "A corner-finding algorithm for chain-coded curves," IEEE Trans. Comput. **C-26**, 297–303 (1977); and [O'Gorman, 1988].

Program

`fitcrit`

> fits straight lines between critical points (curvature maxima) in the segments of pixel chains; corners are located at the location of intersection of their two boundary lines; curves are located by two straight lines drawn from curve onsets (transition points) to the middle point in the arc of the curve.

USAGE: `fitcrit infile outimg [-l FIT_LENGTH] [-r CORNER_ARC] [-c] [-L]`

ARGUMENTS:
- `infile`: input file name (PCC)
- `outimg`: output image filename (TIF)

OPTIONS:
- `-l FIT_LENGTH`: arc length of the fit along the data segments; the longer this length, the greater the degree of smoothing, but the more limited the ability to fit small features; this is usually chosen as the minimum arc length between critical points. (default = 13 [connected pixels]. minimum is 5.)

- `-r CORNER_ARC`: maximum arc length of a corner. Since a corner feature will usually be rounded, there is a maximum arc length preceding the actual corner over which that feature is considered a curve and not a corner.

- `-c`: when set, print out (x,y) coordinates of the straight-line fits between critical points.

- `-L`: print Software License for this module

5.5 Straight-Line Fitting

Typical Application(s) – detection and parametrization of lines, especially in human- or machine-drawn diagrams in which straight lines are intended (e.g., engineering drawings).

Key Words – line fitting, straight-line fitting, least-squares fit, regression fit, eigenvector line fitting, principal axis line fitting.

Related Topics – Hough transform (Section 4.10), polygonalization (Section 5.3), critical point detection (Section 5.4).

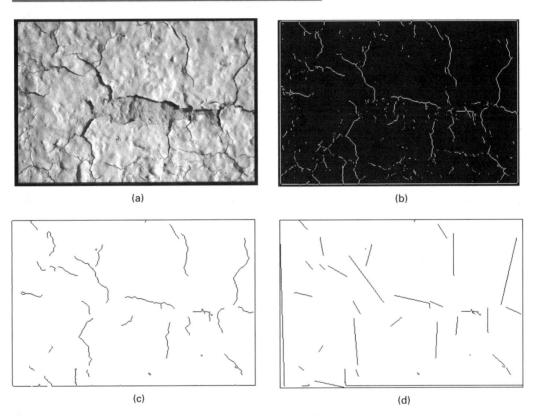

(a)   (b)

(c)   (d)

Figure 5.5.1. Pictorial Example. Illustration of straight-line fitting: (a) original image containing ragged lines (cracks), (b) binarized and thinned image, (c) noise-reduced lines, (d) straight-line fits by means of eigenvector method. (This product/publication includes images from Corel Stock Photo Library, which are protected by the copyright laws of the U.S., Canada, and elsewhere. Used under license.)

A common task in image analysis (and a classic problem in statistics) calls for a determination of the straight line that best fits a chain of points. This line yields a concise representation of the set of points in terms of two parameters, such as its end points, its slope and intercept, or its corresponding polar parameters ρ and θ. The linear parametrization greatly facilitates such operations as template matching (Section 3.8)

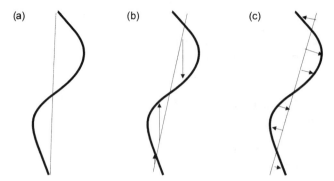

Figure 5.5.2. Image curves and three line different types of line fits using different measures of the deviation ("error") between the curve and fit: (a) end-point fit; (b) least-squares error fit, minimizing y-axis error (note that the y error is particularly poor in this example in which the resultant line is closer to vertical than horizontal – a better choice would have been to minimize the x-error); (c) eigenvector fit in which the error is measured perpendicular to the orientation of the resultant line fit.

and object recognition (Section 4.3). For example, matching of a straight-line template to a chain of points is now reduced simply to comparing slopes and intercepts between two lines. Similarly, it is much easier to recognize simple shapes such as triangles and squares by a simple evaluation of a small number of parameters such as the number of straight-line sections of the contour and subtended angles.

The task addressed in this section is that of finding the optimal straight-line fit to a given group of points. It is assumed that these points have been previously extracted from the image by thinning, contour analysis, or as time samples of a moving object, so they already are connected in a chain.

The most direct strategy to fit a straight line to a set of points is simply to connect end points. However, more sophisticated methods are required to obtain the best fit using all points in the given chain – not just the end points (see Fig. 5.5.1).

We describe two methods for such optimal fitting of straight lines to a chain (an ordered set) of points. First, there is the least-squares method that is widely used in statistical analysis when one variable, y, is dependent on the other, x. Second, there is a method that is generally more appropriate to most image analysis problems, which involve x and y not as statistical variables but as locations of points in an image. Figure 5.5.2 illustrates these two fitting methods along with the simple procedure of connecting end points.

The Hough transform method (Section 4.10), also can be used to obtain parameters of lines; however, the Hough transform performs the more general task of determining how many groups of points in an image are collinear, then finding line parameters of the collinear groups. The Hough transform is inherently more time consuming when there are many points in an image and generally yields less reliable parametric fits. In contrast, the line fitting approach we present here is a faster and more local approach since the line points are already identified and connected in a chain. It is reasonable to combine the two methods by first determining groups of collinear points throughout the image by the Hough technique, then obtaining accurate parameter estimations by line fitting.

5.5.1 LEAST-SQUARES LINE FITTING

One way to fit a line to data points is to minimize the error of the fit to the line with respect to one of the line variables, x or y. For statistical plots in which y is the dependent variable and x is the independent variable, this technique minimizes the sum of squares of y distances between the fitted line and the original data points. That is, for N data points, $\{(x_i, y_i), 0 \le i \le N - 1\}$, the error is

$$E_y = \sum_{i=0}^{N-1} (y_i - y)^2,$$

where y denotes the y coordinate of the point on the fitted line with x coordinate x_i. Substituting $y = mx + b$, differentiating E_y with respect to m and b, and setting these equations to zero yields the least-squares solution:

$$m = \frac{N \sum xy - \sum x \sum y}{N \sum x^2 - \sum x \sum x},$$
$$b = \sum y/N - m \sum x/N.$$

If the denominator of the equation for m is zero, m is the vertical slope of the line passing through the average x location, $\sum x/N$.

Minimization with respect to the y error yields a good fit for lines that are not close to vertical. A better fit for close-to-vertical lines is obtained by minimization with respect to the x variable, that is, by minimization of

$$E_x = \sum_{i=0}^{N-1} (x_i - x)^2,$$

where now x denotes the x coordinate of the point on the fitted line with y coordinate y_i. In this case, the solutions for m and b have the form

$$m = \frac{N \sum y^2 - \sum y \sum y}{N \sum xy - \sum x \sum y},$$
$$b = \sum y/N - m \sum x/N.$$

We observe that, analogous to the remark made above in connection with E_y, if a slope is close to horizontal, $m = 0$, it is better to minimize the y error. In general, it is better to minimize with respect to y for lines with slope in the range $(0 \pm 45)°$ and with respect to x for lines with slope in the range $(90 \pm 45)°$. In practice, this $\pm 45°$ range around each axis can be exceeded quite liberally with similar results, that is, $(0 \pm 80)°$ and $(90 \pm 80)°$ should give reasonable results for most applications. To gain some insight as to the expected slope of the optimal straight line, a crude estimate of the slope of the given chain of points may be obtained from the slope of the line that connects the end points.

5.5.2 LINE FITTING BY EVALUATION OF EIGENVECTORS

Although less frequently used, a method that is in fact more appropriate to most line fitting problems encountered in connection with image analysis uses eigenvector evaluation. In contrast to the least-squares optimization based on a measure of error in either the x or the y variable, the eigenvector strategy is to minimize error in the direction perpendicular to the fitted line. This approach avoids the dependence of E_y and E_x on the slope of the optimal fit, with the worst cases at close-to-vertical or close-to-horizontal lines, as described above. An example of eigenvector line fitting is shown in Fig. 5.5.1.

For a set of points $\{(x_i, y_i); 0 \le i \le N - 1\}$, the objective in eigenvector line fitting is to find the equation of a line that minimizes the error $E_\perp$, defined in terms of the perpendicular distance d_i between the point with x coordinate x_i and the optimal line:

$$E_\perp \equiv \sum_{i=0}^{N-1} d_i^2.$$

The derivation of the solution is beyond the scope of this text; we merely list here the steps involved in obtaining the desired optimal fit:

- Calculate averages of x and y coordinates of the given set of points:

$$\bar{x} = \frac{1}{N} \sum x_i,$$

$$\bar{y} = \frac{1}{N} \sum y_i.$$

- Standardize each data point by subtracting the averages just calculated:

$$\hat{x}_i = x_i - \bar{x},$$

$$\hat{y}_i = y_i - \bar{y}.$$

- Determine the symmetric matrix A, defined in terms of the sum of vectors to each point $\mathbf{v}_i = (\hat{x}_i, \hat{y}_i)$; with $a = \hat{x}_i^2, b = \hat{x}_i \hat{y}_i, c = \hat{y}_i^2$:

$$A = \sum \mathbf{v}_i^\perp \mathbf{v}_i = \begin{bmatrix} a & b \\ b & c \end{bmatrix}.$$

- Determine the smaller eigenvalue, λ, and the associated eigenvector, e_λ, of A:

$$\lambda = [(a + c) - \sqrt{(a + c)(a + c) - 4ac + 4bb}]/2,$$

$$e_\lambda = [b/(\lambda - a), 1].$$

- Find the slope of the desired optimal line as that perpendicular to the slope of the eigenvector:

$$m = -b/(\lambda - a),$$

and if $a = \lambda$, the slope is the vertical.

- Find the intercept $y0$ of the desired optimal line from the average of the data set:

$$y0 = \bar{y} - m\bar{x}.$$

References and Further Reading

The derivation of the results in this section may be found in R. O. Duda and P. E. Hart, *Pattern Classification and Scene Analysis* (Wiley-Interscience, New York, 1973), pp. 332–335; linear regression is further described in W. H. Press, S. A. Teukolsky, W. T. Vetterling, and B. P. Flannery, *Numerical Recipes in C*, 2nd ed. (Cambridge U. Press, Cambridge, 1992), pp. 661–681.

Program

fitline

          fits straight lines to pixel chain segments using the eigenvector method.

          NOTE: The fit is performed to each complete line segment; a segment is a chain of points between features, either end-point features or junction features.

          NOTE: This fit method is not appropriate to approximate curve or corner features within segments -- use a critical point fit or polygonalization for this case.

     USAGE: fitline infile outimg [-c] [-L]

ARGUMENTS: infile: input filename (PCC)

          outimg: output image filename (TIF)

  OPTIONS: -c: when set, print line coordinates

            -L: print Software License for this module

5.6 Cubic Spline Fitting

Typical Application(s) – obtaining more visually pleasing contours for interactive (manual) analysis, noise reduction, smoothing.

Key Words – spline fit, B-splines, cardinal splines, approximating splines, interpolating splines, third-order polynomial fit.

Related Topics – polygonalization (Section 5.3), critical point detection (Section 5.4), line fitting (Section 5.5).

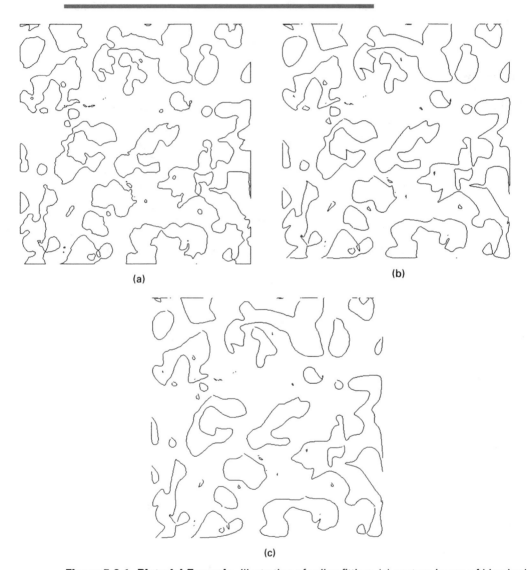

(a)

(b)

(c)

Figure 5.6.1. Pictorial Example. Illustration of spline fitting: (a) contour image of binarized and contour-coded liver tissue image of previous sections, (b) spline fit with small error value, (c) spline fit with larger error value. (Reprinted with Permission from [O'Gorman et al. 85], Copyright © 1986 by IEEE.)

In contrast to the piecewise straight-line approximations in previous sections, the spline fit is a smooth curve fit. As such, it can be used to approximate a chain comprising many different curves. Spline fits approximate a curve by a series of third-order, or cubic, polynomial curve segments (see below). Spline fitting is commonly performed with the results of polygonalization. The polygonal break points are used as nodes of the spline – or guide points – to guide the path of the spline to pass near or through the nodes. The spline is useful in image analysis to aid in visually matching two curves. This is because the smoother curve provides more visual emphasis to its general curvature and less distraction to the smaller, noisy deviations.

We describe two types of splines: interpolating and approximating splines. Interpolating splines must pass through specified node points, while approximating splines pass only close to these nodes. Approximating splines are typically smoother because of this more relaxed requirement. Examples of both of these splines are shown in Fig. 5.6.1.

One could perform higher-order polynomial fits to approximate the lines more closely. However, while a closer fit may be obtained in some cases, the curves may be less well behaved, tending to fit small deviations of undesired noise instead of the overall desired curve. The cubic spline is the lowest-order polynomial fit that has the desired combination of being well behaved and yielding smooth results.

5.6.1   SPLINE FITTING

Given a series of critical points, spline curves are fit to consecutive pairs of points. Each fit requires four consecutive points, the two that are being fit plus one before and one after. This fit is performed to each consecutive point on a curve, such that there are three points of each current fit that overlap the points used in the fit of the previous point. This ensures that the resulting fit is a continuous curve through all points.

In addition to selecting the spline type, the user must also choose the granularity, that is, the number of points to be evaluated between data points. There is the usual trade-off: finer granularity results in a more precise fit, but this is at the expense of more computation time. We will assume uniform spacing of the independent variable x to ensure a consistent fit between points.

To fit an open curve, each end point is specified twice to ensure fitting to each end. To fit a closed curve, the last point is copied to precede the first point, and the first and the second points are copied to follow the last point.

Interpolating Spline

The interpolating spline that we describe here is the cardinal spline. This is used when the resulting curve must pass exactly through a set of nodes. The cardinal spline ensures smoothness by being first-order continuous at its ends, that is, tangents of connected curves have equal slopes.

A cubic spline is specified by four coefficients, and four equations are needed to solve for the optimal polynomial fit. These coefficients are conveniently expressed in matrix form. For the cardinal spline, this coefficient matrix has the form

$$C = 1/2 \begin{bmatrix} -1 & 3 & -3 & 1 \\ 2 & -5 & 4 & -1 \\ -1 & 0 & 1 & 0 \\ 0 & 2 & 0 & 0 \end{bmatrix}.$$

The evaluation of each interpolated point relies on this matrix, as well as on four consecutive data points and on the variable t, $0 \le t < 1$, the latter setting the degree of granularity.

Specifically, x and y coordinates of interpolated points are calculated as follows:

$$x'(i) = 1/2[x(i-1)(-t^3 + 2t^2 - t) + x(i)(3t^3 - 5t^2 + 2)$$
$$+ x(i+1)(-3t^3 + 4t^2 + t) + x(i+2)(t^3 - t^2)],$$
$$y'(i) = 1/6[y(i-1)(-t^3 + 2t^2 - t) + y(i)(3t^3 - 5t^2 + 2)$$
$$+ y(i+1)(-3t^3 + 4t^2 + t) + y(i+2)(t^3 - t^2)].$$

Approximating Spline

The approximating spline we describe here is the B-spline. An approximating spline is not guaranteed to pass through any of the data points; instead, it passes within a guiding polygon specified by the nodes. The B-spline is second-order continuous at its ends, that is, both the first derivative (tangent) and the second derivative of consecutive curves are equal. As a result, the B-spline generally has a smoother appearance than does the cardinal spline.

The coefficient matrix for the B-spline is

$$C = 1/2 \begin{bmatrix} -1 & 3 & -3 & 1 \\ 3 & -6 & 3 & 0 \\ -3 & 0 & 3 & 0 \\ 1 & 4 & 1 & 0 \end{bmatrix}.$$

Expanding and simplifying as for the interpolated spline above, we find expressions for the interpolated points:

$$x'(i) = 1/6[x(i-1)(-t^3 + 3t^2 - 3t + 1) + x(i)(3t^3 - 6t^2 + 4)$$
$$+ x(i+1)(-3t^3 + 3t^2 + 3t + 1) + x(i+2)(t^3)],$$
$$y'(i) = 1/6[y(i-1)(-t^3 + 3t^2 - 3t + 1) + y(i)(3t^3 - 6t^2 + 4)$$
$$+ y(i+1)(-3t^3 + 3t^2 + 3t + 1) + y(i+2)(t^3)].$$

References and Further Reading

[O'Gorman et al., 1985] L. O'Gorman, A. C. Sanderson and K. Preston, Jr., "A system for automated liver tissue damage analysis: methods and results," IEEE Trans. Biomedical Engineering Vol. BME-32(9), pp. 696–706 (Sept. 1985).

Spline polynomials and their extensive use in computer graphics are further discussed in T. Pavlidis, *Algorithms for Graphics and Image Processing* (Computer Science Press, Rockville, MD, 1982), pp. 247–273; J. D. Foley and A. van Dam, *Fundamentals of Computer Graphics* (Addison-Wesley, Reading, MA, 1982), pp. 514–523; and G. Medioni and Y. Yasumoto, "Corner detection and curve representation using cubic B-splines," Comput. Vis. Graph. Image Process. **29**, 267–278 (1987).

Program

`fitspline`

        performs spline fitting to image lines to produce smooth curve approximations.

  USAGE: fit spline infile outimg [-t THRESHOLD] [-g GRANULARITY] [-i] [-L]

ARGUMENTS: infile: input filename (PCC)
                  outimg: output image filename (TIF)

OPTIONS: -t THRESH: threshold on error for spline node determination; nodes are found from polygonalization; (default = 5); the smaller this threshold, the closer the approximation of the original data by the spline fit, but the greater the number of splines required.

      -g GRANULARITY: number of points between each pair of spline nodes; the greater the number of points, the smoother the approximation. (default = 5)

            -i: perform interpolation fitting, that is, ensure that fit runs through node points; a B-spline is used; the default is NOT interpolation, instead it is an approximation spline; a cardinal spline is used; the approximation spline yields a close fit, but one that does not usually pass through the node points; the approximation spline appears smoother than the interpolation spline.

            -L: print Software License for this module

5.7 Morphology and Topology of Line Patterns

Typical Application(s) – direct-space analysis of line patterns to examine pattern morphology and to derive a quantitative description of morphological determinants; identification of topological (point) defects.

Key Words – parallelism, overlap and adjacency; segment clusters and groups; cluster geometry and global descriptors; line pattern topology; branch and end points.

Related Topics – shape features (Section 4.4), convex hull (Section 4.6), thinning (Section 4.7), chain coding (Section 5.1), polygonalization (Section 5.3).

Figure 5.7.1. Pictorial Example. Illustration of morphological and topological analysis applied to a labyrinthine stripe domain pattern observed by polarization microscopy in a thin film of a garnet material in a magnetic field. The top panel displays the result of a decomposition of this globally disordered pattern into clusters of ordered line segments; the bottom panel shows the location of branches and end points in the original pattern along with interconnecting line segments. The spacing between stripe centers is 11 μm. (Reprinted with Permission from [Seul and Wolfe 92B], Copyright © 1992 American Physical Society.)

Globally disordered line patterns are observed in many instances of domain formation in systems such as thin films of liquid crystals, ferroelectric materials, and magnetic garnets. The quantitative description of the complex labyrinthine domain configurations has many similarities to the analysis of fingerprint patterns. This direct-space pattern analysis has the advantage of providing a methodology to identify morphological characteristics as well as structural motifs. This is accomplished by identification of regions of local order within a globally disordered line pattern. As illustrated in Fig. 5.7.1, ordered regions are composed of clusters of parallel, nonintersecting line segments that realize a locally unidirectionally ordered, or lamellar, configuration.

In this section, we introduce a methodology for the decomposition of line patterns into segment clusters and the statistical analysis of segment cluster characteristics such as size, aspect ratio, and average local orientation of lines. In addition, we provide a

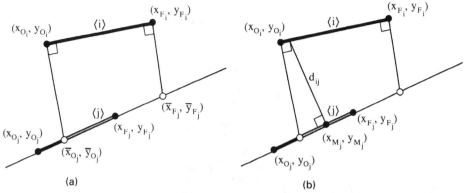

(a)   (b)

Figure 5.7.2. (a) Sketch illustrating definition of overlap parameter in terms of the projection of segment $\langle i \rangle$ onto segment $\langle j \rangle$. In the example, p_{ij} is defined by the points $(\bar{x}_{O_I}, \bar{y}_{O_I})$, (x_{F_J}, y_{F_J}) and indicated by the heavy portion of segment $\langle j \rangle$. (b) Sketch illustrating the definition of adjacency parameter d_{ij}. Coordinates (x_{M_j}, y_{M_j}) identify the midpoint of $p_j^{(i)}$, the overlap of segment $\langle i \rangle$ with segment $\langle j \rangle$, highlighted by the heavy line reproduced from (a); d_{ij} denotes the signed distance of (x_{M_j}, y_{M_j}) from the line containing segment $\langle i \rangle$. (Reprinted with Permission from [Seul et al., 92], Copyright © 1992 Taylor & Francis.)

method to locate characteristic points within a pattern, notably branch points and end points. These correspond to topological defects and represent a fundamental aspect of line pattern disorder.

5.7.1 SEGMENT CLUSTER ANALYSIS

The starting point for the segment cluster analysis is a piecewise linear approximation of a thinned line pattern generated by means of polygonalization. As described in Section 5.3, polygonalization of a line pattern entails the replacement of the original set of vertices, defining the contours of the skeletonized pattern, by a subset of vertices that are connected by linear segments.

For the purpose of the cluster analysis, each segment $\langle i \rangle$ is represented as a data structure of the form $[(x_0, y_0), (x_F, y_F), m_i, i, k]$ where (x_0, y_0) and (x_F, y_F) denote segment end-point coordinates, $m_i \equiv (y_F - y_0)/(x_F - x_0)$ denotes its slope, i represents a segment index, and k is the index of its parent line. The parent line is a portion of the original curvilinear skeleton contour that is bounded by end points or branch points.

Segment Adjacency Lists

Line segments are next sorted into clusters by evaluation, for each pair of segments produced by the polygonalization of the entire thinned image, of a set of three geometrical parameters measuring the degree of parallelism, mutual overlap, and adjacency, as defined in Fig. 5.7.2. The pairwise comparison of all segments in the polygonalized image represents the most significant computational cost of the procedure.

The requisite sorting procedure is accomplished in two steps. First, a segment adjacency list SAL$^{(i)}$ is constructed for each segment $\langle i \rangle$. A segment $\langle j \rangle = \langle i \rangle$ is entered into SAL$^{(i)}$ if it successively satisfies criteria of parallelism, mutual overlap,

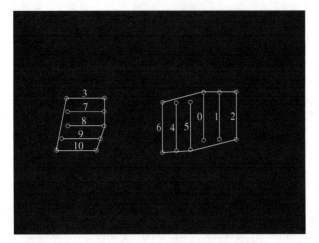

Figure 5.7.3. Illustrative example of segment cluster analysis, performed by invoking **XSGLL** with the -*t* option. A collection of linear segments is first sorted into two segment group lists, as detailed in Tables 5.7.1 and 5.7.2. Based on the end points of each cluster's constituent line segments, a convex hull is constructed to delineate each cluster's boundary. (Reprinted with Permission from [Seul et al., 92], Copyright © 1992 Taylor & Francis.)

and adjacency. These criteria are specified by the user in the form of threshold values for closeness of angular orientation, overlap, and adjacency, respectively, such that $|\tan(\theta_i - \theta_j)| \leq \tan \delta\theta_0$, $p_{ij} \geq p_0$ and $|d_{ij}| < d_0$.

The result of this procedure is a segment adjacency list $\mathrm{SAL}^{(i)}$ for each segment $\langle i \rangle$. Entries are data structures of the form $\{j, d_{ij}, p_{ij}, p_{ji}\}$, listed, for later convenience, in the order of decreasing inverse overlap parameter p_{ji}; the reference segment $\langle i \rangle$ is placed at the top. A typical example, based on the analysis of the simple model pattern in Fig. 5.7.3, is discussed in connection with Table 5.7.1.

Segment Group Lists: Clusters

The array of segment adjacency lists now at hand provides an extensive but highly redundant representation of the spatial correlations between segment pairs. We therefore cast this information in a more consolidated form by contracting segment adjacency lists into segment group lists, SGL, the data structure appropriate for the representation of segment clusters in a given pattern: Each segment may be entered in at most one SGL.

To generate the segment group lists, segment adjacency lists are scanned and segments in these lists are assigned into groups. A segment group list $\mathrm{SGL}^{(i)}$ is complete when the segment adjacency lists $\mathrm{SAL}^{(j)}$ for each segment $\langle j \rangle \in \mathrm{SGL}^{(i)}$ have been scanned and exhausted of active (that is, unassigned) segments. Segments are accepted into a segment group list by application of several criteria based on overlap parameters. A corresponding SGL level can be maintained as segments are assigned. This SGL level can assume three possible values, 0, 1, 2, reflecting the level of indirection in making segment assignments; such a parameter is displayed in Table 5.7.2. An example of a segment group list, based on the continued analysis of Fig. 5.7.3, is shown in Table 5.7.2 [Seul et al., 1992].

Table 5.7.1. Segment Adjacency Lists

Typical Example of Segment Adjacency Lists (SALs), Constructed in the Course of Analyzing the Line Pattern in Fig. 5.7.3. In Each Segment Adjacency List, the Reference Segment is Placed at the Top, and the Following Segments are Inserted in the Order of Decreasing Inverse Overlap Parameter p_{ji}. The Value of 999.0 was Arbitrarily Chosen to Initialize Parameters; in General, $0 \le p_{ij}$, $p_{ji} \le 100$. The Threshold Values for Degree of Parallelism, Overlap, and Adjacency were Set to $10.0°$, 50.0 and 50.0, Respectively. Distance d_{ij}, is Measured in Suitable Units, Here Pixels of a Graphics Display.

SAL⟨0⟩	segment:	0	$d(0,0)$:	0.00	$p(0,0)$:	999.00	$p(0,0)$:	999.00
	segment:	1	$d(0,1)$:	28.00	$p(0,1)$:	100.00	$p(1,0)$:	100.00
	segment:	5	$d(0,5)$:	−23.00	$p(0,5)$:	76.92	$p(5,0)$:	79.21
	segment:	4	$d(0,4)$:	−47.00	$p(0,4)$:	76.92	$p(4,0)$:	79.21
SAL⟨1⟩	segment:	1	$d(1,1)$:	0.00	$p(1,1)$:	999.00	$p(1,1)$:	999.00
	segment:	2	$d(1,2)$:	29.00	$p(1,2)$:	100.00	$p(2,1)$:	100.00
	segment:	0	$d(1,0)$:	−28.00	$p(1,0)$:	100.00	$p(0,1)$:	100.00
SAL⟨2⟩	segment:	2	$d(2,2)$:	0.00	$p(2,2)$:	999.00	$p(2,2)$:	999.00
	segment:	1	$d(2,1)$:	−29.00	$p(2,1)$:	100.00	$p(1,2)$:	100.00
SAL⟨3⟩	segment:	3	$d(3,3)$:	0.00	$p(3,3)$:	999.00	$p(3,3)$:	999.00
	segment:	7	$d(3,7)$:	29.00	$p(3,7)$:	100.00	$p(7,3)$:	100.00
SAL⟨4⟩	segment:	4	$d(4,4)$:	0.00	$p(4,4)$:	999.00	$p(4,4)$:	999.00
	segment:	5	$d(4,5)$:	24.00	$p(4,5)$:	100.00	$p(5,4)$:	100.00
	segment:	6	$d(4,6)$:	−25.00	$p(4,6)$:	97.17	$p(6,4)$:	99.04
	segment:	0	$d(4,0)$:	47.00	$p(4,0)$:	79.21	$p(0,4)$:	76.92
SAL⟨5⟩	segment:	5	$d(5,5)$:	0.00	$p(5,5)$:	999.00	$p(5,5)$:	999.00
	segment:	4	$d(5,4)$:	−24.00	$p(5,4)$:	100.00	$p(4,5)$:	100.00
	segment:	6	$d(5,6)$:	−49.00	$p(5,6)$:	97.17	$p(6,5)$:	99.04
	segment:	0	$d(5,0)$:	23.00	$p(5,0)$:	79.21	$p(0,5)$:	76.92
SAL⟨6⟩	segment:	6	$d(6,6)$:	0.00	$p(6,6)$:	999.00	$p(6,6)$:	999.00
	segment:	5	$d(6,5)$:	49.00	$p(6,5)$:	99.04	$p(5,6)$:	97.17
	segment:	4	$d(6,4)$:	25.00	$p(6,4)$:	99.04	$p(4,6)$:	97.17
SAL⟨7⟩	segment:	7	$d(7,7)$:	0.00	$p(7,7)$:	999.00	$p(7,7)$:	999.00
	segment:	8	$d(7,8)$:	30.00	$p(7,8)$:	100.00	$p(8,7)$:	100.00
	segment:	3	$d(7,3)$:	−29.00	$p(7,3)$:	100.00	$p(3,7)$:	100.00
SAL⟨8⟩	segment:	8	$d(8,8)$:	0.00	$p(8,8)$:	999.00	$p(8,8)$:	999.00
	segment:	7	$d(8,7)$:	−30.00	$p(8,7)$:	100.00	$p(7,8)$:	100.00
	segment:	9	$d(8,9)$:	26.00	$p(8,9)$:	84.29	$p(9,8)$:	89.39
SAL⟨9⟩	segment:	9	$d(9,9)$:	0.00	$p(9,9)$:	999.00	$p(9,9)$:	999.00
	segment:	10	$d(9,10)$:	26.00	$p(9,10)$:	90.00	$p(10,9)$:	90.00
	segment:	8	$d(9,8)$:	−26.00	$p(9,8)$:	89.39	$p(8,9)$:	84.29
SAL⟨10⟩	segment:	10	$d(10,10)$:	0.00	$p(10,10)$:	999.00	$p(10,10)$:	999.00
	segment:	9	$d(10,9)$:	−26.00	$p(10,9)$:	90.00	$p(9,10)$:	90.00

Table 5.7.2. Segment Group Lists

Typical Example of Segment Group Lists (SGLs), Constructed from the Segment Adjacency Lists of Table 5.7.1 and Based on the Analysis of the Line Pattern in Fig. 5.7.3. The Segment Group Lists are Sorted in the Order of Increasing Adjacency Parameter d_{ij}, and the Reference Segment is Assigned the Value 0. Distance d_{ij} is Measured in Suitable Units, Here Pixels of a Graphics Display.

SGL⟨0⟩	segment:	6	$d(0,6)$:	−72.00	$p(0,6)$:	78.22	$p(6,0)$:	74.53	SGL level:	2
	segment:	4	$d(0,4)$:	−47.00	$p(0,4)$:	76.92	$p(4,0)$:	79.21	SGL level:	0
	segment:	5	$d(0,5)$:	−23.00	$p(0,5)$:	76.92	$p(5,0)$:	79.21	SGL level:	0
	segment:	0	$d(0,0)$:	0.00	$p(0,0)$:	999.00	$p(0,0)$:	999.00	SGL level:	0
	segment:	1	$d(0,1)$:	28.00	$p(0,1)$:	100.00	$p(1,0)$:	100.00	SGL level:	0
	segment:	2	$d(0,2)$:	57.00	$p(0,2)$:	100.00	$p(2,0)$:	100.00	SGl level:	2
SGL⟨3⟩	segment:	3	$d(3,3)$:	0.00	$p(3,3)$:	999.00	$p(3,0)$:	999.00	SGL level:	0
	segment:	7	$d(3,7)$:	29.00	$p(3,7)$:	100.00	$p(7,0)$:	100.00	SGL level:	0
	segment:	8	$d(3,8)$:	59.00	$p(3,8)$:	100.00	$p(8,1)$:	100.00	SGL level:	2
	segment:	9	$d(3,9)$:	85.00	$p(3,9)$:	89.39	$p(9,1)$:	84.29	SGL level:	2
	segment:	10	$d(3,10)$:	111.00	$p(3,10)$:	78.79	$p(10,1)$:	74.29	SGl level:	2

To reduce noise and to construct well-defined polygonal contours for the segment clusters identified in the foregoing analysis, merging of segments in a given segment group list may become necessary. This additional step entails the removal of vertices shared by two segments, so that a pair of segments with end points (x_{O_1}, y_{O_1}), (x_{F_1}, y_{F_1}) and (x_{O_2}, y_{O_2}), (x_{F_2}, y_{F_2}) would be replaced by a single segment with end points (x_{O_1}, y_{O_1}), and (x_{F_2}, y_{F_2}) in the event that $x_{F_1} = x_{O_2}$, $y_{F_1} = y_{O_2}$.

The segment group lists faithfully represent the segment clusters in the patterns we have investigated. We note that a further level of global feature classification may be implemented on the basis of the segment group lists by considering lines in their entirety.

5.7.2 STATISTICAL ANALYSIS OF SEGMENT CLUSTERS

Geometrical Aspects

The set of end points of linear segments, contracted by merging segments sharing a common vertex and then suitably ordered as detailed in the previous section, defines a polygonal cluster boundary. To delineate the overall shape of segment clusters, we construct the convex hull of the original polygonal boundary to assign to each segment cluster a uniquely defined, convex, closed bounding contour (Section 4.6). The outcome of such an analysis, performed for a typical labyrinth in a magnetic garnet film, is displayed in Fig. 5.7.4. This figure provides a demonstration of the degree of performance of this procedure in deriving a truthful representation of the pattern morphology. The original labyrinthine structure is effectively decomposed into an assembly of segment clusters whose pertinent geometrical properties and mutual relationships with respect to position and orientation are available for explicit evaluation and inspection.

To derive quantitative information concerning geometrical properties of segment clusters, a number of additional measures may be used. These include their length L,

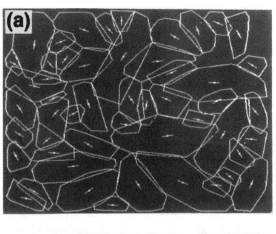

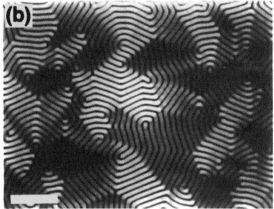

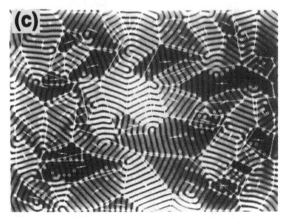

Figure 5.7.4. Segment cluster analysis of a labyrinth pattern in magnetic garnet film: (a) Convex hulls of polygonal segment clusters. A nematic director, attached to the convex hull center, indicates the average normal direction of the segments contained in a given cluster. (b) Original labyrinth pattern following application of low-pass and dilation filters to render visible regions of oriented stripe segments. (c) Superposition of (a) and (b). End points of linear segments, generated by approximation, are also indicated (by ○). (Reprinted with Permission from [Seul et al., 92], Copyright © 1992 Taylor & Francis.)

maximal width W, and area A, as well as their average segment normal $\hat{n}$; the latter is indicated in Fig. 5.7.4 in the form of a line segment attached to the convex hull center.

The area A is computed in the form of the zero-order moment (Section 4.4) from the directed set of segment end points that define a polygonal contour for each segment cluster. Length L and maximal width W for the ith cluster are readily evaluated from the information stored in the corresponding segment group list $SGL^{(i)}$, so that L is given by

$$L = |d_{\text{tail},(i)} - d_{\text{head},(i)}| \simeq (N^{(i)} - 1)d \tag{5.5}$$

where $d_{\text{head},(i)}$ and $d_{\text{tail},(i)}$ respectively represent the adjacency parameters for segments stored at the top and the bottom in segment group list $SGL^{(i)}$, a doubly linked list (Section A.5) containing $N^{(i)}$ segments, in order of increasing adjacency parameter; d denotes the average stripe period of the entire pattern. The maximal cluster width W is simply defined as the maximal length of the segments in $SGL^{(i)}$:

$$W = \max\{l_k\}_{1 \leq k \leq N^{(i)}}. \tag{5.6}$$

The average segment normal may be regarded as the equivalent of a nematic director, a quantity indicating the average orientation assumed by elongated molecules in a nematic liquid crystal. While $\hat{n}$ measures the (azimuthal) orientation of an individual cluster, the corresponding two-dimensional nematic order parameter, $S \equiv \langle [2\cos^2(\phi - \phi_0) - 1]/2 \rangle$, provides an average measure of alignment of the clusters in a set with respect to a preferred direction ϕ_0, where, customarily, $\phi_0 = 0$.

Topological Characteristics: Branch and End Points

Disclinations are topological defects that affect the local rotation symmetry of a given structure. In a unidirectionally periodic (lamellar) array of straight lines, disclinations of the type sketched in Fig. 5.7.5 may occur. We may assign formal charges to such topological defects by noting the rotation of the (unsigned) local layer normal along a contour surrounding the defect. For example, the local layer normal undergoes a $\pm\pi$ rotation along a contour enclosing a $\pm\pi$ disclination (Fig. 5.7.5). The charge of a disclination is then simply defined as the value of the layer-normal rotation in units of 2π: Thus a $\pm\pi$ disclination carries a charge of $\pm1/2$.

We thus recognize end points and branch points in the PCC as disclination defects of the line pattern with respective topological charges of $+1/2$ and $-1/2$. Both types of defects are marked in Fig. 5.7.6. Sorting of the stack data structure underlying the implementation of the PCC analysis (Section 5.1) permits the identification of such disclination pairs and thus the explicit reconstruction from the PCC of the vertices defining the connecting line segment. An example of disclination dipole arrays is shown in Fig. 5.7.6.

To extract quantitative information from the shape of the line segment joining $+1/2$ and $-1/2$ charges, a variety of descriptors may be considered. Of interest among these are characteristic measures of chain configuration introduced in the context of the study of polymer chain statistics and random flights. A simple descriptor is the ratio of end-to-end distance d_{ee} of a given trajectory and its contour length

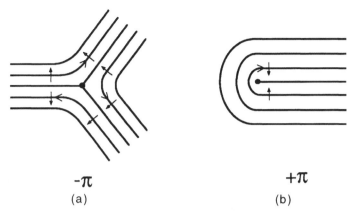

$-\pi$

(a)

$+\pi$

(b)

Figure 5.7.5. Sketch of disclination defects in a layered medium: (a) $-1/2$ disclination, (b) $+1/2$ disclination. A bound pair of oppositely charged disclinations represents a dislocation. (Reprinted with Permission from [Seul et al., 92], Copyright © 1992 Taylor & Francis.)

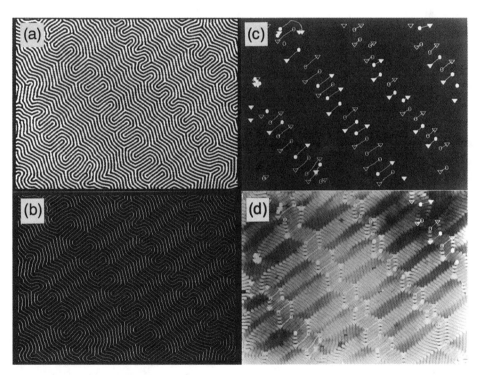

Figure 5.7.6. Topological analysis of a line pattern. The line pattern in (a) was recorded from a magnetic garnet film subjected to a magnetic field; thinning (Section 4.7) yields the pattern in (b). Panel (c) depicts pairs of oppositely charged disclinations, joined by (thinned) linear segments of the original pattern in a dipolar configuration; triangles and circles denote $-1/2$ and $+1/2$ disclinations, respectively. In panel (d), arrays of disclination dipoles are superimposed on a filtered version of the original pattern in (a). (Reprinted with Permission from [Seul and Wolfe 92A], Copyright © 1992 American Physical Society.)

l: $\delta \equiv d_{ee}/l$; given that $l \geq d_{ee}$, it is apparent that $0 \leq \delta \leq 1$. As with other global shape descriptors (Section 4.4), δ yields a highly degenerate classification, with many configurations leading to the same δ. However, δ is readily evaluated from the PCC representation, is insensitive to local pixel noise, and captures the essence of the transformation of straight line segments ($\delta = 1$) into meandering lines ($\delta \ll 1$) that leads to configurations such as those in Fig. 5.7.1.

References and Further Reading

[Seul and Wolfe, 1992A] M. Seul and R. Wolfe, "Evolution of disorder in magnetic stripe domains. I. Transverse instabilities and disclination unbinding in lamellar patterns," Physical Review A **46**, pp. 7519–7533 (1992).

[Seul and Wolfe, 1992B] M. Seul and R. Wolfe, "Evolution of disorder in magnetic stripe domains. II. Hairpins and labyrinth patterns versus branches and comb patterns formed by growing minority component," Physical Review A **46**, pp. 7534–7547 (1992).

[Seul et al., 1992] M. Seul, L. R. Monar, and L. O'Gorman, "Pattern analysis of magnetic stripe domains: morphology and topological defects in the disordered state," Philos. Mag. B **66**, 471–506 (1992).

The methodology used here for the analysis of line segments is described in more detail by L. O'Gorman and G. I. Weil, "An approach toward segmenting contour line regions," in *Proceedings of the Eighth International Conference on Pattern Recognition* (IEEE Computer Society Press Los Alamos, CA, 1986), 254–258, related aspects are discussed in connection with the thin line code by H. V. Jagadish and L. O'Gorman, IEEE Comput. **22**, 33 (1989).

Programs

fitpolyg

```
          performs polygonal line fitting to image lines (see
          also: section 5.3); with -w option set, the program
          produces an output file containing line segment data
          for analysis by xsgll

   USAGE: fitpolyg infile outimg [-t THRESHOLD] [-c] [-w file]
          [-L]

ARGUMENTS:   infile: input filename (PCC)
             outimg: output image filename (TIF)

 OPTIONS: t THRESH: threshold on error for polygonal fit (default
                    = 5); the smaller is this threshold, the
                    closer will be the polygonal fit to the
                    original data, but the more straight line fit
                    segments will be required.
              -c: when set, prints out the (x,y) coordinates of
                    the polygonal fit endlines.
          -w file: save line segment data to a file (.seg) for
                    analysis by xsgll
              -L: print Software License for this module
```

eh_seg

extracts histogram data from file (SEG), generated by
pccseg, containing line segment data

USAGE: eh_seg infile [-d] [-a] [-l] [-nn] [-if] [-ff] [-w] [-L]

ARGUMENTS: infile: input filename (SEG)

OPTIONS: -s: show input data

construct histogram for:
-a: turn angles between consecutive segments
-l: segment lengths

-nn: set number of bins to n (default: 10)
-if: set initial value to (float)f (default: MIN)
-ff: set final value to (float)f (default: MAX)
-w: write output file of type .hdt (hist data)

-L: print Software License for this module

---

xsgll

constructs segment adjacency lists (SAL) and segment
group lists (SGL), given an input file containing line
segment data; writes output data into file and creates
image containing line segment clusters

USAGE: xsgll infile outimg [-t] [-w] [-a ang_thresh]
[-p ovlp_thresh] [-d d_max] [-m] [-o]
[-c] [-h] [-L]

ARGUMENTS: infile: input filename (.seg) (ASCII)
outimg: output image filename (TIF)

OPTIONS: -t: employ test data in test_segm.c
-w: write sgl parameters and stats to file
NOTE: output filename is constructed from input
filename by replacing .seg with .sgl
-a angl_thresh: set angle threshold - default: 10 deg
-p ovlp_thresh: set ovlp_thresh (0 <= pij <= 100) - default: 50
-d d_max: set maximum dist between segments - default: 50
pixels
-m: do not merge segments after SGL constr
-o: overlay topological features onto image
-c: display segment clusters
-h: do not display conv hull, only orig. polygon

-L: print Software License for this module

---

eh_sgl

   extracts histogram data from SGL output files created
   by xsgll

USAGE: eh_sgl infile [-v] [-s] [-a] [-l] [-b] [-h] [-q] [-nn]
           [-if] [-ff] [-w] [L]

ARGUMENTS: infile: input filename (SGL)

OPTIONS: -v: expect vertex coordinates in .sgl file
      -s: show input data

      construct histogram for:
      -s: segment number (per SGL)
      -a: cluster area
      -l: segment length
      -b: segment width
      -h: cluster (convex) hull area
      -q: quotient area/conv hull area

     -nn: set number of bins to n (default: 10)
     -if: set initial value to (float)f (default: MIN)
     -ff: set final value to (float)f (default: MAX)
      -w: write output file of type .hdt (hist data)

      -L: print Software License for this module

6 Analysis of Point Patterns

This chapter contains an introduction to the direct-space analysis of planar patterns composed of point objects. We include in this category patterns in which the original objects are finite in extent, but may be meaningfully replaced by actual points representing, for example, object centroids. Such patterns are encountered in a variety of circumstances spanning a wide range of spatial dimensions, as exemplified by images of atoms, recorded by modern scanning tunneling microscopy or atomic force microscopy; biological cells or lipid vesicles; magnetic bubble domains; or stars. In contrast to such classical methods of structural analysis as x-ray and neutron scattering, direct-space analysis has the advantage of providing a great deal of geometrical and topological information about the pattern configuration. As discussed in the following sections, this is particularly valuable in the analysis of pattern disorder. Strategies of point pattern analysis differ according to the desired information. We address the most common situations in this chapter.

Section Overview

Section 6.1 describes the properties of the Voronoi diagram and possible approaches to its construction. The Voronoi diagram is a planar graph that permits the identification of the set of nearest neighbors (NNs) for each member of a point pattern and provides the basis for direct-space analysis of such patterns.

Section 6.2 discusses the characterization of point patterns in terms of distribution functions that capture the statistics of local distances and orientations in a given pattern.

Section 6.3 presents a discussion of a rather different type of point pattern description, emphasizing geometrical as well as topological features. Such a characterization frequently is desirable to capture the essential features of cellular patterns such as the networks formed by soap froths.

Section 6.4 extends the analysis of point patterns from NN relationships between points to increasingly distant neighbors. This is helpful in ascertaining how geometrical and topological quantities of interest vary as increasingly distant pairs of points are considered. For example, clustering of points would manifest itself in this type of analysis.

6.1 The Voronoi Diagram of Point Patterns

Typical Application(s) – tesselation of planar patterns of point particles in images of atoms; cells adsorbed to surfaces; layers of colloidal spheres in suspension; domain patterns in a wide variety of physical–chemical systems; stars and galaxies.

Key Words – planar graph, tesselation, triangulation, proximity, nearest neighbor (NN).

Related Topics – region peak detection (Section 3.4); binary region detection (Section 4.3), medial axis transform (thinning) (Section 4.7; point pattern analysis (Chap. 6).

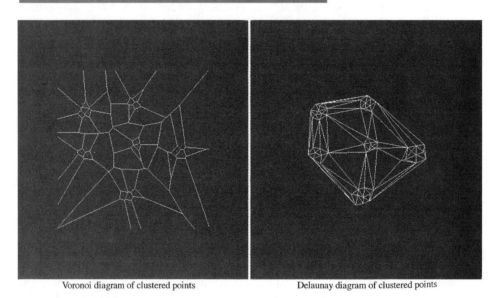

Voronoi diagram of clustered points   Delaunay diagram of clustered points

Figure 6.1.1. Pictorial Example. Point pattern with superimposed Voronoi diagram (left) and Delaunay triangulation (right).

The principal analytical tool to extract statistical information pertaining to the spatial distribution of the points in a planar pattern is the Voronoi diagram. As illustrated in Fig. 6.1.1 and in more detail in Fig. 6.1.2, this construct generates a uniquely defined partition, or tesselation, of the plane into a set of N polygonal cells, also referred to as Voronoi polygons. Each polygon contains exactly one member, $P_i \in \mathbf{P}$, of the given pattern $\mathbf{P} \equiv \{P_i; 1 \leq i \leq N\}$ and delineates the locus of all points in the plane that are closer, in terms of Euclidean distance, to P_i than to any other member of $\mathbf{P}$.

This definition contains a prescription for the actual construction of the desired plane partition. The set of edges defining each Voronoi polygon is in fact generated from the intersection of perpendicular bisectors of the line segments connecting any one point to all its nearest neighbors in $\mathbf{P}$. By construction, the resulting partition is composed of convex polygonal regions, each representing the two-dimensional equivalent of what is familiar in the field of solid-state physics as the Wigner-Seitz cell, introduced in Fig. 6.1.2.

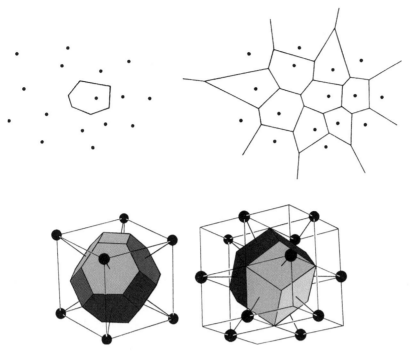

Figure 6.1.2. Top, a simple point pattern (left-hand figure), along with its Voronoi diagram (right-hand figure). The characteristically shaped polygonal region indicated in the left-hand panel for one of the points in the pattern is referred to as that point's Voronoi polygon. It corresponds to the well-known Wigner–Seitz cell, a construct that serves to characterize crystal lattices in two and three dimensions. (Reprinted with Permission from [Preparata and M. Shamos 85], Copyright © 1985 by Springer-Verlag, New York, Inc.) Bottom, Wigner–Seitz cells for a (three-dimensional) body-centered cubic lattice (left-hand figure), and for a face-centered cubic lattice (right-hand figure). (Reprinted with Permission from [Ashcroft and Mermin 76], Copyright © 1976 by Holt, Rinehart and Winston.)

6.1.1 PLANAR TESSELATIONS AND TRIANGULATIONS

The complete set of polygon vertices and edges generated in the Voronoi tesselation defines a planar graph, the Voronoi diagram. As illustrated in Fig. 6.1.2, each node or vertex V of the Voronoi diagram marks the intersection of exactly three polygon edges (with the one exception of the case of a rectangular array of points). As a result, the planar graph obtained, as indicated in Fig. 6.1.3, by connecting with a straight line each pair of sites in **P** whose Voronoi polygons share an edge, represents a triangulation. Accordingly, the resulting Delaunay triangulation is a planar graph whose nodes are formed by the sites of the original point pattern **P** and whose branches define a planar partition of that pattern into a set of triangular regions (Fig. 6.1.4).

The numbers of vertices v, edges e, and disjoint regions or faces f of a planar partition are related by Euler's formula $v - e + f = 2$. For the Delaunay triangulation, this result produces the following valence relations: $3v = 2e$, reflecting the fact that three edges meet at every vertex and each edge links two vertices, and $\sum_n f_n = 2e$, where f_n denotes the number of faces with n edges, embodying the statement that two faces share an edge. These two conditions permit Euler's relation to be recast in the

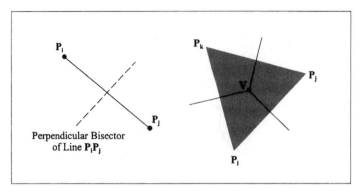

Figure 6.1.3. Graphical Illustration of the definition of the perpendicular bisector of a pair of points P_i and P_j (left-hand figure). The intersection of three perpendicular bisectors defines a Voronoi polygon vertex V (right-hand figure).

form $[6 - (\sum_n f_n)/f] = (6 - \langle n \rangle) = 12/f$, yielding as an immediate corollary for large networks ($f \to \infty$) in two dimensions the classic result $\langle n \rangle = (\sum_n f_n)/f = 6$. These simple valence conditions may also be used to show that a Voronoi diagram on N sites has at most $3N - 6$ edges and $2N - 5$ vertices. More generally, inequalities hold, so that a polygonal network with N vertices of degree ≥ 3 has at most $3N - 6$ edges and $2N - 4$ interior regions.

6.1.2 THE PROXIMITY (NEAREST-NEIGHBOR) PROBLEM

The prescription just given for the Voronoi tesselation relies on the set of perpendicular bisectors of line segments connecting NN's and thus implies, as a requirement for the

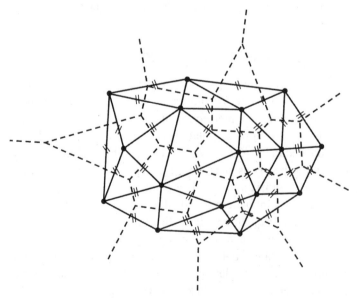

Figure 6.1.4. Illustration of the Delaunay triangulation (solid lines) as the straight-line dual graph of the Voronoi diagram (dashed lines). (Reprinted with Permission from [Preparata and M. Shamos 85], Copyright © 1985 by Springer-Verlag, New York, Inc.)

construction of the Voronoi diagram, the identification of the set of NNs for every member of the given point pattern. Consequently, the Voronoi diagram represents the solution of the NN problem, also known as the proximity problem. This has a variety of important consequences for the algorithmic solution to a class of fundamental problems in the area of computational geometry.

The Voronoi diagram may be generalized in various ways, on the one hand including its construction in higher spatial dimensions and on the other hand pertaining to collections of more complex objects, such as lines or circles of differing radii; the latter problem is known to be equivalent to the consideration of weighted point sets. A close connection exists between the Voronoi diagram and the skeleton (Section 4.7) associated with line patterns and elongated polygonal shapes. The skeleton represents the locus of all internal points of the polygon that are equidistant from at least two distinct points on the boundary. The skeleton may be envisioned to represent the locus of intersection of the half planes that contain, for each polygon edge, the points in the plane closer to that edge than to any other edge.

6.1.3 CONSTRUCTION OF THE VORONOI DIAGRAM

As we have discussed, the construction of the Voronoi diagram entails for each site $P_i \in \mathbf{P}$, $1 \leq i \leq N$, the identification, of the set of edges forming the Voronoi polygon associated with that site. Each edge is defined as a straight-line segment connecting two successive vertices of the polygon.

Algorithmic Approaches

A first approach to the algorithmic solution of this problem would be to implement the above-mentioned prescription of determining the intersection of bisectors, proceeding by means of the successive construction of individual polygons and their addition to the diagram. While this procedure has the advantage of conceptual simplicity and relative ease of implementation, it is unfortunately excessively slow.

A second, more sophisticated approach achieves far higher computational efficiency by invoking a recursive strategy. The given point pattern is first subdivided and the Voronoi diagram is constructed for each of the resulting subsets of sites; in the second step, the resulting partial diagrams are merged. While strategies are known that permit the efficient execution of this merge operation, its details are complex and are further discussed in the technical literature [Preparata and Shamos, 1985].

Fortune's Algorithm

A third distinct procedure, introduced by Fortune [Fortune, 1987], is the one implemented here. It uses a sweep-line or plane-sweep algorithm in combination with a geometric mapping of the Voronoi diagram. As the nomenclature suggests, a plane-sweep algorithm samples a given collection of objects in the plane (in the present situation the sites of a point pattern) by sweeping a horizontal line across it, from bottom to top, say, while registering all intersections of that line with the objects of interest.

The difficulty of this approach when applied to the construction of the Voronoi diagram arises from the fact that, as is apparent from Fig. 6.1.5, the sweep line will

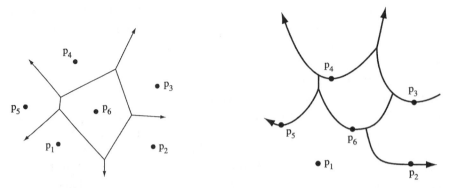

Figure 6.1.5. Geometric mapping of a Voronoi diagram of six points (left-hand figure) into a transformed version (right-hand figure) in which the original sites represent the lowermost points on the region boundary (see text). (Reprinted with Permission from [Fortune 87], Copyright © 1987 by Springer Verlag, New York, Inc.)

generally intersect the interior of the Voronoi polygon, or region, associated with a specific site before encountering the site itself, and this makes it very difficult to maintain the schedule of intersection events. The problem may be circumvented by a geometric rearrangement of the Voronoi diagram, invented by Fortune. As illustrated in Fig. 6.1.5, the original diagram is transformed so as to ensure that the site of each Voronoi polygon becomes the lowermost point of a (transformed) Voronoi region, now no longer polygonal. In a bottom-to-top sweep, these extremal points are thus encountered first as the sweep intersects a new polygon in its upward progression.

Both the Voronoi diagram and its straight-line dual graph, the Delaunay triangulation, may be generated from the transformed graph in a manner that is efficient in terms of computational performance as well as storage. Performance is improved by taking advantage of hashing techniques.

References and Further Reading

[Ashcroft and Mermin, 1976] N. W. Ashcroft and N. D. Mermin, "Solid State Physics," Saunders College, 1976.

[Fortune, 1987] S. J. Fortune, "A sweepline algorithm for Voronoi diagrams," Algorithmica **2**, 153–174 (1987).

[Preparata and Shamos, 1985] F. P. Preparata and M. I. Shamos, *Computational Geometry*, 2nd ed. (Springer-Verlag, New York, 1985).

The details of the mapping underlying Fortune's algorithm are discussed in the original work in which a useful geometric interpretation is also offered that is further discussed in J. O'Rourke, *Computational Geometry in C* (Cambridge U. Press, New York, 1993).

A review of theoretical results for the Voronoi diagram and an authoritative discussion of the fundamental significance of the Voronoi diagram to classes of problems in computational geometry is in [Preparata and Shamos, 1985]; a review of the properties of the Voronoi diagram that are of particular interest to applications in physics and materials science is in F. Aurenhammer, Assoc. Comput. Mach. Comput. Surv. **23**, 345–405 (1991).

Programs

spp

scans a point pattern in a given image and generates
data for subsequent Vornoi analysis.

USAGE: spp inimg [-d] [-w file] [-L]

ARGUMENTS: inimg: gray-scale or binary image filename (TIF)

OPTIONS: -d: scan pattern of pointlike objects marked in
max_index-1 (254); data are displayed in order
of increasing y-coordinates

-w file: write file (.vin) to disk

-L: print Software License for this module

xvor

constructs Voronoi diagram or Delaunay triangulation
for a set of points in the plane by using Fortune's
algorithm [Fortune 1987]

USAGE: xvor infile outimg [-w file] [-s] [-t] [-g] [-L]

ARGUMENTS: infile: input data file of type *.vin, produced by spp
or xal; assumes y-sorted site coordinates unless
option -s invoked

outimg: ouput image filename (TIF)

OPTIONS: -w file: write output to file (.vdt) as well as stdout

-s: y-sort data in input file

-t: perform Delaunay triangulation

-g: debug mode

-L: print Software License for this module

6.2 Spatial Statistics of Point Patterns: Distribution Functions

Typical Application(s) – evaluation of distribution functions for pair distances and coordination numbers to assess degree of randomness.

Key Words – distance statistics, angle statistics, site coordination; random, clustered, ordered patterns.

Related Topics – point detection (Section 3.4), binary region detection (Section 4.3).

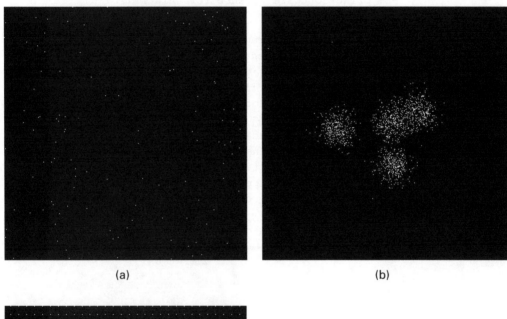

(a)   (b)

Figure 6.2.1. Pictorial Example. A random point pattern, exhibiting (a) Poisson statistics is contrasted with patterns exhibiting (b) site clustering and (c) site ordering. (right panel).

(c)

In this section, we show how to examine a pattern's configurational statistics to assess deviations from randomness in the spatial distribution of pattern sites. Quite generally, it is useful to distinguish random and nonrandom point patterns; in the latter category, ordered arrays and clustered patterns represent opposite limiting cases (see Fig. 6.2.1).

In nature, regular arrays, or lattices, which exhibit translational and certain types of rotational symmetries, may arise in the presence of a repulsive interaction between constituent particles. In contrast, patterns composed of clusters of particles are indicative of a (net) attractive interaction between constituents. The statistics of the particles' spatial distribution may reveal quantitative information about the pertinent interaction, including its range, and may even provide bounds on the functional form of its distance dependence.

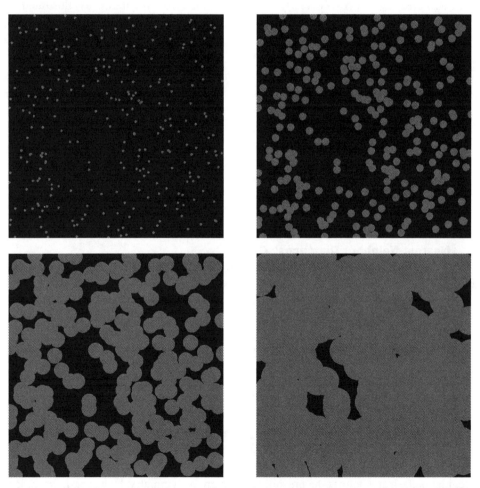

Figure 6.2.2. Four intermediate states in the course of constructing $N(t)$, the number of connected regions associated with the random point pattern in Fig. 6.2.1. The patterns correspond to: $N/N_O = 0.93822$ (top left); $N/N_O = 0.6139$ (top right); $N/N_O = 0.05405$ (bottom left); $N/N_O = 0.00386$ (bottom right).

6.2.1 DISTANCE STATISTICS

To characterize the spatial distribution of sites in a given point pattern, we introduce here simple functions that capture aspects of a pattern's distance statistics.

The Number of Connected Regions: $N(t)$

An intuitive characterization of a pattern's distance statistics invokes the following simple construction (see Fig. 6.2.2). On each of the N_0 sites in the pattern, place a circle of radius t. As t is increased, circles of neighboring sites will begin to overlap so that the total number of connected regions $N(t)$ will decrease.

To construct $N(t)/N_0$, simply count the number of distinct regions remaining for given t. Here, the normalized distance variable $t \equiv r/r_0$ depends on the actual distance variable r as follows: let S denote the total area available for the placement of sites so that S/N_0 represents the average area per site; then $r_0 \equiv (S/N_0)^{1/2}$. Plots of the fraction of connected regions N/N_0 clearly reveal deviations from regularity, as illustrated in Fig. 6.2.3.

Fraction of Connected Pixels: $P(t)$

An additional quantity reflecting the configuration of points in a set is the fraction of connected pixels within an area of interest containing all the sites in the set. As increasingly larger solid circles, placed on each pattern site in the manner just described for the construction of $N(t)$, begin to overlap, the fraction of connected pixels continues to increase, eventually approaching unity.

To evaluate $P(t)$, determine, at each stage, the total number of pixels within connected regions relative to the total number of pixels within the chosen original area of interest. Examples in Fig. 6.2.3 illustrate the behavior of $P(t)$ for random and nonrandom patterns.

Nearest-Neighbor Distances: $Q(t)$

A sensitive measure of the spatial distribution of a set of N_0 points in a pattern is available in the form of the cumulative distribution of NN distances, based on the identification of NNs in the Voronoi diagram.

To evaluate $Q(t)$, determine, for each site in the pattern, the set of NNs from the Voronoi diagram. Next, determine the distribution of all NN distances r_2 in the pattern. Now $Q(t)$ represents the fraction of all NN distances for which $r_2/r_0 < t$.

6.2.2 SITE COORDINATION STATISTICS

The Variance of the Nearest-Neighbor Distribution

The local site coordination simply is defined as the number of NNs associated with each site. By definition, this coordination number n is the same for each site in a regular pattern and defines the prevailing point group symmetry. However, in irregular, disordered patterns, frequent deviations in coordination number do occur, and this provides the basis for a measure of disorder.

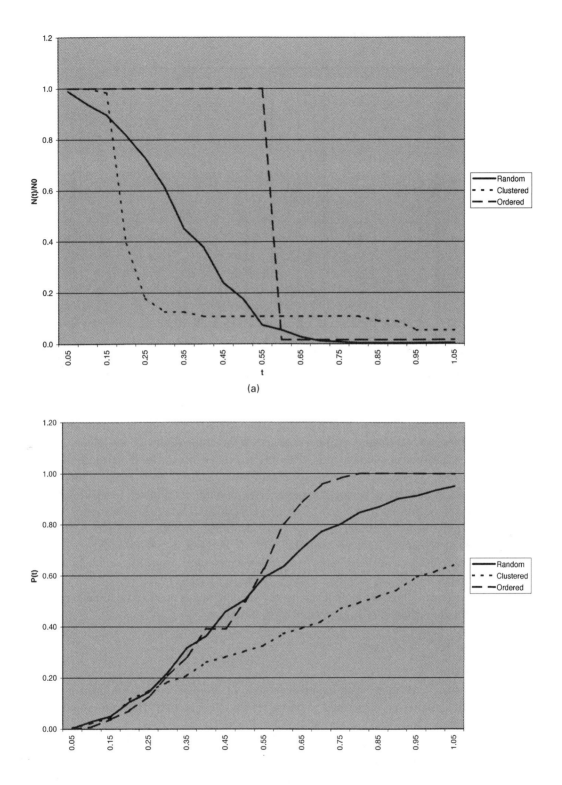

Figure 6.2.3. Plots of (a) $N(t)/N_0$ and (b) $P(t)$, for random (solid curves), clustered (dotted curves), and ordered (dashed curves) point patterns, illustrating the sensitivity of each function to deviations from randomness.

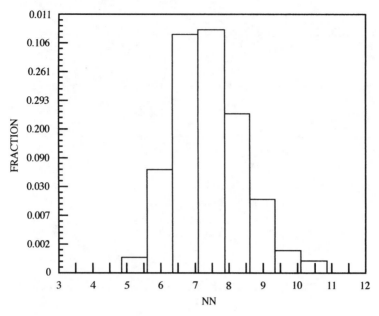

Figure 6.2.4. Site coordination histogram for a random (Poisson) point pattern.

Specifically, the distribution of nearest neighbors $f(n)$, evaluated for the entire collection of sites, provides a global measure of the range of deviations in coordination number. A fundamental theorem, due to Euler, states that in two dimensions the mean of $f(n)$ is 6 (Section 6.1). We may regard the quantity $C \equiv n - 6$ as a topological charge, in analogy to electrical charge. Thus, fivefold coordinated $(-)$ and sevenfold coordinated $(+)$ sites in a sea of sixfold coordinated sites have a tendency to form pairs (dipoles; see Section 5.7). With this identification, Euler's theorem is the statement of topological charge neutrality, $\langle C \rangle = 0$. Thus, if only fivefold coordinated and sevenfold coordinated sites exist in a pattern, their respective numbers must be identical.

In general, higher topological charges may of course exist, and a suitable measure for the spread about $\langle C \rangle$ is provided by the variance

$$\mu_2 \equiv \sum_{n \geq 3} n^2 f(n).$$

This parameter, along with the concentration of droplets with $C = 0$, provides a first useful classifier for disordered patterns (see also Section 6.3). The corresponding histogram for the Poisson pattern is shown in Fig. 6.2.4.

NN Bond Angles

Closely related to the histogram of site coordinations is the distribution of angles subtended by successive NN connecting line segments or bonds: This is simply the distribution of angles ϕ characterizing the pattern's Delaunay triangulation. Thus the condition $\langle n \rangle = 6$ implies that $\langle \phi \rangle = \pi/3$. The variance of this distribution of bond angles may thus be used to signal transitions between ordered and disordered patterns that alter the degree of orientational order.

References and Further Reading

General methods of point pattern statistics are described in Chap. 8 of B. D. Ripley, *Spatial Statistics* (Wiley, New York, 1980). Different possible definitions of randomness in point patterns and algorithms to generate corresponding configurations of points are discussed in K. Gotoh, "Comparison of random configurations of equal disks," Phys. Rev. E **47**, 316–318 (1993), which also cites the closed form of the distribution of NN distances $dQ(t)/dt$ for patterns obeying a Poisson distribution. Further results for random Voronoi polygons, including the distribution of nearest neighbors shown in Fig. 6.2.4, are discussed in R. E. M. Moore and I. O. Angell, "Voronoi Polygons and Polyhedra," J. Computational Physics **105**, pp. 301–305 (1993).

An illustration of the methodology in analyzing domain patterns observed in certain polymer systems is given by H. Tanaka, T. Hayashi, and T. Nishi, "Digital image analysis of droplet patterns in polymer systems: point patterns," J. Appl. Phys. **65**, 4480–4495 (1989).

Programs

dpp

```
            draws random, clustered, or ordered point pattern in
            a given area, with preselected coverage.
     USAGE: dpp outimg [-p t] [-c f] [-L]
 ARGUMENTS: outimg: image filename (TIF)
   OPTIONS:    -p t: select pattern type;
                     t=r: random (default)
                     t=c: clustered
                     t=o: ordered
               -c f: coverage; 0 <= f <= 100; default: f=5
                     Note: f denotes the percentage of "ON" pixels
                           in the selected area of interest
                 -L: print Software License for this module
```

xptstats

```
            constructs the distribution functions N(t)/N_0 and P(t)
            for a given point pattern
     USAGE: xptstats infile [-n nsteps] [-w f] [-L]
 ARGUMENTS:    infile: input data file of type .vin, produced by
                       spp or xah
   OPTIONS: -n nsteps: set number of steps (default: 20)
                 -w f: write data to file
                   -L: print Software License for this module
```

xvora

extracts statistics of a given Voronoi diagram from the representation stored in a file of type .vdt; a histogram may be contructed for various quantities

USAGE: xvora infile outimg [-d] [-k S] [-p a] [-b] [-e] [-a] [-l]
                          [-n nbins] [-i f] [-f f] [-s] [-w f]
                          [-L]

ARGUMENTS: infile: input data file (ASCII) produced by xvor;
                      .vdt (Voronoi data)
                   or .ddt (Delaunay data)
   outimg: ouput image filename (TIF)
   OPTIONS: -d: show input data
             -k S: construct k-NN shells, k<=S (default: S=2)
             -p a: enable labeling; a is a string, as follows:
                 a=nall: enable numbering of all sites
          a=n57: enable numbering of 5-fold, 7-fold sites
          a=n57a: ->in addition, mark active AOI
             -b: no border strip

           construct histogram for:
            -e: edge number (per Voronoi polygon)
            -a: polygon area
            -l: nn-distance (edge bisector length)
          -n n: set number of bins to n (default=10)
          -i f: set initial value to f(float)
          -f f: set final value to f(float)
            -s: evaluate nn-distance statistics function Q = Q(t)

          -w f: write data to file
                f=h: hist data --> file of type .hdt
                f=s: site topol data --> file of type .std
                f=p: poly topol data --> file of type .ptd

            -L: print Software License for this module

6.3 Topology and Geometry of Cellular Patterns

Typical Application(s) – statistics of cellular patterns and polygonal networks; analysis of pattern-coarsening dynamics.

Key Words – coordination number, topological charge, Lewis law, Aboav–Weaire law, charge compensation.

Related Topics – k-NN shells (Section 6.4).

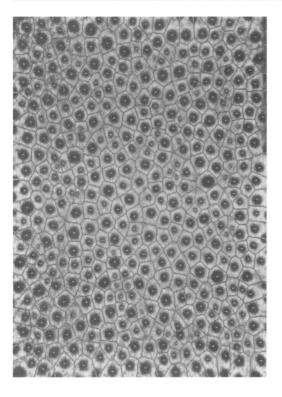

Figure 6.3.1. Pictorial Example. Pattern of (dark) circular droplet domains (bubbles) in a monomolecular film of organic molecules floating on water, which were made visible by inclusion in the film of a fluorescent dye that is excluded from the domains. Superimposed on the domain pattern is the Voronoi diagram constructed from the set of domain centroids, which are also indicated. The vertical dimension of the field of view is 1500 μm [Sire and Seul, 1995]. (Reprinted with Permission from [Sire and Seul 95.])

The Voronoi diagram and the distribution and correlation functions introduced in the preceding sections of this chapter provide methods of characterizing the spatial configuration of point patterns based solely on the position of the points in the set. In this section we extend our view to include the analysis of interrelated topological and geometrical quantities of polygonal networks. Such networks, also referred to as cellular patterns, arise in many contexts. Examples are foams, soap bubble froths, and metallurgical grains (Fig. 6.3.1). Cellular patterns and point patterns are intimately related: As discussed in Section 6.1, every point pattern generates a polygonal network in the form of its Voronoi diagram. Polygonal cells are characterized not only by their position relative to their neighbors in the pattern, but also by properties such as their area and their number of sides.

6.3.1 GEOMETRICAL AND TOPOLOGICAL ASPECTS OF PATTERNS

A prominent phenomenon for which the connection between geometry and topology is of fundamental importance is the coarsening of foams and soap froths that are composed of a polygonal network of cells or bubbles. As foams age, large bubbles grow larger while small bubbles shrink and disappear. The kinetics of coarsening of (planar) froths is described by an equation, derived long ago by von Neumann, which states that the rate of change dA/dt of the area A of any given n-sided cell in the froth is proportional to topological charge $C \equiv n - 6$ (see Section 6.2.2); that is, $dA/dt = kC$. According to this equation, cells with more than six sides will grow while cells with less than six sides will shrink and in fact will eventually collapse. This cell collapse is accompanied by rearrangements of the entire network.

In characterizing the statistical properties of cellular patterns, certain interdependent geometrical and topological quantities are of particular interest, and these are described below.

6.3.2 NN INDEXING OF GEOMETRICAL PROPERTIES OF DOMAINS

In approaching the statistical analysis of cellular patterns, it is useful to correlate geometrical quantities of NN cells or domains. That is, to construct the Voronoi diagram of a set of domains of finite size, the domain pattern must first be converted into a suitable point pattern; data derived in the geometrical analysis of domains such as position, area, shape, and moments (Sections 4.4–4.6) must be preserved and must be indexed in such a way that, for any given cell, NNs may be assigned on the basis of the subsequent topological analysis.

The requisite steps may be summarized as follows.

- Apply a suitable filling algorithm (Sections 4.3 and 4.4) to convert the set of domains into a representative point pattern, for example in the form of domain centroids. In this step, geometrical quantities are accumulated in the order in which domains are encountered in the course of a raster scan of the pattern.
- Next, use the set of domain centroids as input for the construction of the Voronoi diagram. This graph may be viewed as a cellular pattern that is uniquely associated with the original domain pattern.
- Finally, merge data from the geometrical and topological analyses by matching centroid coordinates and extracting geometrical and topological quantities of interest into an appropriate data structure.

6.3.3 STATISTICAL ANALYSIS OF CELLULAR PATTERNS

Domain Area and Coordination Number: Lewis Law

An object of fundamental interest in the analysis of domain and cellular domain patterns is the joint probability distribution $p(n, A)$ of coordination number n and area A. It describes the interdependence of area and coordination number of any domain in the planar pattern: the larger a given domain, the more likely it is to have a large number of NNs; $p(n, A)$ gives the quantitative dependence of these two quantities.

We may construct $p(n, A)$ by merging two types of data sets, one giving $A_i \equiv A(x_i, y_i)$, the other giving n for the site (x_i, y_i), as derived from the Voronoi diagram of domain centroids. From this merged data set we construct the distribution of domain areas, $p(A_n)$ of domains of given n. The moments of $p(A_n)$ are directly related to the joint distribution; for example,

$$\langle A_n \rangle = \sum_{i=0}^{N-1} A_i \, p(n, A_i).$$

An important relation first conjectured by Lewis for cellular patterns in plant leaves and since confirmed in cellular patterns of many varieties relates the geometrical quantity $\langle A_n \rangle$ normalized to the overall average area $\langle A \rangle$, to the topological charge $C \equiv (n - 6)$:

$$\langle a_n \rangle \equiv \langle A_n \rangle / \langle A \rangle = 1 + \lambda C.$$

Here, λ is an adjustable parameter and serves to classify patterns. The Lewis law has been shown to follow from very general arguments. An example, extracted from Fig. 6.3.1, is shown in Fig. 6.3.2(a).

Compensation of Topological Charge: Aboav–Weaire Law

A further relation, surmised by Aboav and later correctly stated by Weaire [Weaire and Rivier, 1984], addresses the compensation of topological charge at the NN level. Guided by the analogy to electrical charge (Section 6.2.2), one might guess that the topological charge of a given domain may be completely screened by clustering around it domains of opposite charge. For an n-fold coordinated domain, complete compensation would imply that $C + n\overline{C_{NN}} = 0$, where $\overline{C_{NN}}$ denotes the average topological charge of the NNs of the domain of interest.

However, topological charges must obey the sum rule $\langle C + n\overline{C_{NN}} \rangle = \mu_2$, where μ_2 is the variance in the distribution of cell coordination numbers encountered in the pattern (see also Section 6.2). This sum rule precludes exact compensation of topological charge at the NN level. The degree to which this global balance of topological charge is approached by individual NN clusters is expressed in the Aboav–Weaire law,

$$C + n\overline{C_{NN}} = (1 - a)C + \mu_2.$$

Note that the Euler law $\langle C \rangle = 0$ ensures validity of Weaire's sum rule. Charge compensation is indicated by values $a \sim 1$ of the parameter a, implying domain configurations in which domains of high coordination number (and large area) tend to be surrounded by domains of low coordination number (and small area), and vice versa. Applied to the pattern in Fig. 6.3.1, this analysis yields the plot in Fig. 6.3.2(b).

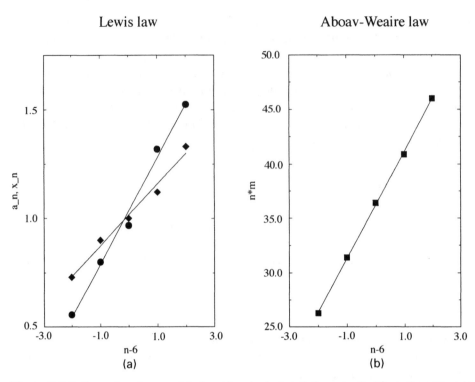

Figure 6.3.2. Examples of geometrical and topological statistics extracted from the pattern in Fig. 6.3.1. Plots (with linear fits) show the validity of the Lewis law (a) for the droplet domain (solid circles) and for the corresponding Voronoi cell (diamonds) and (b) the Aboav–Weaire law. Here, $x_n \equiv \langle A_n \rangle / \langle A \rangle$ and $a_n \equiv \langle A_{Vn} \rangle / \langle A_V \rangle$ respectively represent the average area of an n-fold coordinated droplet domain, normalized by the mean domain area $\langle A \rangle$, and the average area of an n-sided Voronoi cell, normalized by the mean cell area $\langle A_V \rangle$; n and m respectively denote the droplet coordination number and the average coordination number of droplets in the NN shell of an n-fold droplet [Sire and Seul, 1995]. (Reprinted with Permission from [Sire and Seul 95.])

In view of the Lewis law, the Aboav–Weaire law implies an inverse correlation between the area of a domain or cell and the (average) area of its NNs. This inverse correlation, made explicit in Fig. 6.3.3, expresses the strong tendency of disordered cellular patterns observed in nature to assume configurations in which large cells are surrounded by small cells (and vice versa).

Along with the previously introduced quantities μ_2 and λ, the adjustable parameter a represents a third quantity facilitating the classification of cellular patterns.

Boundary Separation of Adjacent Domains

Additional information may be extracted from the merged data sets. For example, it is often informative to examine the distance of closest approach between adjacent circular bubbles. Defining this boundary separation as s_{ij} and the distance between nearest neighbor sites i and j in the Voronoi diagram as d_{ij}, one immediately has $s_{ij} = d_{ij} - (A_i^{1/2} + A_j^{1/2})$. The quantity s_{ij} is of interest in cases in which droplet

Area Correlations between NN Domains

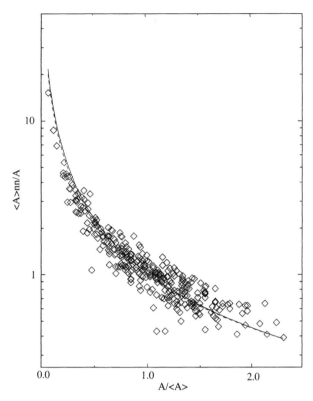

Figure 6.3.3. Scatter plot for a droplet domain pattern such as that in Fig. 6.3.1, showing the dependence between $A/\langle A \rangle$ and the average area of its NNs, $\langle A \rangle_{nn}$. A pronounced inverse correlation is apparent: when $A/\langle A \rangle$ exceeds unity, the corresponding value of $\langle A \rangle_{nn}/A$ falls below unity (and vice versa). The solid and the dashed curves reflect predictions derived from simple theoretical models based on the Lewis and the Aboav–Weaire laws [Sire and Seul, 1995]. (Reprinted with Permission from [Sire and Seul 95.])

domains interact explicitly by means of interactions of long range: In such a case, bubbles will prefer spatial arrangements with correspondingly optimal values of s_{ij}, and this optimal value may be recovered even if the bubble pattern looks spatially disordered, as in the example of Fig. 6.3.1.

References and Further Reading

[Sire and Seul, 1995] C. Sire and M. Seul, "Maximum entropy analysis of disordered droplet patterns," J. Phys. I **5**, 97–109.

[Weaire and Rivier, 1984] D. Weaire and N. Rivier, "Soap, cells and statistics – random patterns in two dimensions," Contemp. Phys. **25**, 59–99 (1984).

Reviews of structure and coarsening dynamics of cellular patterns are available in Weaire and Rivier (1984) and in J. Stavans, "The evolution of cellular structures," Rep. Prog. Phys. **56**, 733–789 (1993).

Program

xsgt

merges geometry and topology data for droplet patterns based on previous analysis by xah and xvora

USAGE: xsgt infile [-k S] [-n n] [-e] [-w] [-L]

ARGUMENTS: infile: input file name of type .xxx (current default)
NOTE: full names of the two required input files, infile.gdt and infile.std, and optionally an output file, are generated after stripping the .xxx suffix; input data file types:
.gdt, generated by xah
.std, generated by xvora

OPTIONS: -k S: construct k-NN shells, k<=S (default: S=2)
-n n: set number of bins to n (default: 10)
-w: write output file of type .sgt (site geometry topology)

-L: print Software License for this module

6.4 The *k*-Nearest-Neighbor (*k*-NN) Problem

Typical Application(s) – partitioning of a given point set into clusters; range finding, particle tracking; evaluation of pattern statistics as a function of increasing index *k*.

Key Words – nearest-neighbor (NN) shell, fractal measure, range finding, clustering.

Related Topics – proximity problem (Section 6.1), cellular patterns (Section 6.3).

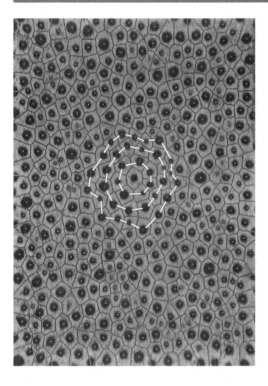

Figure 6.4.1. Pictorial Example. Pattern of droplet domains analyzed in Section 6.4 with superimposed Voronoi diagram (see also Pictorial Example in Section 6.1) with three successive *k*th nearest neighbor shells (*k* = 1, 2, 3) for a randomly chosen site in the pattern. (Reprinted with Permission from [Sire and Seul 95.])

The proximity problem introduced in Section 6.1 frequently presents itself in a more general form, requiring the identification of increasingly distant neighbors, up to the *k*th nearest, of a given site in a point pattern. The solution to the *k*-NN problem, for a selected pattern site, is a catalog of pattern points in the order of increasing *k*. In this section, we distinguish a **topological** and a **geometrical** variant of the problem (see Fig. 6.4.1).

6.4.1 TOPOLOGICAL AND GEOMETRICAL *k*-NN PROBLEM

Topological *k*-NN Problem

As discussed in Section 6.1, the Voronoi diagram embodies the solution of the topological proximity problem. Specifically, the Delaunay triangulation, the planar graph dual of the Voronoi diagram, uniquely identifies the set of NNs for each member of the point set. Here, k serves to count successive NN shells surrounding a given site in a pattern and thus represents topological distance; the standard NN problem corresponds to the special case $k = 1$. For example, in connection with the analysis of cellular patterns (Section 6.3), it may be necessary to check how many NN shells are affected by the presence of a local defect in the pattern.

Another quantity of interest in this context is the fractal measure. This is defined as the proportionality constant, c, in the expression $N_k \equiv ck^2$, relating the number, N_k of sites contained within the k-th NN shell of a given "root" site to the square of the shell index. That is, c represents the topological analog of π, while the shell index, k, represents the radius of a topological circle and N_k its area. When averaged over a set of points forming an ordered lattice, the fractal measure reflects the local lattice symmetry, so that, for example, for a triangular lattice, $c = 3$.

Generalizing this concept, we may plot the sequence of N_k, the number of neighbors in the kth shell, as a function of distance d_k from the root site. When averaged over the set of points, this sequence represents the discrete two-dimensional analog of the radial distribution function, widely used in materials characterization of ordered and amorphous solids.

Geometrical k-NN Problem: Range Finding

Naturally, simple Euclidean distance, a geometrical quantity, may also be invoked to generate an ordered catalog of sites. The solution to the geometrical k-NN problem proceeds directly from a set of object coordinates and invokes sorting and subsequent searching strategies.

This is a task that is commonly called for in connection with range finding. It is often desirable to restrict a search to points that lie within a prescribed maximum range from a reference site. For example, if point objects are free to move about during the time interval between successive snapshots, their velocity limits the distance traveled and thus the range of the search step in object tracking.

A prototypical application calling for the identification of Euclidean k-NNs arises in the task of constructing clusters and partitions of a point pattern. Spatial non-uniformities in a pattern of point objects frequently suggest an intuitive partitioning of the pattern into clusters (see Fig. 6.2.1). A global measure of the degree of clustering was introduced in the form of cumulative distribution functions in Section 6.2. When clustering is indicated by such global measures, an explicit partitioning may be constructed by evaluation of pairwise Euclidean distances, which may be grouped according to a suitable criterion.

6.4.2 IDENTIFICATION OF kTH NEAREST NEIGHBORS

Topological kth Nearest Neighbors

While generalizations of the Voronoi diagram, referred to as order-k diagrams, have been designed to address this problem, we solve it here by a simple extension of the standard proximity problem.

The Voronoi diagram described in Section 6.1 supplies, for any member $P_i \equiv P(x_i, y_i)$ in a point pattern, the set of (first) NN sites ($k = 1$). Accordingly, one may construct the second NN shell by looking up the nearest neighbors of each member of the first NN shell of P_i, and this process may be iterated to successively larger NN shells: Care must of course be taken to ensure that each site is uniquely assigned to one k-NN shell (for each root site).

A suitable data structure in which to collect the members of successive NN shells is an array of lists for each site in the set. The array is dimensioned to accommodate the desired maximal k, k_{max}, and contains, for each k, $0 \leq k \leq k_{max}$, a list of indices designating those sites in the set comprising the k-NN shell.

Euclidean _k_-Nearest Neighbors

In many applications it is not the topological NN relationships that are of primary interest, but rather the simple Euclidean distance from a specific point of interest. That is, one would like to determine the kth NNs of a specific point in the plane by sorting other points in order of increasing Euclidean distance. For example, the task of tracking mobile particles requires the assignment of NNs on the basis of a measurement of Euclidean distance traveled by particles between snapshots. The desired number of neighbors k, usually in the range $0 < k \leq 5$ for most applications, is set by the user. NNs would be identified by selecting $k = 1$; groups of three most adjacent points are obtained for $k = 2$.

Most often, the choice for k is obvious, as in the examples of pairs or triplets just mentioned. However, there are trade-offs in setting k, specifically in choosing the value to be larger or smaller than the obvious choice. For instance, if the desired grouping consists of three elements, the obvious choice for k would be 2, thereby adding two neighbors to each element to construct a group of 3. However, we might want to increase k by 1, $k = 3$, so that even if there is one element of errant noise that is mistaken as a NN, still the true neighbors will be found. In many cases, it is advisable to perform noise reduction (Section 4.2) before region detection and centroid construction (Section 4.3) to reduce the number of erroneous neighbors.

Alternatively, we might be interested in decreasing k by 1, so that $k = 1$. In this case, we will not obtain all elements in the group, but the incomplete sampling may be sufficient and will likely contain fewer erroneous neighbors. Of course, the price we pay for making k larger is both the computational expense and the undesired inclusions into groups. Conversely, making k smaller results in computational savings, but incomplete groupings.

Each neighbor-to-neighbor pair is called a connection, and this connection has two features, illustrated in Fig. 6.4.2. One is the Euclidean distance d_{ij} between elements i and j of the set. The other is the angle ϕ_{ij} measured with respect to some fixed axis, usually the horizontal axis of the image (see also Section 6.2). Having determined the k-NNs for each element along with the distance and angle features for each connection, we can make global measurements. For example, in an image of this page, each text character may be viewed as an element whose neighbors to the left and the right are in the same text line. If we perform a NN analysis with $k = 2$, most connections would have angles corresponding to a full text line and we are able to determine the text line orientation from this information. Performing a NN analysis with $k = 4$ would result in many connections within text lines, but also those between adjacent text lines. These

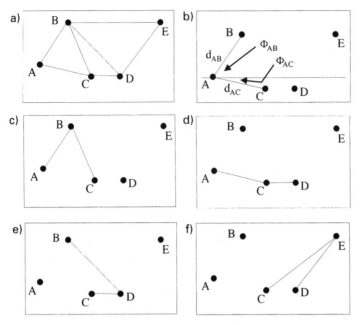

Figure 6.4.2. Example illustrating Euclidean k-NN analysis among five elements, with $k = 2$. In (a), elements A–E are shown with connections; in (b), connections are shown for element A; also shown are interelement distances d_{AB} and d_{AC} and angles with respect to the horizontal axis ϕ_{AB} and ϕ_{AC}. Panels (c)–(f) display connections for elements B, C, D, and E.

connections between text lines will have longer distances and angles differing by 90° from those of connections within the same text line.

The brute force approach to the construction of Euclidean NNs would require evaluation of $N(N - 1)/2$ pairwise distances $d(i, j)$ for an N-member point set and the sorting of this array with respect to $d(i, j)$ for each member in the set. Here, Euclidean distance is defined in the usual way, i.e., $d_{ij} = [(x_i - x_j)^2 + (y_i - y_j)^2]^{1/2}$.

To make this process more efficient, we first order the elements according to either their x or their y coordinate value (whichever has the wider spread of element locations), then determine the distances from each element to its ordered neighbors. If this distance to an element l is less than that for a previously determined NN m, $d_{il} < d_{im}$, element l is inserted in the list of NNs for element i; otherwise, provided that k neighbors have not already been chosen, l is appended to the end of the neighbor list for element i. Neighbors are tested up until the next neighbor's distance along the ordered axis exceeds the kth NN distance already found.

The result of the Euclidean NN analysis is a list of points or image elements, and for each of these an ordered list of the k-NNs. For each connection, the distance has already been calculated and stored. Following the NN analysis, the angle of each connection is evaluated as the angle between neighboring elements and the horizontal axis that defines the angle 0°. The range of angles is 0°–180°, that is, the angle is not a directed one (in which case the range would be 0°–360°), so $d_{ij} = d_{ji}$.

The distances and the angles are accumulated in two histograms to aid in analysis. A peak on the distance plot indicates a common distance between neighbors. Two peaks on the distance plot indicate two predominant distances. A peak on the angle plot indicates a common orientation of the neighboring elements with respect to one another. For example, a predominating sixfold symmetry in the configuration of points in the plane would manifest itself in the form of peaks at 60° in the angle plot. These features are useful to describe average global relationships between image elements (see also Section 6.2).

References and Further Reading

[Sire and Seul, 1995] C. Sire and M. Seul, "Maximum entropy analysis of disordered droplet patterns," J. Phys. I **5**, 97–109.

Generalizations of the planar Voronoi diagram are described in F. P. Preparata and M. I. Shamos, *Computational Geometry*, 2nd ed. (Springer-Verlag, New York, 1985). An interesting discussion of structural characterization from a geometrical point of view is presented in the review by R. Zallen, in: E. W. Montroll and J. L. Lebowitz, *Fluctuation Phenomena* (Elsevier Science Publishers, North-Holland, Amsterdam, 1987), Chap. 3.

Programs

xvora

> with option -k set, determines topological k-nearest neighbors (in kth shell of root site) based on Voronoi analysis; for details of usage, see section 6.2.

xknn

> finds Euclidean k-nearest neighbors for each point in a list with given x–y coordinates

USAGE: xknn infile outimg nn [-l] [-L]

ARGUMENTS: infile: input file with list of (x,y) pairs (ASCII)
outimg: output image filename (TIF)
nn: number of nearest neighbors to calculate (int)

OPTIONS: -l: print (x,y) point with neighbor list

-L: print Software License for this module

Frequency Domain Analysis

It is in the spatial domain that we observe images. Individual objects have size and shape while collections of objects are arranged in space in relation to one another. Chapters 4, 5, and 6 focus on methods of analyzing individual objects and details of the arrangements of objects in the spatial domain.

In many applications, the primary interest of analysis is not in features of individual objects or even in the details of the arrangement of objects, but rather in the image texture or the degree of regularity governing a pattern formed by multiple identical objects. These properties are most conveniently analyzed not in the spatial domain but in the spatial frequency domain. For example, a fine texture or pattern composed of many small, densely arranged elements is said to have high spatial frequency characteristics; conversely, a coarse texture or pattern composed of a small number of larger elements is said to have low spatial frequency characteristics. The following examples illustrate the connection between spatial domain and frequency domain. Corresponding figures are shown in Section 7.1.

In pictures of foliage, a characterization based on the analysis of individual leaves would not only be very tedious, but would likely miss the principal features of the images that emerge from the collection of leaves. More concise and more pertinent is an analysis based on the examination of the leaves as a group. For instance, one tree may have many, narrow, sharply pointed leaves, while another has larger, broad leaves. In this case, the former is classified on the basis of its high frequency features, and the latter by lower frequency features (see Fig. 7.1.6).

Regularity in the arrangement of objects manifests itself in the degree of order or randomness in the corresponding pattern (for a related discussion, see Section 6.2). Thus, windows in an office building, bricks in a wall, or tiles on a floor usually produce regular, highly structured patterns. In contrast, leaves on a tree, pebbles on a beach, or water droplets on a window typically display a more random spatial arrangement (see Fig. 7.1.7).

Order and disorder are concepts of fundamental importance in the physical sciences. Materials adopting an ordered arrangement of constituents display physical and chemical characteristics that are altered or lost when the ordered state is destroyed. The transformation between ordered and disordered states is induced by external stimuli. For example, order in a crystal is destroyed by melting. On the other hand, a magnetic field can induce order in a magnet by aligning spins in a common direction. Similarly, lateral pressure applied to a two-dimensional film of rod-shaped molecules tends to induce tighter molecular packing as well as molecular ordering. Such films also exhibit spontaneously formed stripe domain patterns whose morphology can be altered by changes in pressure and temperature. A disordered configuration of stripes will rearrange in response to such external stimuli to a pattern in which stripes are aligned in a common direction (see Fig. 7.1.8).

Section Overview

Section 7.1 introduces the transformation of images from their natural representation in the spatial domain to the frequency domain by means of the Fourier transform (FT). The relation between textures and patterns in the spatial domain and their respective transformations in the frequency domain are described.

Section 7.2 focuses on the practical aspects of applying the discrete two-dimensional (2D) FT to the analysis of images. We place particular emphasis on the application of the transform to image filtering. Thus we show how to perform filtering in the frequency domain to extract, enhance, or suppress certain frequency features. This implementation of filtering in the frequency domain will be compared with the implementation of the analogous operations in the spatial domain, discussed in Sections 3.2–3.4.

7.1 The 2D Discrete Fourier Transform

Typical Application(s) – segmentation by pattern or texture, recognition by pattern or texture; filtering, convolution.

Key Words – discrete Fourier transform (DFT), fast Fourier transform (FFT), inverse Fourier transform; correlation, power spectrum; sampling rate, resolution, maximum frequency, minimum size; aliasing, "jaggies," windowing.

Related Topics – convolution (Section 3.1), subsampling (Section 3.7), multiresolution (Section 3.8); Fourier descriptors (Section 4.5).

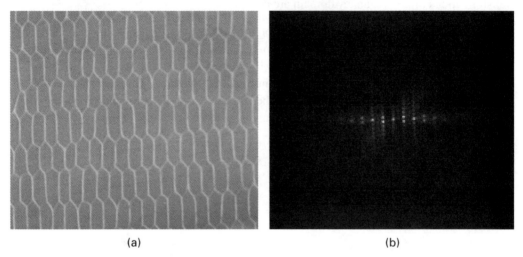

(a)   (b)

Figure 7.1.1. Pictorial Example. (a) Image of a hexagonal mesh structure observed in adult bovine lens fiber cells, (b) corresponding power spectrum, produced by Fourier transformation, with linearly scaled intensities. The multiple periodicities and the hexagonal symmetry of the pattern are reflected in the set of diffraction peaks displayed in the power spectrum. (Panel (a) Reprinted with permission [Costello 97] – Copyright © 1997 by Dr. J. Costello.)

This section introduces Fourier transformation as a tool for frequency domain analysis of images and provides the standard fast algorithm to evaluate the FT. The basis for this analytical approach is the interpretation of an image as an intensity function of two spatial variables (see Fig. 7.1.1).

In its analysis of time-dependent functions, or waveforms, digital signal processing relies in a fundamental way on the application of the FT (and other transformations) to determine the frequency content of the waveform. In analogous fashion, the 2D FT facilitates the interconversion between the spatial domain representation and the spatial frequency domain representation of images. Given its intimate connection to diffraction experiments probing the atomic or molecular structure of materials, the FT also plays an important role in materials science and in physics (see Section 6.3). Properties of the FT are reviewed in Section A.1.

7.1.1 THE DISCRETE FOURIER TRANSFORM

Relevant to image analysis is a discrete version of this transform, defined in the context of a discrete grid of image pixels, $I = I(m, n)$. As described in Section A.4, these discrete functions are sampled representations of continuous signals; that is, the set of image pixels represents a discrete sample of the intensity distribution of the scene recorded in the image. Analogously, the continuous FT has a discrete analog, the discrete Fourier transform (DFT).

The 2D DFT is defined as

$$F(u, v) = \frac{1}{MN} \sum_{m=0}^{m=M-1} \sum_{n=0}^{n=N-1} I(m, n) \exp[-i2\pi(um/M + vn/N)],$$

where the image $I(m, n)$ is transformed to its frequency domain representation $F(u, v)$ by means of the FT, designated by $\mathcal{F} : I(m, n) \xrightarrow{\mathcal{F}} F(u, v)$. Whereas m and n are spatial samples, u and v are corresponding frequency samples along two corresponding axes. The range in the two domains is the same, $0 \leq m \leq M - 1$, $0 \leq n \leq N - 1$ and $0 \leq u \leq M - 1$, $0 \leq v \leq N - 1$. Also in this equation, $i \equiv \sqrt{-1}$, and $\exp[\ ] \equiv e^{[\ ]}$.

The size of the 2D DFT is equal to that of the original image; that is, the transform of an image of size $M \times N$ also has dimensions $M \times N$. The lowest-frequency component is that at the origin of the transform, that is, at $(u, v) = (0, 0)$. This component reflects the average pixel value in the image. In the context of electrical engineering and signal processing, this component is sometimes referred to as the dc (direct current) component. The highest-frequency component is located at $u = \pm M/2$, $v = \pm N/2$. This component corresponds to the smallest resolvable length of a periodic signal in the original image, which is two pixels in length.

One can gain intuition for the nature of this transform by examining an image of a one-dimensional (1D) wave pattern, written mathematically as a sinusoidal function, $I(n) = A \sin kn = A[\exp(ikn) - \exp(-ikn)]/2i$. Fourier transformation of this equation yields the important result,

$$I(n) \xrightarrow{\mathcal{F}} F(u) = \begin{cases} A & \text{for } u = \pm k \\ 0 & \text{otherwise} \end{cases}.$$

The image and its transform are shown in Fig. 7.1.2. Note that a sinusoidal function of a single frequency k produces a transform with peaks at both $+u$ and $-u$. Although negative frequency has no meaning for real images, it is always true that the transform will contain symmetrically located, positive and negative peaks corresponding to each frequency component.

The FT is represented here in a related form, known as the power spectrum, to which we will sometimes refer as the transform. The power spectrum is a representation of the power (or quantity) of signal at each spatial frequency. At the center of the spectrum, the frequency equals zero, and it increases radially from this center to maximum frequency at the spectrum boundaries. The power spectrum display characteristics are discussed more fully below.

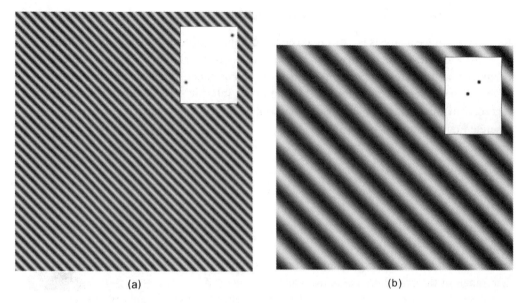

(a) (b)

Figure 7.1.2. Power spectra of periodic waveforms: (a) image of sinusoidal waveform with a frequency of 10 pixels, Inset: power spectrum of (a) showing two symmetric peaks indicating strong single frequency, (b) image of sinusoidal waveform with frequency of 40 pixels, Inset: power spectrum of (b) showing two symmetric peaks of lower frequency [closer to (0, 0)] than that in (a).

Some properties of the 2D Fourier transform are illustrated in Figs. 7.1.2–7.1.4. Figure 7.1.2 shows the contrast between lower and higher frequencies. Figure 7.1.2(a) contains higher-frequency sinusoids than those of Fig. 7.1.2(b). Figure 7.1.2(c) shows that the peaks in the transform of Fig. 7.1.2(a) are located further from (0, 0) than those of Fig. 7.1.2(d), indicating higher frequency for the former and lower frequency for

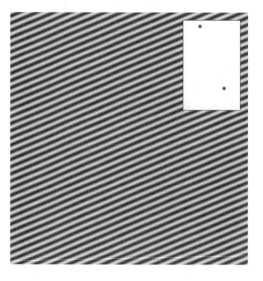

Figure 7.1.3. Power spectrum of rotated, periodic waveform: image of sinusoidal waveform of the same frequency as that of Fig. 7.1.2(a) but rotated +67 degrees; Inset: power spectrum of (a) showing two symmetric peaks that are rotated 67° with respect to those in the inset of Fig. 7.1.2(a).

the latter (Note: peaks in the power spectra are located symmetrically about the origin which is not shown).

Figure 7.1.3 shows the effect of orientation on the frequency transform. The figure displays a sinusoid with the same frequency as that of Fig. 7.1.2(a), but rotated by $+67°$. The peaks in the power spectrum of Fig. 7.1.3 are correspondingly rotated by the same angle with respect to those in Fig. 7.1.2. Note that the peaks in the power spectrum indicate the angle of the sinusoidal wave front.

Figure 7.1.4. Power spectrum of two sinusoids: image displaying two superimposed sinusoidal waveforms, those of Figs. 7.1.2(b) and 7.1.3; Inset: power spectrum showing two pairs of peaks indicating the two sinusoids of different frequencies and orientations.

Figure 7.1.4 shows the effects of two (or more) sinusoids in the same image. Figure 7.1.4 contains the superposition of the waveforms in Figs. 7.1.2(b) and 7.1.3. The power spectrum of Fig. 7.1.4 shows four peaks, corresponding to the summation of the individual power spectra of each of the waveforms.

These examples illustrate the fact that the FT identifies the presence of sinusoidal components of different frequencies in the input image and so yields a convenient description of the spatial frequency content of that image. Fourier analysis is based on the fundamental assumption that any image may be approximated in the form of a superposition of sinusoidal wave patterns of increasing frequencies with suitably chosen amplitudes. Proceeding from that assumption, we may invoke the FT to determine the frequencies present in a general image.

For example, it can be shown mathematically that a stripe pattern consists of a summation of sinusoids whose frequencies are odd multiples of the fundamental (stripe) frequency. This is observed in Fig. 7.1.5. Figure 7.1.5(a) shows a square wave and Fig. 7.1.5(b) shows a series of peaks in the power spectrum of this image. The positions of these peaks are given by $u = k, 3k, 5k, \ldots$, where k is a constant representing the fundamental frequency of the stripe. In Fig. 7.1.5 and for some following spectra, we also display two cross sections of the power spectra. These are useful to display peaks along certain directions in the spectra. The top cross section displays power in the spectrum as a function of radius (summed over all angles), from low frequency $(0, 0)$ in the middle of the plot to high frequencies at the left and the right. The lower cross section displays power as a function of azimuthal angle (summed over all radii), from $0°$ in the middle of the plot to $+180°$ at the right and $-180°$ at the left.

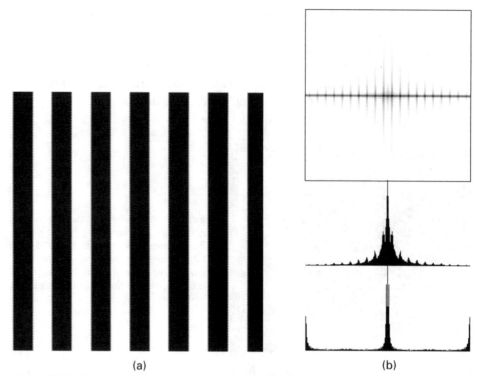

(a)   (b)

Figure 7.1.5. Square wave and its power spectrum: (a) image containing square wave pattern, (b) power spectrum of (a) showing series of frequency peaks. Figure (b) has three portions. The top is the power spectrum. The middle shows the power as a function of radius in the power spectrum, where radius = 0 in the center of the plot. The lower portion shows the power as a function of angle in the power spectrum, where angle = 0 is in the center of the plot.

Figures 7.1.6–7.1.8 show power spectra of additional example images. Figure 7.1.6 contrasts high- and low-frequency images. It also shows how object orientation is evident in the power spectrum. Figure 7.1.7 shows how constructed objects with regular patterns can be distinguished from natural objects with no regularity in pattern. Figure 7.1.8 shows how the oriented patterns can be distinguished from random patterns in the frequency domain.

7.1.2 THE FAST FOURIER TRANSFORM (FFT)

The DFT as defined above requires a large amount of computation. An efficient implementation of this transform, described by Cooley and Tukey (1965), the fast Fourier transform (FFT), has been essential in the development of the FT into a practical tool of enormous impact in the analysis of signals. The gain in computational efficiency is considerable: Whereas straightforward evaluation of the DFT requires of the order of N^2 operations, the FFT requires of the order of only $N \log_2 N$ operations. For an image of 1024×1024 pixels, this translates into a computational advantage of 10:1 in favor of the FFT.

(a)

(c)

(b)

(d)

Figure 7.1.6. Pair of images illustrate low- and high-frequency objects, in this case leaves. In the power spectra shown in this and other figures in this chapter, the origin (0,0) is at the center of the image: (a) cluster of large, long leaves; (b) cluster of small, rounder leaves; (c) power spectrum of (a) [top] and plots showing the relatively wider low-frequency peak and some directionality in the leaf orientation at approximately $-10°$; (d) power spectrum of (b) [top] and plots showing the relatively narrower low-frequency peak, and no directionality in the leaves. (This product/publication includes images from Corel Stock Photo Library, which are protected by the copyright laws of the U.S., Canada, and elsewhere. Used under license.)

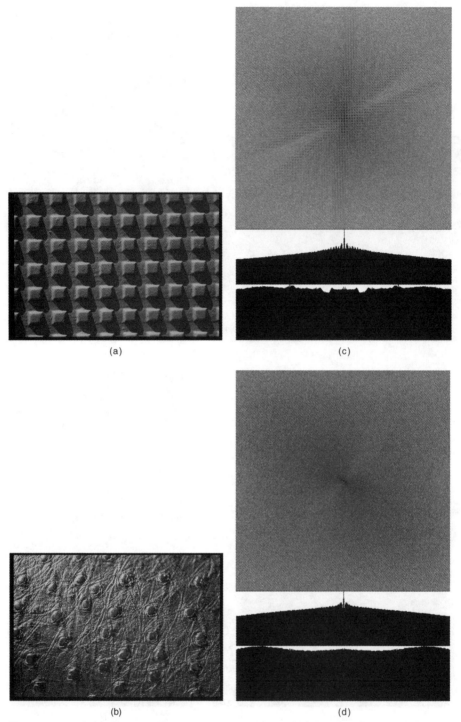

(a)

(b)

(c)

(d)

Figure 7.1.7. Pair of images illustrate structured (constructed) and unstructured (natural) objects: (a) very structured image containing aligned, square bumps; (b) unstructured image of striations and randomly located round bumps; (c) power spectrum of (a) shows structure, the cross section of radial power shows periodicity of bumps, and the cross section of angular power shows some directionality at ~90°; (d) power spectrum of (b) shows little structure, but there is some directionality (of the striations) at ~135°. (This product/publication includes images from Corel Stock Photo Library, which are protected by the copyright laws of the U.S., Canada, and elsewhere. Used under license.)

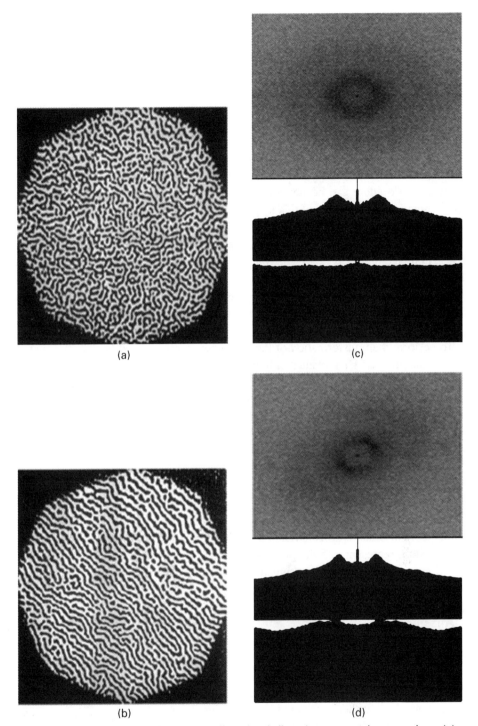

(a)

(c)

(b)

(d)

Figure 7.1.8. Pair of images illustrate unaligned and aligned structure or image regions: (a) unaligned regions in image; (b) aligned regions in image; (c) power spectrum of (a) and radial power plot show a frequency peak due to the common size of the regions, but there is no alignment: the power spectrum merely displays a circular ring instead of individual peaks; (d) power spectrum of (b) shows same frequency peak, but the shape of the power spectrum now indicates the existence of a preferred direction: distinct peaks replace the circular ring in (c) with corresponding changes in the azimuthal plot of the respective power spectra (see [Seul and Chen, 1992]). (Reprinted with Permission from [Seul and Chen 93], Copyright © 1993 American Physical Society.)

The essential idea behind the efficiency gained because of the FFT is its divide-and-conquer strategy. It is computationally more efficient to evaluate the FT for shorter rather than longer sequences, given the N^2 requisite operations, and subsequently to recombine the partial transforms. Accordingly, in the FFT algorithm, the original sequence is subdivided into shorter sequences and transformed; partial transforms are then recombined. The subdivisions are recursive, that is, the original is divided into two equal-length sequences, and those each into two, etc. Consequently, there is a requirement that the signal size be a power of 2, $N = 2^n$. This ensures that the subdivision procedure can be iterated down to minimal sequences of length 2; we discuss practical implications of this length restriction in Subsection 7.1.3.

An additional reason for the efficiency of the FFT is that it is a separable transform (see also Sections 3.1 and A.2). That is, instead of the FFT's being performed on the entire 2D image of size $N \times M$, the transform can instead be applied individually to each row and column. When row transformation is performed first, the intermediate result has the form of M rows of Fourier transforms, each of length N. Next, column transformation is performed on each of the N columns of length M to produce the final result.

To perform the transformation of the frequency domain signal back to the spatial domain, the inverse FFT is evaluated in the same manner as the forward FFT.

The most common application of the FFT to filtering – discussed in detail in Section 7.2 – does in fact rely on the successive application of the forward FFT and the inverse FFT. That is, the image is first transformed to the frequency domain by use of the FFT. The frequency signal is modified, or filtered, as desired. The modified signal is inverse transformed back to the spatial domain to obtain the modified image.

7.1.3   PRACTICAL CONSIDERATIONS

Practical use of the FFT requires additional considerations beyond FT theory. Thus, an image is a discrete representation of an underlying continuous intensity distribution $I(x, y)$, and it represents a finite portion of a scene that in reality extends beyond the field of view. The following four considerations apply to the use of the FFT for image analysis.

Aliasing

The sampling rate for images corresponds to the (spatial) density of pixels. This rate must be sufficiently high to ensure that the smallest features of interest are retained. When an image is digitized, a sampling rate (spatial pixel density) is either imposed (by CCD cameras, for example) or must be chosen (by scanners, for example). Sampling and resolution are further considered in Subsection 7.1.4.

Too high a sampling rate will yield too many pixels, increasing the cost of computation and storage. However, too low a sampling rate can introduce serious ill effects. The primary concern – aside from a loss of features – is aliasing, a general term applied to image artifacts that arise when a sampling rate is too low to capture the high frequencies in a signal: Rather than simply disappearing, high-frequency features reappear in undesired locations of the image.

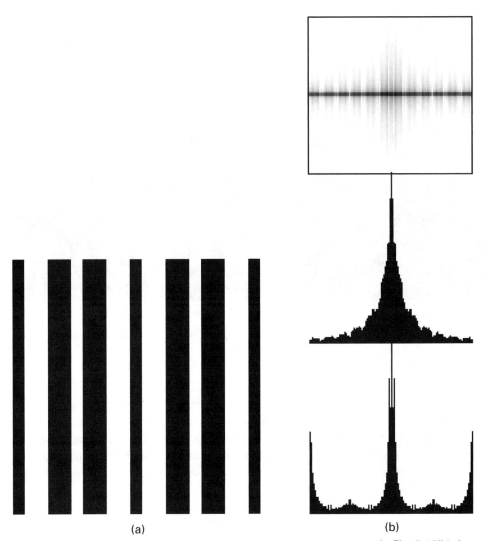

Figure 7.1.9. Aliasing due to undersampling: (a) same square wave as in Fig. 7.1.5(b), but because of undersampling, stripes have nonuniform widths; (b) power spectrum of (a) shows interference of higher frequencies above sampling rate folding over into the lower-frequency range.

This phenomenon is best understood by invoking the frequency domain. In the FT of a sampled signal, frequencies are limited to $u = M/2$, $v = N/2$, where M and N represent the effective sampling rates in two spatial directions. Higher frequencies are cut off, but these appear instead within these limits, introducing spurious interference in the lower frequencies – thus the term aliasing. Figure 7.1.9(a) shows an undersampled square wave [the same as that in Fig. 7.1.5(a)]; undersampling causes the irregular stripe pattern seen in this image. The power spectrum shown in Fig. 7.1.9(b) shows much interference [compared with the power spectrum in Fig. 7.1.5(b)] caused by aliasing. High-frequency peaks that would manifest themselves at spatial frequencies exceeding the limits set by the inadequate sampling rate have effectively wrapped around or

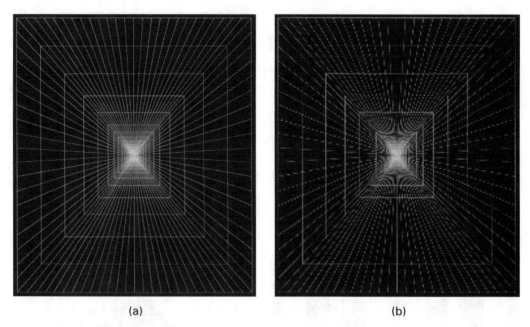

(a)   (b)

Figure 7.1.10. Aliasing manifested as moiré pattern in line image: (a) original image, (b) undersampled image of (a) resulting in moiré pattern.

Two specific examples of aliasing are the following. A periodic signal that is sampled just below its Nyquist rate – a rate corresponding to twice the highest frequency in the spectrum of the signal (Section A.4) – will have the appearance of a beating pattern, reflecting a repetition of the pattern at a frequency much lower than its true frequency. Figure 7.1.10 illustrates this effect, which normally occurs with high-frequency images such as the line pattern in this figure.

A temporal version of this phenomenon is best known in the motion picture domain, specifically in cowboy movies, in which the revolution of the wagon wheel becomes too high with respect to the frame rate of the video camera, thus causing the wheel to appear to begin to rotate backwards. Another example of aliasing is often seen in computer animation and graphics. A diagonal edge in these hard-edged pictures that is sampled at too low a rate will have the appearance of stair steps rather than that of a smooth straight edge. These extraneous stair steps are sometimes called "jaggies" (Fig. 7.1.11).

Aliasing is avoided in either of two ways. One is to ensure that the sampling rate is greater than or equal to the Nyquist rate (see Section A.4). The other is to reduce or eliminate, before sampling, those frequency components exceeding half the intended sampling rate. This is accomplished by low-pass filtering, described in Section 3.2 as a process of smoothing in the spatial domain. In Sections 3.7 and 3.8, low-pass filtering is performed before image size is reduced – a process that is equivalent to reducing the sampling rate – precisely to avoid aliasing. Filtering may also be accomplished in the frequency domain, and that is the topic of Section 7.2.

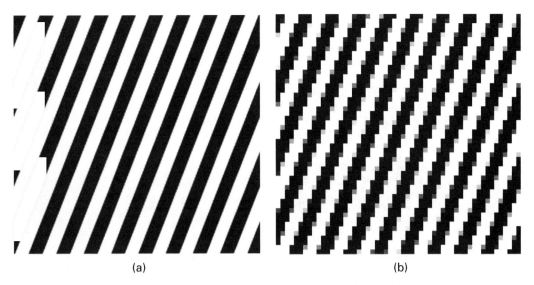

(a)   (b)

Figure 7.1.11. Aliasing manifested as "jaggies" on diagonal hard edges: (a) image of square wave pattern, (b) jaggies caused by undersampling of image in (a).

Windowing

Before a FT is performed, a window function is applied to each row and column of the image to taper the image to zero at the four boundaries. The otherwise abrupt transition at the boundaries would introduce high-frequency components into the transform and would introduce ringing in the inverse transform (see also Section 3.1).

One can best understand the origin of ringing by thinking of the image as an infinite function, representing the actual scene, that is truncated at the edges of the field of view. That is, the image $I(m, n)$ can be thought of as the product of an ideal, infinitely extended image, $I_{\text{ideal}}(m, n)$, $-\infty < (m, n) < \infty$ and a 2D box function, defined to assume the value 1 over the range in which the image is defined and 0 outside of that range. Fourier transformation of this product introduces high-frequency components that arise in the transform of the box function, that is, the ringing is introduced, by means of convolution in the frequency domain, by the box truncation of the original image (see Section A.1). To minimize the effect, truncation by the box function is replaced by multiplication of the image with a window function designed to gradually reduce pixel values to zero at the image boundaries. This discussion is in fact entirely equivalent to the consideration of filter roll-off in Section 3.1.

The choice of a suitable windowing function is guided by the desire to minimize high-frequency components of the window function while retaining as much of the original image as possible. That is, one seeks a window whose central peak is as wide as possible while still providing a smooth transition to zero at the boundaries. Several window functions are used in practice, including Hamming window, Gaussian window, Kaiser window, and others. While their relative merits are debated in the literature, a popular choice is the Hamming window, which serves in most practical situations. A windowed image is shown in Fig. 7.1.12.

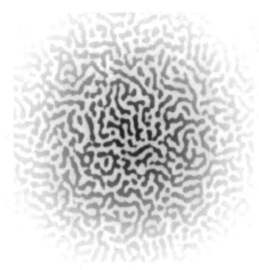

Figure 7.1.12. Hamming windowed image. The image of Fig. 7.1.8(a) is windowed to taper intensities to zero at image boundaries.

Padding

Given its divide-and-conquer methodology, the FFT operates only on signals whose length is an integer power of 2. Therefore, if the image row or column lengths M and N, respectively, are not a power of 2, the image must be extended by padding with zero values to ensure that $M = 2^m$ and $N = 2^n$, where m and n are integers. Alternatively, the size can be reduced to the lower power of 2. The advantage of padding is that the full signal is retained, but this is at the expense of greater computation and memory costs. Reduction of image size results in faster analysis, but portions of the image are lost.

Note that padding does not eliminate the need for windowing. In fact, the windowing operation should be applied before padding so that the signal tapers to the padded zeros. Note also that, provided that windowing and padding are properly applied before the forward transform, no further preprocessing need be applied before the inverse FFT is invoked.

Region Isolation

The FFT is inherently a global transformation, that is, it operates on the entire input image to derive its frequency content. Very often, it is only a portion of the image that contains the pattern or texture of interest, and transformation of the whole image will add contributions in addition to the areas of interest. This may cause confusion in locating particular peaks of interest among the others in the frequency domain. In this case, a purer frequency domain signal can be obtained first by isolation of the area of interest in the spatial domain by manual or automatic means and extraction this region; bear in mind the foregoing considerations as to image size.

Display

The FT produces a set of coefficients in the form of complex numbers, each consisting of real and imaginary parts: $c(u, v) = a + ib$. The most useful form for viewing the transform is the power spectrum; this is a real representation that we obtain by taking the magnitude of the complex Fourier coefficients, $|c(u, v)| = (a^2 + b^2)^{1/2}$. Therefore, a plot of the frequency domain is, in this book, displayed in the form of the power spectrum.

To accommodate the large potential range of coefficients – including the dc component representing the average value of pixels in the entire image as well as possibly very small coefficients for higher-frequency components – the power spectrum is frequently displayed on a logarithmic scale; that is, $P \rightarrow \log_2(P + 1)$, where the addition of 1 eliminates the undefined condition, $\log_2(0)$.

As a final point regarding the power spectrum display, we have already mentioned that each frequency in the real image generates a pair of coefficients corresponding to $\pm(u, v)$. The transform is therefore usually displayed so as to place the origin $(0, 0)$ in the center rather than in the top-left corner of the display. This representation readily highlights the symmetry of the transform around $(0, 0)$.

7.1.4 SAMPLING RANGE, SAMPLING FREQUENCY, AND SAMPLE SIZE

With reference to Fig. 7.1.13 and to Table 7.1.1, we now discuss the relationship between parameters in the spatial and the frequency domains.

Consider an image containing a regular stripe pattern – the simplest periodic signal. To maintain the integrity of the stripes in a sampled image, the sampling rate must ensure that there is at least one pixel over each stripe and one pixel between adjacent stripes; otherwise stripes would simply merge because of undersampling (see Fig. 7.1.9).

In more general terms, if the minimum feature size is the stripe width L_{min} [units] (where [units] is usually mm or inches) and stripes are spaced apart by L_{min} [units], the period of the stripes is $P_{min} = 2L_{min}$ [units]. Then the distance between samples, or spatial resolution, is $R \leq L_{min}$ [units] and the sampling rate is $r = 1/R$ [units^{-1}]. In terms of the minimum spatial feature size, $r \geq 1/L_{min}$ [samples/unit distance] or simply $r \geq 1/L_{min}$ [units^{-1}] (see also Section A.4).

To obtain a spatial resolution of $R = L_{min}$ [units], the image of side length S [units] is sampled by M samples such that $R = S/M$ [units]. This sampling resolution dictates the minimum resolvable signal size. For an object of size L_{min} [units], its signal period is $P_{min} = 2L_{min}$ [units] or, in terms of resolution, $P_{min} = 2S/M$ [units]. This minimum period signal corresponds to the maximum frequency $f_{max} = 1/P_{min} = M/2S$ [units^{-1}].

The largest feature size will produce a low-frequency component that may overlap the zero-frequency component $(u, v) = (0, 0)$ if the frequency domain resolution is not sufficiently high. This frequency domain resolution is dictated by the initial image samples, M. The lowest frequency is $f_{min} = f_{max}/(M/2) = 1/S$. Therefore the largest feature whose spatial frequency signature is resolvable from the dc component must have size $L_{max} \leq S$ [units].

To illustrate the significance of these relationships, we consider three examples corresponding to typical cases of known and unknown parameters:

- **Example 1** – Resolution of digitizing device is known: Assume a resolution of a digitizing device of 120 pixels/cm. That is, $r = 120$ cm^{-1} and $R = 1/120$ cm.

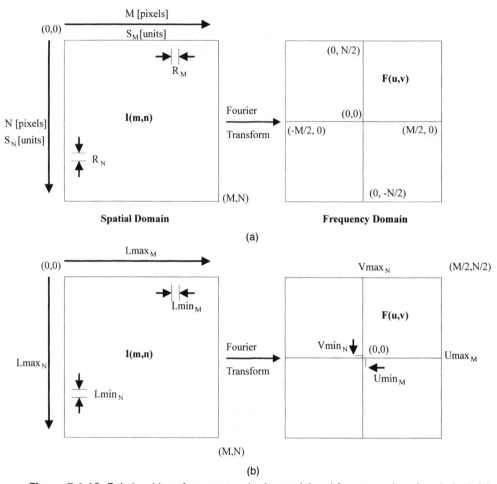

Figure 7.1.13. Relationships of parameters in the spatial and frequency domains. In both (a) and (b), the spatial domain plot is on the left and the frequency domain plot is on the right (see also Table 7.1.1.). (a) Relationships among image size and sampling length parameters, (b) relationships among object size and frequency parameters.

Consequently, the minimum resolvable feature size is $L_{\min} = R = 1/120$ cm. The corresponding periodic signal component has the period $P_{\min} = 2R = 1/60$ cm.

- **Example 2** – Minimum feature size is known: Assume a minimum feature size of $L_{\min} = 0.25$ cm or, equivalently, assume a periodic signal component of $P = 0.5$ cm. Then the resolution must be $R = L_{\min} = 0.25$ cm or smaller to retain this feature. The corresponding requisite sampling rate is at least $r = 1/R = 4$ cm^{-1}.

- **Example 3** – Minimum and maximum feature sizes are known: Assume a known range of feature sizes $L_{\min} = 0.025$, $L_{\max} = 12.5$ cm. To retain the smaller feature size, the requisite resolution is $R = 0.025$ cm and the required sampling rate is at least $r = 1/R = 1/0.025 = 40$ cm^{-1}. To distinguish (from zero) the

Table 7.1.1. Frequency and Spatial Domain Parameters

This Table Summarizes the Relationships Between Features and Requirements in Spatial and Frequency Domains for 1D Signals, with the Understanding that all Expressions Apply to Both x and y axes and, by Extension, to Two Dimensions. Figure 7.1.13 Provides a 2D Illustration of the Parameters (See Also Section A.1).

The Sample Length of a Given Spatial Signal and that of its Frequency Domain Representation are Equal and Given Here by M Pixels. If the Side Length of the Image is S [units], the Sampling Resolution is $S = L/M$ [units], while the Sampling Rate is $r = 1/R = M/S$ [unit^{-1}]. The Maximum Resolvable Frequency is $1/2$ of the Sampling Rate, $f_{max} = r/2 = 1/2R$ [unit^{-1}]. The Size of the Smallest Resolvable Feature or the Minimum Period of a Periodic Signal is $L_{min} = 1/f_{max} = 2R$ [units].

RELATIONSHIP	SPATIAL DOMAIN [UNITS]	FREQUENCY DOMAIN [UNITS^{-1}]
Image size and resolution	Image side length S Number of samples M Spatial resolution $R = S/M$	Minimum frequency resolution $f_{min} = 1/S$ Number of frequency samples M Frequency resolution $r = M/S$ Maximum frequency $f_{max} = r/2 = M/(2S)$
Object size, period, and frequency	Minimum object size L_{min} Minimum object period $P_{min} = 2L_{min}$ Maximum object size L_{max}	Maximum object frequency $f_{max} = 1/(2L_{min})$ Minimum object frequency $f_{min} = 1/L_{max}$
Required sampling	Sampling resolution $R \leq L_{min}$	Sampling frequency $r \geq 1/L_{min}$ Sampling frequency $r \geq 2f_{max}$

frequency component corresponding to the larger feature size requires sampling of a sufficiently large portion of the signal to include L_{max}. The resulting frequency resolution is $M = L_{max}/R$, and the number of pixels is at least $M = L/R = 12.5/0.025 = 500$ pixels.

References and Further Reading

[Cooley and Tukey, 1965] J. W. Cooley and J. W. Tukey, "An algorithm for the machine calculation of complex Fourier series," Math. Comput. **19**, 297–301 (1965).

[Costello, 1997] J. Costello, North Carolina State University, unpublished.

[Seul and Chen, 1992] M. Seul and V. S. Chen, "Isotropic and aligned stripe phases in a monomolecular organic film," Phys. Rev. Lett. **70**, 1658–1661 (1992).

An introductory discussion of the role of the Fourier transform in signal processing is presented by R. N. Bracewell, *The Fourier Transform and Its Applications* (McGraw-Hill, New York, 1978); the analogy of image processing and digital signal processing is pursued by J. S. Lim, *Two-Dimensional Signal and Image Processing* (Prentice-Hall, Englewood Cliffs, NJ, 1990).

Program

powspec

performs 2D FFT to produce image of power spectrum.

USAGE: powspec inimg outimg [-l] [-p] [-w] [-c] [-L]

ARGUMENTS: inimg: gray-scale input image filename (TIF)
               outimg: output image filename (TIF)

OPTIONS: -l: if set, displays linear scale for power in
                 power spectrum, instead of logarithmic default.
             -p: if set, and image row or column is not a power
                 of 2, image size is increased by zero padding;
                 otherwise image size is decreased (default).
             -w: if set, no smoothing window is applied
                 to initial image; default is to apply window.
             -c: display cross section of power spectrum
                 of horizontal line through (0,0).

             -L: print Software License for this module

7.2 Frequency Domain Filtering

Typical Application(s) – smoothing, edge detection, texture segmentation, pattern segmentation.

Key Words – low-pass filtering, high-pass filtering, bandpass filtering, band-stop filtering, cutoff frequency, Gaussian filter, Butterworth filter.

Related Topics – convolution (Section 3.1), noise reduction (Section 3.2), edge enhancement (Section 3.3), subsampling (Section 3.6), multiresolution (Section 3.7).

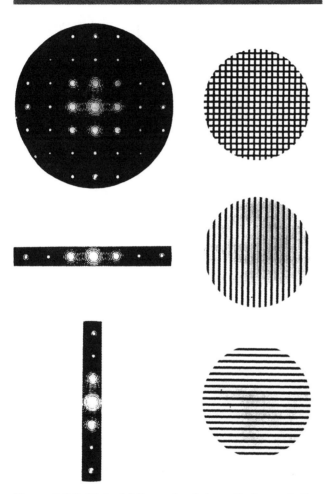

Figure 7.2.1. Pictorial Example. A dramatic demonstration of the effects of filtering in the spatial frequency (Fourier) domain on images in the spatial domain, first given by Abbe in his investigation of image formation in optical microscopes [Goodman, 1968]. Horizontal and vertical slits, placed along respective axes over the central portion of the power spectrum, selectively transmit (pass) only the central row and column of peaks, respectively. Image features related to the blocked peaks are thereby completely eliminated. The action of the slits is equivalent to that of directional low-pass filters. (Material Reproduced with Permission of The McGraw-Hill Companies from [Goodman 68], Copyright © 1968 McGraw-Hill, Inc.)

In this section, we discuss the design of suitable filters to isolate and extract image features such as those illustrated and discussed in the previous section. In Section 3.1, we discussed the topic of filtering from the point of view of performing a convolution operation in the spatial domain. Here, we revisit the topic by taking the point of view of filtering in the transform domain. The spatial frequency representation greatly facilitates the design of filters: We specify the desired filter performance by identifying bands of spatial frequencies in the original image as either retained (passed) or rejected (stopped). Fig. 7.2.1 illustrates the effects of spatial filtering on the corresponding image.

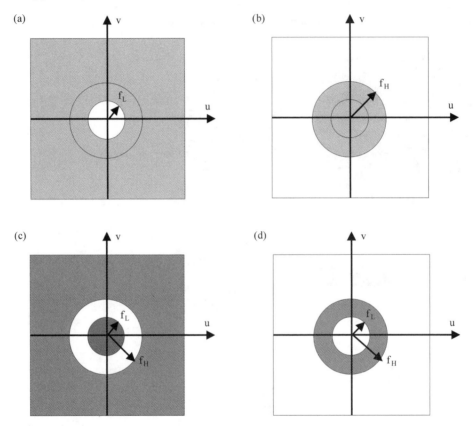

Figure 7.2.2. Generic filter types, characterized by respective frequency ranges that are passed (white regions) and those that are reduced (gray regions): (a) low pass, (b) high pass, (c) band-pass, (d) band stop.

Figure 7.2.2 depicts the frequency bands extracted by several generic types of filters. These bands have the form of circular or annular regions surrounding the origin (zero-frequency point) at the center of the power spectrum (see also Section 3.1). For example, a low-pass filter passes image frequencies below a preselected cutoff frequency f_L; a high-pass filter passes all frequencies above a preselected frequency f_H. A bandpass filter passes all frequencies within a band delimited by f_L and f_H; a band-stop filter passes all frequencies except those between f_L and f_H.

While the illustrations in Fig. 7.2.2 suggest the notion of a precise cutoff frequency, this is an idealized concept. The adverse side effects of using idealized filters with sharply delimited passbands and stop bands – particularly ringing (Sections 3.1 and 7.1) – make the use of the idealized filters with step-shaped or box-shaped transfer functions impractical. Thus the cutoff frequency represents a point in a gradual transition from passed to stopped frequencies. We now consider the design of masks to implement filters for extraction, reduction, and enhancement of specific features in the frequency domain.

7.2.1   DESIGN OF A LOW-PASS FILTER

The objective in low-pass filtering is to retain the low-frequency band while reducing components in higher-frequency bands. To minimize ringing, the idealized low-pass filter (Section 3.1) is approximated by a filter with gradual roll-off. A mathematical function with suitable characteristics is chosen as a model for the desired filter transfer function (see Section A.2), sampled and applied to the (transformed) input image.

Among the many possible filter transfer functions, we consider the two most common, namely the Gaussian filter and the Butterworth filters (Fig. 7.2.3). Gaussian filters, already described in Sections 3.2 and 3.4, will be examined in Subsection 7.2.3 from the point of view of symmetry in spatial and frequency domain filter design. Butterworth filters provide added versatility (Section 3.1).

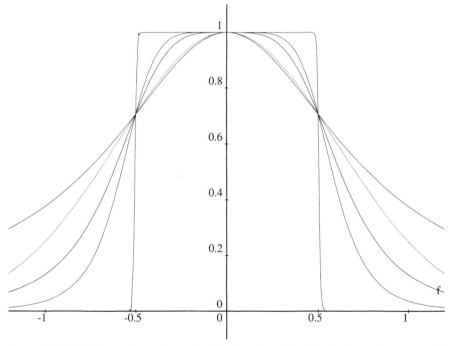

Figure 7.2.3. Butterworth and Gaussian filter functions. The 1D plot shows the cross section of low-pass filters with the cutoff frequency at $f_L = 0.5$. In order from the most gradual tapering to the sharpest cutoff, the curves are Butterworth of the order of 1, Gaussian, Butterworth of the order of 2, Butterworth of the order of 3, and Butterworth of the order of 50.

Gaussian Filters

The Gaussian filter assumes the form,

$$G[f(u, v)] = \exp\left[-\frac{1}{2\sigma^2} f^2(u, v)\right],$$

where f is the radial frequency value, $f(u, v) = \sqrt{u^2 + v^2}$. To apply the filter, perform a forward FFT, multiply (pixel by pixel) the resulting frequency domain representation by this Gaussian filter function, and transform back to the spatial domain by application of the inverse FFT.

The standard deviation σ sets the width of the low-frequency band that is passed by a Gaussian filter and thus determines the cutoff frequency. The choice of the cutoff frequency, and hence σ, must be based on the knowledge of the frequency characteristics of the objects of interest, or on the inspection of the signal in the frequency domain (Section 7.1).

Here, the accepted low-pass cutoff frequency f_L is taken to be the frequency at which attenuation reaches $1/\sqrt{2}$ of the peak amplitude or, alternatively, $1/2$ of the peak power; this is also known as the -3-dB point, because attenuation by -3 dB corresponds to a reduction of $1/2$ in peak power. The standard deviation σ is directly related to f_L by $\sigma = (f_L/\sqrt{\ln 2})$.

Butterworth Filter

The Butterworth filter assumes the form

$$B(f) = \frac{1}{1 + [\sqrt{2} - 1][f(u, v)/f_L]^{2n}},$$

where the newly introduced parameter n is the order of the filter. As with the Gaussian, the Butterworth filter function provides a gradual taper, but provides added versatility in the form of an n-tunable shape (see also Section 3.1). That is, n can be varied from 1 to higher values to increase the sharpness of the filter. The sharper the filter, the more pronounced the ringing. Figure 7.2.3 shows the Butterworth filter shape for different values of n. An example of low-pass filtering by application of a Butterworth function is shown in Fig. 7.2.4.

Many other filter functions have been investigated in the fields of signal and image processing. For most practical situations, the Butterworth filter is a good choice that provides simple adjustment capability. A common mode of using the filter (spatial or frequency domain) is to test it on a representative image and to tune parameters to optimize results. As a rule of thumb, typical values of the Butterworth filter parameters would be f_L equal to half the original frequency bandwidth and $n = 1$.

7.2.2 DESIGN OF HIGH-PASS, BANDPASS, AND BAND-STOP FILTERS

While high-pass, bandpass, and band-stop filters can be designed in a manner analogous to that just described for the low-pass filter, we introduce here an alternative design

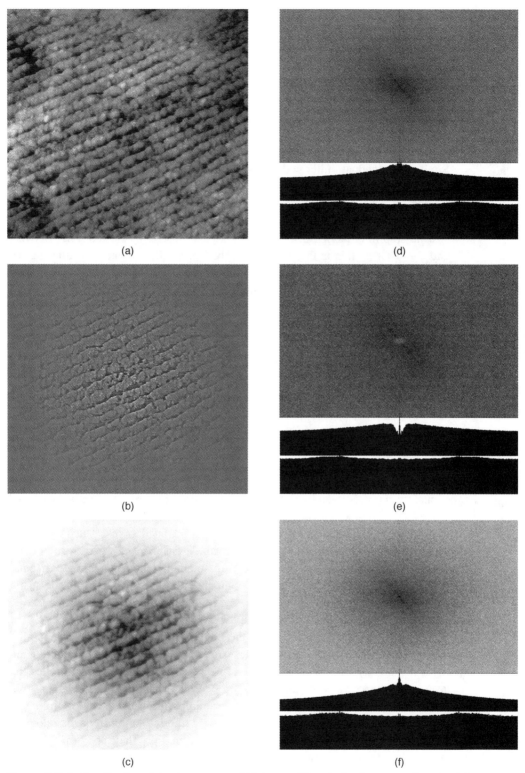

(a)

(d)

(b)

(e)

(c)

(f)

Figure 7.2.4. Results of applying low-pass and high-pass filters in the frequency domain: (a) original image (see also: Fig. 2.4.1(a)); (b) low-pass-filtered image (shown with window superimposed); (c) high-pass-filtered image (shown with window superimposed); (d)–(f) power spectra corresponding to (a)–(c), respectively.

strategy. This entails additive and subtractive combinations of the original frequency domain image with the low-pass result already described.

Thus, to design a high-pass filter with a specified cutoff frequency, the low-pass filter is first designed with that same cutoff and applied to the frequency domain signal. Then this low-pass-filtered signal is subtracted from the original frequency domain signal to yield the high-pass frequency signal (see also Section 3.3). Inverse transformation yields the high-pass signal in the spatial domain. An example is shown in Fig. 7.2.4.

Similarly, a bandpass filter with a passband delimited by f_L and f_H is obtained as follows. The original image is low-pass filtered with cutoff f_L and high-pass filtered with cutoff f_H. The resulting two images are subtracted from the original image to yield a bandpass image whose frequencies are concentrated between f_L and f_H.

A band-stop filter passes all frequencies except those in a selected range between f_L and f_H. That is, this is the inverse of a bandpass filter. Band-stop filtering can be performed similarly to the procedure given above. The original image is low-pass filtered with cutoff f_L and high-pass filtered with cutoff f_H. These two images are added to yield a band-stop filter result in which the frequencies in the middle stop band are reduced or eliminated.

7.2.3 FILTER DESIGN IN SPATIAL AND FREQUENCY DOMAINS

As we have seen, filtering may be performed in the spatial domain (Sections 3.2–3.4) or in the frequency domain. If spatial domain and frequency domain implementations of a specific filtering operation are to produce equivalent results, the design of the respective filters must reflect certain relationships, or symmetries, in their pertinent parameters such as length and width (see also Table A.1 in Section A.1).

These are summarized here for the Gaussian filter, a convenient case because its functional form is preserved by Fourier transformation. In particular, the inverse transform of a Gaussian function with standard deviation σ in the frequency domain yields a Gaussian function with standard deviation $1/\sigma$ in the spatial domain. Stated intuitively, a broad Gaussian, with a large low-pass bandwidth, transforms to a narrow filter in the spatial domain. Conversely, a narrow Gaussian, retaining only a small low-pass band, transforms to a broad filter in the spatial domain. From the perspective of the spatial domain, a larger filter yields a smoother image, while a narrow filter produces less smoothing and retains higher-frequency features.

For example, the inverse transform of the single-sample filter, the smallest possible frequency domain filter, produces the largest smoothing window in the spatial domain, simply a rectangular filter window of the same size as the image. This implies that the average intensity value of an image can be either computed by averaging over all pixel values in the spatial domain or by examining the value at the origin, $(0, 0)$, of the frequency domain. To perform averaging by filtering in the frequency domain, transform the image, center this filter at $(0, 0)$ and multiply the transform to extract the peak value (which corresponds to the average image intensity), and then inverse transform.

Conversely, the largest size of a frequency domain filter corresponds to the full range of the frequency domain; such a filter would leave the frequency distribution of the signal unaltered. Inverse transformation of this largest frequency domain filter produces the smallest spatial domain filter, a single pixel in width. A one-pixel spatial filter produces no smoothing and so leaves the image unaltered as well.

Spatial Domain Filtering Versus Frequency Domain Filtering

Although equivalent results can be achieved by means of filtering by direct convolution in the spatial domain (Section 3.1) or by Fourier transformation and multiplication in the frequency domain, there are computational trade-offs. The cost of filtering in the frequency domain is fixed – that is, it is not dependent on the size of the equivalent spatial domain convolution filter. Conversely, in the spatial domain, the computational cost of filtering depends on the filter size. In general, when the spatial domain filter size is greater than some number, then it is more efficient to filter in the frequency domain. In most practical cases, it is more efficient to perform filtering in the spatial domain.

In the spatial domain, a separable low-pass filter of size $k \times k$ requires $(k + k)$ multiplications for each pixel and thus $2kN^2$ multiplications for an image of size $N \times N$. Frequency domain processing requires the following number of multiplications: N^2 for windowing, $2N^2 \log_2 N$ for the forward 2D FFT, N^2 for filtering, and $2N^2 \log_2 N$ for the inverse 2D FFT, yielding a total of $N^2(4 \log_2 N + 2)$ multiplications. For example, with $N = 256$, the total is $34N^2$; for $N = 512$, the total is $38N^2$, and for $N = 1024$ the total is $42N^2$. For these cases, frequency domain filtering is the preferred approach only when k exceeds 17, 19, and 21, respectively, for the three considered image sizes. These are unusually large values of k, whose typical range is $3 \leq k \leq 9$, and in the absence of special mitigating circumstances – notably, dedicated hardware capable of performing the FFT – filtering is thus most efficiently performed in the spatial domain.

References and Further Reading

[Goodman, 1968] J. Goodman, *Introduction to Fourier Optics* (McGraw-Hill, San Francisco, 1968), Chap. 7.

An introduction of the Fourier transform as a tool for image processing can be found in R. C. Gonzalez and R. E. Woods, *Digital Image Processing*, 2nd ed. (Addison-Wesley, Reading, MA, 1992), pp. 201–213; and, in a somewhat more formal treatment, in W. K. Pratt, *Digital Image Processing* (Wiley, New York, 1978), pp. 279–303. A reference to methods of digital signal processing is L. R. Rabiner and B. Gold, *Theory and Application of Digital Signal Processing* (Prentice-Hall, Englewood Cliffs, NJ, 1975), pp. 75–204; Practical considerations of the performance characteristics of transforms for image processing are discussed in L. O'Gorman and A. C. Sanderson, "A comparison of methods and computation for multi-resolution low- and band-pass transforms for image processing," Comput. Vis. Graph. Image Process. **35**, 276–292 (1986).

Program

fltrfreq

          performs frequency domain filtering

    USAGE: fltrfreq inimg outimg [-l f1] [-h f1] [-b f1 f2]

                               [-s f1 f2] [-n ORDER] [-g] [-p] [-L]

ARGUMENTS: inimg: gray-scale input image filename (TIF)

           outimg: output image filename (TIF)

  OPTIONS: -l f1: LOW-PASS FILTER passing frequencies below
                 cutoff, f1.

        -h f1: HIGH-PASS FILTER passing frequencies above
                 cutoff, f1.

     -b f1 f2: BANDPASS FILTER passing frequencies between
                 f1-f2.

     -s f1 f2: STOP-BANDPASS FILTER passing frequencies not
                 f1-f2.

     -n ORDER: order of Butterworth filter, default = 1.

           -g: flag to perform Gaussian filter, default is
                 Butterworth.

           -p: if set, and image row or column is not a power
                 of 2, image size is increased by zero padding;
                 otherwise, image size is decreased (default).

          NOTE: the frequencies (f1,f2) should be
                 expressed as a number 0 to 1.0; this is a
                 fractional frequency value of the full
                 original passband. For example, a
                 low-pass filter with cutoff frequency of
                 0.5 will reduce the bandwidth by half.
                 Where the image is not square, the
                 fractional frequency value is relative
                 to the higher frequency corresponding to
                 the longer x or y axis.

          -L: print Software License for this module

Program Descriptions

8.1 Introduction

This chapter provides a reference to all of the programs described in this book and available for use in the software package. A listing by book chapters and sections is provided in Section 8.2 to give the reader detailed explanation of each technique. Following this in Section 8.3 is an alphabetical listing of all programs with program synopsis, interactions with other programs, and notes. The purpose is to help give the user a basic understanding of what a program does and how it may fit in a sequence of processing steps with other programs. All program options and parameter values are not discussed here; instead, these are described in the program usage notes in the text of this book, in the program interface, or by running the program at the command line without parameters. The user is encouraged to also gain an understanding of the strengths, weaknesses, and tradeoffs of particular programs by trying them and observing the results for different parameter values and for different image types.

8.2 Programs by Chapter

CHAPTER 2 – GLOBAL IMAGE ANALYSIS

Section 2.1 Intensity Histogram: Global Features
histstats

Section 2.2 Histogram Transformations: Global Enhancement
histex, histramp, histexx

Section 2.3 Combining Images
imgarith, imgbool, combine, inv

Section 2.4 Geometric Image Transformations
xscale, imgrotate

Section 2.5 Color Image Transformations
rgb2gray

CHAPTER 3 – GRAY-SCALE IMAGE ANALYSIS

Section 3.1 Local Image Operations: Convolution
xconv

Section 3.2 Noise Reduction
lpfltr, medfltr

Section 3.3 Edge Enhancement and Flat Fielding
bc

Section 3.4 Edge and Peak Point Detection
xedgefilter, peak

Section 3.5 Advanced Edge Detection
bcd

Section 3.6 Subsampling
subsample

Section 3.7 Multiresolution Analysis
multires

Section 3.8 Template Matching
xcorr

Section 3.9 Gabor Wavelet Analysis
gabor

Section 3.10 Binarization
binarize, threshm, thresho, threshe, threshk, threshc

CHAPTER 4 – BINARY IMAGE ANALYSIS

Section 4.1 Morphological and Cellular Processing
cellog, morph

Section 4.2 Binary Noise Removal
kfill

Section 4.3 Region Detection
contour, xcp, xah, xcc, xrg

Section 4.4 Shape Analysis: Geometrical Features and Moments
xpm, xfm

Section 4.5 Advanced Shape Analysis: Fourier Descriptors
xbdy

Section 4.6 Convex Hull of Polygons
xph

Section 4.7 Thinning
thin

Section 4.8 Linewidth Determination
thinw

Section 4.9 Global Features and Image Profiles
globalfeats, profile, imggrid

Section 4.10 Hough Transform
hough

CHAPTER 5 – ANALYSIS OF LINES AND LINE PATTERNS

Section 5.1 Chain Coding
pcc, pccde, pccdump, pccfeat

Section 5.2 Line Features and Noise Reduction
linerid, linefeat, linexy, structrid, structfeat

Section 5.3 Polygonalization
fitpolyg

Section 5.4 Critical Point Detection
fitcrit

Section 5.5 Straight-Line Fitting
fitline

Section 5.6 Cubic Spline Fitting
fitspline

Section 5.7 Morphology and Topology of Line Patterns
eh_seg, xsgll, eh_sgl

CHAPTER 6 – ANALYSIS OF POINT PATTERNS

Section 6.1 The Voronoi Diagram of Point Patterns
spp, xvor

Section 6.2 Spatial Statistics of Point Patterns: Distribution Functions
dpp, xptstats, xvora

Section 6.3 Topology and Geometry of Cellular Patterns
xsgt

Section 6.4 The k-Nearest Neighbor (k-NN) Problem
xknn

CHAPTER 7 – FREQUENCY DOMAIN ANALYSIS

Section 7.1 The 2D Discrete Fourier Transform
powspec

Section 7.2 Frequency Domain Filtering
fltrfreq

8.3 Alphabetical Listing of Programs

When refering to similar programs beginning with the same letters, we use a "*" to indicate the variable part of the program name, e.g., **fit*** includes the programs **fitcrit**, **fitline**, **fitpolyg**, and **fitspline**. We frequently refer to the "line analysis programs of Chapter 5." These are: **pccde**, **pccdump**, **pccfeat**, **linerid**, **linefeat**, **linexy**, **structrid**, **structfeat**, **fitpolyg**, **fitcrit**, **fitline**, and **fitspline**.

bc

Background Correction (in Edge Enhancement and Flat Fielding, Section 3.3)

Synopsis – The program removes a known background bias from a gray-scale image to produce an output image whose background is uniformly flat. This correction may be needed due to non-uniform light distortion in the optical system capturing the image.

Interactions – Background correction is applied at the beginning of processing, directly after image capture if the optical capture device has a known non-uniform background.

Notes:

* A common procedure to determine if this is needed is to capture several images of uniform background (that is with no objects of other image components) at several lighting levels. These images are inspected for non-uniformity (non-flatness). If this exists, the reference image of the lighting level to be used for subsequent capture is used for the background correction program.

bcd

Boie-Cox Edge Detection Filter (in Advanced Edge Detection, Section 3.5)

Synopsis – The program performs optimal edge detection on a gray-scale input image, producing a gray-scale or binary output image in which intensity values are larger for stronger edges. This program implements the Boie-Cox optimum edge detection algorithm.

Interactions – Edge detection is often performed on images after noise reduction for the purpose of segmenting image objects. This optimal edge detection algorithm has some degree of noise reduction capabilities already, but noise reduction will still be a preferred preprocessing step in many cases since the amount and type of noise reduction can be controlled explicitly with a separate program.

Notes:

* This program gives the option of following optimal edge detection by binarizing the edges with respect to a chosen threshold, thus producing a binary image.
* If the user desires a more rudimentary edge detection program, choose **xedgefilter**.

binarize

Binarize (in Binarization, Section 3.10)

Synopsis – The program binarizes a gray-scale input image, producing a binary output image in which pixel intensity values that are above a chosen threshold are set to ON and otherwise are set to OFF.

Interactions – This program may be preceded by some gray-scale processing steps, especially noise reduction (**lpfltr**, **medfltr**). After binarization, the binary image processing programs may be used (in Chapter 4 programs).

Notes:

- This binarization program is the most rudimentary of the thresholding programs in this package.
- The user chooses the threshold value for this program, whereas the other thresholding programs determine a threshold by some optimization calculation (**thresh***).

cellog

Cellular Logic (in Morphological and Cellular Processing, Section 4.1)

Synopsis – The program performs cellular logic on a binary input image, producing a binary output image. Cellular logic operates iteratively upon an image, changing pixel values from ON to OFF or vice versa depending upon the number and configuration of pixel values around each pixel.

Interactions – This program is preceded by binarization if the image is not already binary (**binarize**, **thresh***). There are two common cases where cellular logic is most often used. One is for noise reduction, after which the image is inspected or further processing is performed. The other is for identifying an object of certain characteristics that the cellular logic filtering can detect.

Notes:

- A common noise reduction purpose of cellular logic is to reduce isolated, small dots and holes. If this is the purpose, another option is to use the **kfill** program, which was specifically designed to reduce "salt-and-pepper" binary noise.
- Cellular logic processing has similar functionality to morphological processing (**morph**), however there are differences in how these operations perform, so the user may want to test both on the images of interest to see which yields the preferred results.
- In contrast to morphological processing (**morph**), cellular logic has a dedicated parameter for specifying connectivity so for instance, an operation can be specified that does not break a single image region into disconnected regions.

combine

Combine (in Combining Images, Section 2.3)

Synopsis – The program combines two input images by adding their corresponding intensities and clipping values that exceed a given value.

Interactions – The image combination result of this program can be used for other programs.

Notes:

- This program is similar to **imgarith** for addition however, this program does not adjust all the resultant intensity values if one or more is out of range; instead it simply clips the value, i.e., sets it to a maximum value of 255 or a chosen clipping value less than 255.
- This program is used when no automatic adjustment of intensities is desired and, either the image has already been adjusted to prevent clipping, or clipping is tolerated.
- This program is a relatively basic way to perform a Boolean combination of images. The program performs no adjustment for any scale, rotational or translational differences between input images.

contour

Contour Region Detection (in Region Detection, Section 4.3)

Synopsis – The program finds the contours of regions in a binary input image, producing a binary output image containing these contours. The program can also find and show the centroids of regions found.

Interactions – This program is preceded by binarization (**binarize, thresh***) if the image is not already binary. Line processing (using the line analysis programs of Chapter 5) can follow contour detection to determine features of the contours found.

Notes:

- Contours describe the edges of binary regions.
- Noise reduction is incorporated in this program through a user-selectable parameter value whereby regions with small contours can be omitted.

dpp

Draw Point Patterns (in Spatial Statistics of Point Patterns: Distribution Functions, Section 6.2)

Synopsis – The program takes no input image; it produces a binary image with a choice of a random, clustered, or ordered point pattern with selected density of coverage.

Interactions – This program can be used to produce synthetically generated point images for testing programs **spp**, **xvor**, or **xptstats**.

eh_seg

Histograms of Segments (in Morphology and Topology of Line Patterns, Section 5.7)

Synopsis – The program takes a file containing line segment endpoints from **fitpolyg** and produces a list of histogram data of the line segments. The histogram displays either angles between polygonal line segment fits or segment lengths.

Interactions – This program is used after PCC coding (**pcc**) and after **fitpolyg**. The resulting histogram data is often the last step of analysis.

Notes:

- Line noise reduction (**linerid**, **structrid**) is usually a preprocessing step before this program so that the resulting histogram contains only lines of interest.

eh_sgl

Histograms of the Segment Group List (in Morphology and Topology of Line Patterns, Section 5.7)

Synopsis – The program takes a file containing segment group lists from **xsgll** and produces an output image containing histogram data of the segment group list. The histogram displays: segment number, cluster area, segment length, segment width, cluster convex hull area, or the ratio of area to convex hull area.

Interactions – This program is used after PCC coding (**pcc**) and after **fitpolyg**. The resulting histogram data is often the last step of analysis.

Notes:

- Line noise reduction (**linerid**, **structrid**) is usually a preprocessing step before this program so that the resulting clusters contain only lines of interest.

fitcrit

Critical Point Detection (in Critical Point Detection, Section 5.4)

Synopsis – The program takes an input PCC file and produces a binary image containing connected line segment fits between critical points – curvature maxima and corners. Optionally, the program also prints out the coordinates of the line fits.

Interactions – This program is used after PCC coding (**pcc**) and often after line noise reduction (**linerid**, **structrid**). This is sometimes the final step of analysis where the objective is to precisely find curvature maxima, or intended corners in the input image.

Notes:

- This program is useful to obtain corners of rectilinear line images, such as engineering drawings. When these are digitized, then binarized and thinned, the corners often end up rounded.
- If the intention is to just fit straight lines to a line image, the program **fitpolyg** may be preferable to this because less complexity also reduces chance for error.
- Two user parameters regulate the closeness of the fit to the input image lines, where the tradeoff is between precision of critical point location and the number of line fits.

fitline

Fit Line (in Straight-Line Fitting, Section 5.5)

Synopsis – The program takes an input PCC file and produces a binary image containing a line fit for each chain between feature points (endpoints or junctions) in the image.

Interactions – This program is used after PCC coding (**pcc**) and often after line noise reduction (**linerid, structrid**). This is sometimes the final step of analysis where the objective is to fit a line to each chain.

Notes:

- This program does not maintain connectivity of chains at junctions. To fit straight lines while maintaining connectivity of structures, use **fitpolyg** or **fitcrit**.
- This program attempts to fit the closest line to each full chain, so endpoints will usually not be close to intended critical points (corners). To fit endpoints of lines precisely, use **fitcrit**.
- If a more precise curved fit is desired to a line fit, choose **fitspline**.

fitpolyg

Fit Polygonal Lines (in Polygonalization, Section 5.3)

Synopsis – The program takes an input PCC file and produces a binary image containing connected line segment fits to the connected input chains. Optionally, the program also prints out the coordinates of the line fits.

Interactions – This program is used after PCC coding (**pcc**) and often after line noise reduction (**linerid, structrid**). This is sometimes the final step of analysis where the objective is to fit straight lines to image curves for the purpose of finding features or compressing chains to a sequence of line endpoints.

Notes:

- If a more precise location of critical points (corners and curve boundaries) is desired, then the program, **fitcrit** may be preferable to this.
- Although not as precise as **fitcrit**, this program is a fast and reliable way to simplify a line image into one containing just line segments.
- The term "polygonalization" does not imply closed curves as output; rather it means that the line segment fits to the open or closed curves of the input image are connected.
- A user parameter regulates the closeness of the fit to the input image lines, where the tradeoff is between closeness and number of fits.

fitspline

Fit Spline (in Cubic Spline Fitting, Section 5.6)

Synopsis – The program takes an input PCC file and produces a binary image containing a cubic spline fit for each chain between feature points (endpoints or junctions) in the image.

Interactions – This program is used after PCC coding (**pcc**) and often after line noise reduction (**linerid**, **structrid**). This is sometimes the final step of analysis where the objective is to fit smooth curves to each chain.

Notes:

- This program does not maintain connectivity of chains at junctions. To fit straight lines while maintaining connectivity of structures, use **fitpolyg** or **fitcrit**.
- This program attempts to fit the closest line to each full chain, so endpoints will usually not be close to intended critical points (corners). To fit endpoints of lines precisely, use **fitcrit**.
- Three user parameters regulate the closeness of the fit to the input image lines, where the tradeoff is between smoothness of the spline, closeness of the fit, and number of fits.
- If only line fits are required rather than splines, choose **fitline**.

fltrfreq

Filter Frequency (in Frequency Domain Filtering, Section 7.2)

Synopsis – The program takes a gray-scale input image and filters it in the frequency domain to produce an output gray-scale image. The user chooses filtering parameters of low-, band-, or high-pass filters, along with values of frequency bounds.

Interactions – This program is used to reduce noise or to reduce or retain objects in chosen frequency bands. Results can be inspected by looking at the output image of this program or by viewing the frequency-domain power spectrum with **powspec**.

Notes:

- Alternatively to frequency domain analysis, filtering can be performed in the spatial domain using **lpfltr** or **xconv** with appropriate filter masks.
- Filtering usually only *reduces* image content in chosen bands, it does not totally eliminate it.
- A low-pass filter retains frequency information below a chosen frequency and reduces it above. A high-pass filter retains frequency information above a chosen frequency and reduces it below. A band-pass filter retains frequency information between chosen low and high frequencies and reduces it below and above these.

gabor

Gabor Wavelet Filtering (in Gabor Wavelet Analysis, Section 3.9)

Synopsis – The program performs Gabor wavelet filtering on an input gray-scale image, producing two output gray-scale images. Each image, contains results of different orientations and scales of Gabor filtering. The two images are for real – or symmetrical – filtering, and imaginary – or asymmetrical – filtering.

Interactions – This program may be performed as an exploratory step before future processing. The results of this program indicate how well objects or regions of particular orientation and scale can be detected. With this information, a region of interest can be identified for future processing such as for low-pass, band-pass, high-pass, or matched filtering.

Notes:

- Gabor filtering is often used to detect and segment textures or patterns.
- The sub-image that shows most clearly the object or region of interest, has the scale and orientation properties best matching those of the region of interest.
- The only difference between the real and imaginary image outputs is the symmetry of the Gabor waveform: symmetric for real, or anti-symmetric for imaginary. Often for natural images, there is no perfect symmetry or anti-symmetry, so both real and imaginary results are similar, thus the user need only look at one result, usually the real image.

globalfeats

Global Features (in Global Features and Image Profiles, Section 4.9)

Synopsis – The program takes a binary input image and produces a list of global features: image size, total number of pixels, number of ON-valued pixels, percentage of ON pixels in image, 1st and 2nd moments of x,y locations of ON pixels, number of edge pixels of regions, and percentage of edge pixels in image.

Interactions – This program is often the last analysis step where an original gray-scale image may have been noise-reduced, binarized, perhaps noise-reduced again, and then processed with this program. These features then enable a global image comparison between images processed by different programs and with different parameters.

Notes:

- Use as few or as many global feature values as is necessary for your purpose. For instance, consider a large filled circle in one image A, and the same circle filled with dots in another image B. These two images will have similar 1st and 2nd moments, but have very different number of pixels, edge pixels, etc. If moments are important to you, then these images are globally similar; if the number of regions, these images are very different.
- Other programs, such as **xah**, **xcc**, and **xrg**, find feature values for individual regions in an image, and **xpm**, **xfm**, and **xbdy**, find features of a single regions in and image.

histex

Histogram Expansion (in Histogram Transformations: Global Enhancement, Section 2.2)

Synopsis – The program takes a gray-scale input image and produces a gray-scale output image with the range of pixel values expanded to the entire gray-scale range for the purpose of maximizing visible contrast.

Interactions – This program is often run on an image before other processing so that the human viewer can best see details and judge results of subsequent processing.

Notes:

- This program only changes the *visible* contrast of the image. There is no gain or loss of information content.

histexx

Histogram Expansion (in Histogram Transformations: Global Enhancement, Section 2.2)

Synopsis – The program takes a gray-scale input image and produces a gray-scale output image with intensity values expanded around a chosen intensity value.

Interactions – This program is often run on an image before other processing so that the human viewer can best see details in an intensity range of interest.

Notes:

- There is a potential loss of information (entropy) in the resultant image, so the result should be used only for visible inspection of the image, not for subsequent processing.
- This program is useful to better view regions of the image with intensities around a certain value.

histramp

Histogram Ramp (in Histogram Transformations: Global Enhancement, Section 2.2)

Synopsis – The program takes a gray-scale input image and produces a gray-scale output image with intensity values expanded in low or high intensity values, or uniformly throughout the intensity range such as to increase visible contrast by expanding that range.

Interactions – This program is often run on an image before other processing so that the human viewer can best see details in an intensity range of interest.

Notes:

- There is a potential loss of information (entropy) in the resultant image, so the result should be used only for visible inspection of the image, not for subsequent processing.
- Dark parts of the image are contrast-enhanced when the slope is less than zero. Light parts of the image are enhanced when the slope is greater than zero. When the slope is zero, all intensities are contrast-enhanced equally – this operation is called *histogram equalization*.

histstats

Histogram Statistics (in Intensity Histogram: Global Features, Section 2.1)

Synopsis – The program takes a gray-scale image and produces an image containing the input image histogram and also writes to the standard output a list of global image features: range of occupied bins, maximum bin height, average bin location, standard deviation, and number of occupied bins. Optionally, a file can be specified to receive the histogram bin values.

Interactions – This program can be used on an image before or after other processing. However, the histogram output is not used for another program, it is just useful for viewing.

Notes:

- As with any histogram analysis, this produces *global* results across the whole image, so it is useful to obtain statistics on the image or a single object alone in the image, but not to obtain statistics on one of many objects in the image. To do this, the area of the object of interest can be extracted and the histogram analysis run on that sub-image.

hough

Hough Transform (in Hough Transform, Section 4.10)

Synopsis – The program transforms an input binary image to an output gray-scale image showing accumulated polar coordinate values. Peaks in the output image indicate lines in the input image, and their polar coordinate locations indicate the line angle and distance from the origin.

Interactions – Preceding Hough processing, common processing includes noise reduction (**lpfltr**, **medfltr**, **kfill**) and binarization (**binarize**, **thresh***). However, expected lines in this binary image may be broken up. It is this type of image where the Hough transform is useful to detect these lines.

Notes:

- The Hough transform is useful for detection of broken lines. If image lines do not have breaks, then a more direct detection scheme is via binarization (**binarize**, **thresh***), thinning (**thin**), and PCC coding (**pcc**), followed by the line analysis programs of Chapter 5. This is in part because peak detection on the Hough transform can be difficult and unreliable.
- Although a Hough transform procedure can be used to detect higher order geometric shapes such as circles and ellipses, each higher order requires an additional transform axis (2 for lines, 3 for circles, 4 for ellipses). Peak detection is less reliable and computationally expensive as the number of axes increases. For circle detection, an alternative is the program, **xcc**. This package offers only Hough line detection.

imgarith

Image Arithmetic (in Combining Images, Section 2.3)

Synopsis – The program takes two gray-scale (or binary) input images and performs arithmetic operations – addition, subtraction, or multiplication – to produce a gray-scale output image.

Interactions – There are a number of examples of use of this program with other programs. For instance, two programs – or the same program with different parameter values – may be run on a single image, then the two resultant images added together to combine results. A program may be run on an image to identify an object or area, then the resultant image subtracted from the original image. An image may be multiplied by an image with a ramped increase in intensity from one side to another to brighten the image side corresponding to the higher side of the ramp.

Notes:

- This program is a relatively basic way to arithmetically combine images. The program performs no adjustment for any scale, rotational, or translational differences between input images.
- If any resultant intensity values fall outside of the valid intensity range (0-255), then all resultant values are scaled so that they are within the range.

imgbool

Image Boolean (in Combining Images, Section 2.3)

Synopsis – The program takes two binary or gray-scale input images and performs Boolean operations – OR, AND, or XOR – to produce a binary or gray-scale output image.

Interactions – This program is often used to either combine or rid objects from binary images before subsequent processing.

Notes:

- This program is a relatively basic way to perform a Boolean combination of images. The program performs no adjustment for any scale, rotational or translational differences between input images.
- This program is also useful when one input is binary and the other is gray-scale. In this case, the binary image may be used to isolate a portion of the gray-scale image (by ANDing the gray-scale image with an image containing one or ON values at the desired image location) or to add a binary gray-scale image by ORing or XORing them.

imggrid

Image Grid (in Global Features and Image Profiles, Section 4.9)

Synopsis – The program takes an input gray-scale or binary image and simply overlays a grid of chosen spacing upon it to produce a gray-scale or binary output image.

Interactions – Grids have been used since the beginning of scientific measurement. A simple use is to use as an overlay to help estimate the pixel lengths of objects in an image. Another use is to manually count objects in some sampling of the grids to estimate total objects in an image.

Notes:

- The grid intensity is binary, but is not one value; instead the binary value takes the furthest value (ON or OFF) of the input image pixel value at that location.
- There are other programs to automatically determine object sizes and other features mentioned in Chapter 4. However, this manual program may be useful as an exploratory step to determine which automatic program to use or what parameter values (size, etc.) to set.

imgrotate

Image Rotate (in Geometric Image Transformations, Section 2.4)

Synopsis – The program rotates a gray-scale image to a desired angle, yielding the rotated output gray-scale image.

Interactions – This program can be used before other analysis programs that expect a particular oriented image. For instance if matching is performed by the rudimentary method of subtracting images, then the candidate image orientation should be oriented to align with the image to be matched.

Notes:

- The pixels in the output image are interpolated during rotation to produce a smooth result, however this reduces the image resolution somewhat from the original.
- The program enables the user to rotate with respect to a chosen point in the image (the default is the center). This feature enables the user to perform the rotation with respect to an object in the image by specifying the center of that object.
- The rotation angle is measured counter-clockwise from zero degrees along the positive x-axis.

inv

Invert (in Combining Images, Section 2.3)

Synopsis – The program inverts all intensity values on an input gray-scale image to produce an output gray-scale image. Each new intensity value is 255 minus the original intensity value.

Interactions – The result of this program can be used for other programs.

Notes:

- An example use of this program is before another program that detects light (or dark) objects such as **binarize** or **thresh***. If the input image contains these objects of interest, but they are dark (or light), then intensity inversion will convert them to the value range of interest.
- This program produces what is commonly called a *negative* image, inverting dark and light intensities. (Or if the negative image is the starting point, it produces the positive image.)

kfill

K-Fill Binary Noise Reduction Filter (in Binary Noise Removal, Section 4.2)

Synopsis – The program reduces noise from a binary input image, producing a binary output image. Noise reduction consists of filling holes, removing isolated, small regions of ON pixels, and smoothing contours or sides of binary regions.

Interactions – This program is preceded by binarization if the image is not already binary (**binarize, thresh***). This noise reduction step is often used before further binary processing such as contour detection (**contour**) or thinning (**thin**), and region detection or line analysis (line analysis programs of Chapter 5).

Notes:

- This program performs similarly to cellular logic (**cellog**) and morphological processing (**morph**), in which it operates iteratively on the image and operates with respect to pixels and their neighbors. Unlike these other two programs, the dedicated purpose of **kfill** is noise reduction.
- The smallest mask size of three will give the smoothest noise reduction. Larger mask sizes give coarser results.

linefeat

Line Features (in Line Features and Noise Reduction, Section 5.2)

Synopsis – The program takes an input PCC file and prints an output of line features including: type of line, length, start and end coordinates, bounding box, and centroid; and also gives a line summary containing number of lines and average statistics.

Interactions – This program is used after PCC coding (**pcc**) and often after line noise reduction (**linerid**, **structrid**).

Notes:

- This program gives a more formatted and informative listing of line features than **linexy**, which simply dumps line coordinates. However, **linexy** may be more easily read into another program.
- A *line* feature is a single line segment, whether connected to other lines or not. If structures containing many connected lines are of interest, these are *line structures*, and the program **structfeat** lists features for these.

linerid

Line Rid (in Line Features and Noise Reduction, Section 5.2)

Synopsis – The program takes an input PCC file and filters out small lines – either isolated lines or branch lines – producing an output PCC file.

Interactions – This program is used after PCC coding (**pcc**) and often is the first step of line noise reduction.

Notes:

- This program rids only lines containing endpoints (i.e., not bridge lines) so the number of line structures remains the same after processing.
- This program and **structrid** are the main filtering programs for PCC code.

linexy

Line XY Coordinates (in Line Features and Noise Reduction, Section 5.2)

Synopsis – The program takes an input PCC file, and lists the line end coordinates ($x1$, $y1$), ($x2$, $y2$), ...

Interactions – This program is used after PCC coding (**pcc**) and sometimes follows line fitting (PCC coding (**pcc**) must follow line fitting) using any of **fitpolyg**, **fitcrit**, **fitline**, and **fitspline**.

Notes:

- This program gives a raw listing of line end coordinates. A more formatted and informative listing is created from **linefeat**.

lpfltr

Low-Pass Filter (in Noise Reduction, Section 3.2)

Synopsis – The program performs low-pass filtering on an input gray-scale image, producing a low-passed (or smoothed) gray-scale output image.

Interactions – Noise reduction is one of the most commonly applied programs before subsequent processing.

Notes:

- Low-pass filtering reduces noise by smoothing the image, but does this indiscriminantly, in which is both noise and non-noise image features are smoothed. The user must choose the amount of filtering that trades off noise-reduction and feature smoothing.
- An alternative noise reduction filter is the median filter **medfltr**, which blurs some images less for the same amount of noise reduction.
- The low-pass bandwidth is chosen between 0 and 0.5, with lower values passing less image information, i.e., smoothing more.
- The filter size is chosen as an odd number from three to higher values. The higher the value, the more precisely the low-pass bandwidth can be specified, but with more computation time.

medfltr

Median Filter (in Noise Reduction, Section 3.2)

Synopsis – The program performs median filtering on an input gray-scale image, producing a noise-reduced output gray-scale image. One type of noise that the median filter best reduces is shot or speckle noise, which is isolated pixels of higher or lower intensity than the local area.

Interactions – Noise reduction is one of the most commonly applied programs before subsequent processing.

Notes:

- Median filtering is often chosen when low-pass filtering (**lpfltr**) undesirably smooths the important features of the image. Because a median filter performs a non-linear operation as opposed to the low-pass filter, it can better retain important features while reducing some types of noise. However, the types of noise are limited as just described.
- The filter size is chosen as an odd number from three to higher values. The higher the value, the more the noise reduction, but the more the computation time and the greater tendency to reduce desirable (non-noise) image features.

morph

Morphological Processing (in Morphological and Cellular Processing, Section 4.1)

Synopsis – The program performs morphological processing on a binary input image, producing a binary output image. Morphological processing operates iteratively upon an image, changing pixel values from ON to OFF or vice versa depending upon the number and configuration of pixel values around each pixel.

Interactions – This program is preceded by binarization if the image is not already binary (**binarize**, **thresh***). There are two common cases where morphological processing is most often used. One is for noise reduction, after which the image is inspected or further processing is performed. The other is for identifying an object of certain qualities that the morphological filtering can detect.

Notes:

- A common noise reduction purpose of morphological processing is to reduce isolated, small dots and holes. If this is the purpose, another option is to use the **kfill** program, which is specifically designed to reduce binary noise.
- Morphological processing has similar functionality to cellular logic (**cellog**), however there are differences in how these operations perform, so the user might want to test both on the images of interest to see which yields preferred results.
- One difference of morphological processing from cellular logic (**cellog**) is the concept of a definable structuring element, that is the binary filter mask. This program limits selection of the structuring element to a square shape of chosen (odd side-length) size.

multires

Multiresolution Analysis (in Multiresolution Analysis, Section 3.7)

Synopsis – The program processes a single input gray-scale image to produce a multiresolution pyramid of subsampled gray-scale output images. Each pyramid image is half the size of the preceding image, starting from the original (at the base of the pyramid) up to a four-pixel image (at the top of the pyramid).

Interactions – Since multiresolution analysis gives a range of subsampled images, this single operation can help a viewer see which degree of subsampling is appropriate for an image before using the program **subsample** with a single chosen subsample rate.

Notes:

- Each smaller image in the multiresolution pyramid has less high frequency (fine detail) information, so the user should look at the images from large to smaller sizes to see which level there has an acceptable degree of loss.

pcc

Primitives Chain Code (in Chain Coding, Section 5.1)

Synopsis – The program transforms a binary image containing thinned lines to produce an output file containing the Primitives Chain Code (PCC) of these lines.

Interactions – This program is most often used directly after **thin**. This program must be used before all line analysis programs of Chapter 5: **pccde**, **pccdump**, **pccfeat**, **linerid**, **linefeat**, **linexy**, **structrid**, **structfeat**, **fitpolyg**, **fitcrit**, **fitline**, and **fitspline**.

Notes:

- A common mistake is to apply this program to a non-thinned image. Even if an image looks thinned, application of **thin** reduces the image to contain minimally eight-connected lines expected by **pcc**.
- Another mistake is to apply PCC to an inverted line image, that contains white lines on a black background. If this is the case, invert the image using **inv**.

pccde

Primitives Chain Decode (in Chain Coding, Section 5.1)

Synopsis – The program reverse transforms an input PCC file producing a binary image containing lines.

Interactions – This program is used after performing line processing, such as noise reduction, on a PCC file by programs: **linerid**, **linexy**, and **structrid**.

Notes:

- This program is used to see the results of line processing in the PCC domain.

pccdump

Primitives Chain Code Dump (in Chain Coding, Section 5.1)

Synopsis – The program takes an input PCC file and prints out the raw PCC codes and feature codes and types.

Interactions – This program is used after PCC coding (**pcc**) and often after processing in the PCC domain such as by: **linerid**, **linexy**, and **structrid**.

Notes:

- This program gives a full, raw dump of PCC codes, and can be used to debug or see the finest detail of coding. An alternative that gives more readable text and image output is **pccfeat**, and higher level programs **linefeat** and **structfeat**.

pccfeat

Primitives Chain Code Features (in Chain Coding, Section 5.1)

Synopsis – The program takes an input PCC file and produces a binary image with superimposed squares over features, or prints out the PCC codes and feature codes and types.

Interactions – This program is used after PCC coding (**pcc**) and often after processing in the PCC domain such as by: **linerid**, **linexy**, and **structrid**.

Notes:

- The program **pccdump** gives a full, raw dump of the PCC codes and features, and may be preferred over this program for debugging, for instance.

peak

Peak Detection Filter (in Edge and Peak Point Detection, Section 3.4)

Synopsis – The program attempts to find the dominant peak in a gray-scale input image and to produce an output image with that peak location indicated.

Interactions – An example use of this program is as a last step of analysis (after noise reduction, smoothing, etc.) for the purpose of locating a peak of interest.

Notes:

- This program finds a single peak location. If there are multiple peaks (or objects) in the image, then the user should first find an approximate square area that includes only one peak of interest and have the program start for this region.
- The peak location may be thrown off by noise, image objects, or regions with complex (far from oval) shapes.

powspec

Power Spectrum (in The 2D Discrete Fourier Transform, Section 7.1)

Synopsis – The program takes a gray-scale input image and produces the a gray-scale output image of the power spectrum.

Interactions – This program may be used with noise reduction and filtering programs such as **lpfltr** and **fltrfreq**.

Notes:

- This program gives a visual display of power spectrum in the image, therefore it is commonly used to inspect the results of a frequency-altering programs such as noise removal or low-, band-, and high-pass filtering.
- Power is displayed on the vertical axis with a logarithmic (default) scale, and frequency is displayed along the horizontal axis with zero (or dc) in the middle, positive frequency to the right and negative to the left.

profile

Image Profile (in Global Features and Image Profiles, Section 4.9)

Synopsis – The program takes a binary input image and produces a binary output image containing the horizontal or vertical profile. A profile is a histogram showing the summation of the number of ON-values for each column or each row in the image.

Interactions – One use of this program is to determine the location of the highest density of objects in an image. If both vertical and horizontal profiles are taken and peaks are found of each, the intersection of peak locations on the x- and y-axes will indicate a high density image area. Another use of this program is to determine vertical or horizontal alignment of objects in an image. Well-defined profile peaks indicate object alignment for that profile. After alignment has been ascertained, subsequent processing or feature determination might be performed accordingly.

Notes:

- If this program is used for determining alignment, it should be noted that it is most effective for only vertically or horizontally aligned image objects. If objects are more diagonally aligned, the image can be rotated using **imgrotate**.
- The reduction of image information from two dimensions to one necessarily reduces information, so there will be many images containing complex distributions of objects where this profile program is not useful.

rgb2gray

Color to Gray-Scale Image Conversion (in Color Image Transformations, Section 2.5)

Synopsis – The program converts a color input image to a gray-scale output image. The user chooses which information the output image contains. This can be the red, green, or blue component of the color image (the color image is a combination of these three basis images). Or it can be an intensity, hue, or saturation image, which contains that portion of information from the color image.

Interactions – Since no other program in this analysis package processes color images, a color image must be converted to gray-scale before other processing. This is the common way color image processing is performed, that is by separating one or more of the basis images of interest from the color image, and then processing these as for any other gray-scale image.

Notes:

- The most common color image conversion before other processing is to an intensity image, which contains information on darkness-lightness.
- Less frequently a color image is converted to a hue image, which contains color information, or a saturation image, which contains a measure of the pureness of color.
- Conversion to a red, green, or blue image may be confusing. Since real-life pictures (as opposed to graphics images) rarely contain a pure color of one of these bases, just converting to these separate images will rarely separate red from green from blue objects in the image.

spp

Scan Point Pattern (in The Voronoi Diagram of Point Patterns, Section 6.1)

Synopsis – The program takes an input binary or gray-scale image and produces a file containing data for subsequent Voronoi analysis.

Interactions – This program is used to produce input for the program **xvor**.

structfeat

Structure Features (in Line Features and Noise Reduction, Section 5.2)

Synopsis – The program takes an input PCC file and prints an output of line structure features including: number of lines per structure, total line length, bounding box, and centroid; and also gives a structure summary over the image containing the number of line structures and average statistics.

Interactions – This program is used after PCC coding (**pcc**) and often after line noise reduction (**linerid**, **structrid**). This is often a final step of line analysis to see what structures remain after processing, and their features and statistics.

Notes:

- A line structure is a single connected entity that contains one or more lines. To see only line features, use **linefeat**.

structrid

Structure Rid (in Line Features and Noise Reduction, Section 5.2)

Synopsis – The program takes an input PCC file and rids (filters out) line structures by thresholds in line length and bounding area, and produces a PCC file containing the remaining (unfiltered) structures.

Interactions – This program is often used after PCC coding (**pcc**) to reduce line structures that are not of interest, for instance noisy structures that are small or large structures that are simply greater than the size bounds of interest.

Notes:

- This program and **linerid** are the main filtering programs for PCC code.

subsample

Subsample (in Subsampling, Section 3.6)

Synopsis – The program performs subsampling on a gray-scale image, producing a smaller gray-scale output image. Low-pass filtering is performed during this operation to reduce aliasing (high frequency noise) in the smaller image.

Interactions – Subsampling can be used before other processing steps if it is desired to reduce the image size for faster processing and if information of interest is of sufficiently large size or resolution that it will not be lost in subsampling. This program is also used simply to reduce an image size to fit a space or for smaller memory use.

Notes:

- Since there is information loss in subsampling, it should only be used if the information of interest is not reduced using this step. For instance, it may be acceptable to subsample an image containing large objects of little fine detail since little will be lost in this process. However if an image contains small objects of interest or large objects with fine detail of interest, the image should not be subsampled before object detection.

thin

Thin (in Thinning, Section 4.7)

Synopsis – The program thins lines in a binary input image containing lines, producing a binary output image containing lines of minimal – single-pixel – thickness, while maintaining connectivity.

Interactions – This program may be used before the line analysis programs of Chapter 5, which require that lines be thinned to minimal thickness.

Notes:

- To obtain expected results, the input image should contain only thin, elongated regions, i.e., what we would refer to as lines. The program will thin non-elongated blobs in any binary image, however processing on these types of regions might yield unexpected and non-useful results.
- This program processes the image iteratively. If the program goes through many iterations, this will take some computing time and perhaps the image is not suitable for thinning (due to having rounded blobs versus elongated regions).
- The number of image iterations can be limited by specifying this as a program parameter value, however the final result is not guaranteed to have only one-pixel width lines in this case.
- This program thins the foreground (black) regions. If the user desires the background to be thinned, invert the image first using **inv**.
- The program enables the user to choose larger mask sizes than the default three to potentially reduce the number of iterations, however larger mask values may produce coarser thin line results.

thinw

Thin Width (in Linewidth Determination, Section 4.8)

Synopsis – The program thins lines in a binary input image containing lines, producing a gray-scale output image (note this result is not binary) containing lines whose value at each line pixel is the width of the line at that point.

Interactions – This program is often the last analysis step where an original gray-scale image may have been noise-reduced, binarized, perhaps noise-reduced again, and then processed with this program.

Notes:

- To obtain expected results, the input image should contain only thin, elongated regions, i.e., what we would refer to as lines. The program will thin non-elongated blobs in any binary image however, processing on these types of regions might yield unexpected and non-useful results.
- This program processes the image iteratively. If the program goes through many iterations, this will take some computing time, and perhaps the image is not suitable for thinning (due to having rounded blobs versus elongated regions).

- The number of image iterations can be limited by specifying this as a program parameter value, however the final result is not guaranteed to have only one-pixel width lines in this case.
- This program thins the foreground (black) regions. If the user desires the background to be thinned, invert the image first using **inv**.
- The program enables the user to choose larger mask sizes than the default three to potentially reduce the number of iterations, however larger mask values may produce coarser thin line results.

threshc

Threshold Using Connectivity-Preservation Method (in Binarization, Section 3.10)

Synopsis – The program thresholds a gray-scale input image, producing one or more binary output images with respect to one or more determined thresholds. The one or more threshold values are determined by an algorithm that seeks to preserve connected regions in the image intensities. This is a multi-thresholding algorithm, that is, it seeks to find as many thresholds as are required for multi-level (not just binary) images, however it defaults to a binarization method if two levels are found in the image.

Interactions – This program may be preceded by gray-scale processing steps, especially noise reduction (**lpfltr, medfltr**). After thresholding, binary image processing programs may be used (Chapter 4). If multi-level images are found, then each is stored as a separate binary image.

Notes:

- This thresholding program is one of many offered in this package, different ones appropriate for different images and purposes. See the text of Section 3.10 to get an indication of which will be appropriate for particular image types.
- This connectivity-preserving method works best for non-natural images such as text and graphics images.
- The user can try different thresholding programs on their image type of interest and choose the one that produces the best results (**binarize, thresh***).

threshe

Threshold Using Maximum Entropy Method (in Binarization, Section 3.10)

Synopsis – The program binarizes a gray-scale input image, producing a binary output image whose pixel intensity values that are above a threshold are set to ON and otherwise are set to OFF. The threshold value is determined by an algorithm that seeks to maximize entropy of image intensities for the chosen threshold.

Interactions – This program may be preceded by gray-scale processing steps, especially noise reduction (**lpfltr, medfltr**). After thresholding, the binary image processing programs may be used (Chapter 4).

Notes:

- This thresholding program is one of many offered in this package, different ones appropriate for different images and purposes. See the text of Section 3.10 to get an indication of which will be appropriate for particular image types.
- The user can try different thresholding programs on their image type of interest and choose the one that produces the best results (**binarize, thresh***).

threshk

Threshold Using Kittler's Minimum Error Method (in Binarization, Section 3.10)

Synopsis – The program binarizes a gray-scale input image, producing a binary output image whose pixel intensity values that are above a threshold are set to ON and otherwise are set to OFF. The threshold value is determined by an algorithm first described by Kittler that seeks to minimize a measure of error for the chosen threshold.

Interactions – This program may be preceeded by gray-scale processing steps, especially noise reduction (**lpfltr, medfltr**). After thresholding, the binary image processing programs may be used (Chapter 4).

Notes:

- This thresholding program is one of many offered in this package, different ones appropriate for different images and purposes. See the text of Section 3.10 to get an indication of which will be appropriate for particular image types.
- The user can try different thresholding programs on their image type of interest and choose the one that produces the best results (**binarize, thresh***).

threshm

Threshold Using Moment-Preservation Method (in Binarization, Section 3.10)

Synopsis – The program binarizes a gray-scale input image, producing a binary output image whose pixel intensity values that are above a threshold are set to ON and otherwise are set to OFF. The threshold value is determined by an algorithm that seeks to preserve a statistical measure of moment of image intensities (a different measure than for **thresho**).

Interactions – This program may be preceeded by gray-scale processing steps, especially noise reduction (**lpfltr, medfltr**). After thresholding, the binary image processing programs may be used (in programs of Chapter 4).

Notes:

- This thresholding program is one of many offered in this package, different ones appropriate for different images and purposes. See the text of Section 3.10 to get an indication of which will be appropriate for particular image types.

- The user can try different thresholding programs on their image type of interest and choose the one that produces the best results (**binarize, thresh***).

thresho

Threshold Using Otsu's Moment-Preservation Method (in Binarization, Section 3.10)

Synopsis – The program binarizes a gray-scale input image, producing a binary output image whose pixel intensity values that are above a threshold are set to ON and otherwise are set to OFF. The threshold value is determined by an algorithm that seeks to preserve a statistical measure of moment of image intensities (a different measure than for **threshm**).

Interactions – This program may be preceded by gray-scale processing steps, especially noise reduction (**lpfltr, medfltr**). After thresholding, the binary image processing programs may be used (Chapter 4).

Notes:

- This thresholding program is one of many offered in this package, different ones appropriate for different images and purposes. See the text of Section 3.10 to get an indication of which will be appropriate for particular image types.
- The user can try different thresholding programs on their image type of interest and choose the one that produces the best results (**binarize, thresh***).

xah

Image Binary Area Histogram and Centroids (in Region Detection, Section 4.3)

Synopsis – The program finds the centroids and area histogram of blobs in a binary input image and yields a binary output image with centroids displayed. The area histogram is written to standard output showing the number of ON-pixels per blob, first shown in order of placement in the image, then in sorted order from least to most.

Interactions – This program is preceded by binarization if the image is not already binary. The output centroid image can be used for subsequent programs such as *k*-nearest neighbor analysis (**knn**). The output histogram information is in numeric form to view or be used by a data analysis program (not in this package).

Notes:

- The number and size of blobs will be very dependent upon noise reduction and other processing steps leading up to the blob image.

xbdy

Image Region Boundary Coding (in Advanced Shape Analysis: Fourier Descriptors, Section 4.5)

Synopsis – The program determines the convex boundaries of an input binary image containing a single region and produces an output image containing the convex boundary and Fourier descriptors (optionally).

Interactions – This program is preceded by binarization (**binarize, thresh***) if the image is not already binary. This is often the final analysis step, where one or more of these feature values describe the region of interest.

Notes:

- This program is applied to an image with a single region of interest.

xcc

Image Centers of Circular Regions (in Region Detection, Section 4.3)

Synopsis – The program estimates the centers of what are expected to be circular objects of known radii in a binary input image and produces a binary output image with these center locations indicated by dots.

Interactions – This program is preceded by binarization (**binarize, thresh***) if the image is not already binary. The output image can be used for recreating geometrically perfect circles around the found centers to inspect the deviation of the true shapes from circular.

Notes:

- Some deviation from circular shapes and expected radius size is allowed in the input image however, large deviation in shapes or radius will cause the resultant center locations to deviate from the ideal.

xconv

Convolution (in Local Image Operations: Convolution, Section 3.1)

Synopsis – The program convolves a gray-scale input image with a chosen kernel (or filter) to produce a filtered gray-scale output image.

Interactions – Convolution is one of the most common image processing operations. When convolution is performed toward the beginning of processing, it may be for the purpose of reducing noise by low-pass filtering, or for performing high-pass filtering to retain only high-frequency image features. When it is done toward the end of processing, it may be done in an attempt to detect an object in the image closely matching the image in the chosen convolution kernel (an operation called template matching or correlation, **xcorr**).

Notes:

- This program has one built-in kernel, a Gaussian-shaped kernel, for which the user can choose the filter size. Convolution of an image with a Gaussian-shaped kernel produces low-pass filtering. An alternative spatial domain filtering program only for low-pass filtering is **lpfltr**.
- The program also enables the user to convolve with a user-chosen kernel, which can be used for any other type of filtering: low-pass, band-pass, high-pass, or matched filtering.
- Alternatively to spatial domain convolution, filtering can be performed in the frequency domain using **freqfltr**.

xcorr

Correlation, or Template Matching (in Template Matching, Section 3.8)

Synopsis – The program correlates an input gray-scale image with a filter containing a template, or sub-image of interest, for the purpose of identifying the presence and locations of that template in the output image.

Interactions – This is often done as a final step after noise reduction and other processing steps for the purpose of finding objects of interest.

Notes:

- This template matching does not adapt for rotation or scale variations between the image objects and the filter template. One way to deal with this is to perform template matching multiple times with each possible rotation or scale represented in different template filters.

xcp

Image Contour Coding using Fourier Descriptors (in Region Detection, Section 4.3)

Synopsis – The program finds the contours of regions in a binary input image, producing a list of curvature points and Fourier descriptors that concisely represents the curve.

Interactions – This program is preceded by binarization (**binarize, thresh***) if the image is not already binary. The result is an efficient representation of the contour that a user may want to store or use other processing methods to analyze however, there are no additional programs in this package that accept output of this program.

Notes:

- If further analysis is desired on a contour image, use **contour** followed by the line analysis programs of Chapter 5.

xedgefilter

Image Edge Detection Filter (in Edge and Peak Point Detection, Section 3.4)

Synopsis – The program performs simple edge detection on a gray-scale input image, producing a gray-scale output image that has intensity values larger for stronger edges.

Interactions – Edge detection is often performed on images after noise reduction for the purpose of segmenting image objects.

Notes:

- This program convolves user-chosen masks across the image. If the user wishes to choose the edge detection mask, this program is appropriate.
- If the user desires a more sophisticated edge detection program, choose **bcd**.

xfm

Image Region Fast Moment Calculation (in Shape Analysis: Geometrical Features and Moments, Section 4.4)

Synopsis – The program determines shape features of a single region in a binary input image, producing a list of moments up to order three in x and y.

Interactions – This program is preceded by binarization (**binarize**, **thresh***) if the image is not already binary. This is often the final analysis step, where one or more of these feature values describe the region of interest.

Notes:

- This program is applied to an image with a single region of interest.

xknn

k-Nearest Neighbors (in The k-Nearest Neighbor (k-NN) Problem, Section 6.4)

Synopsis – The program takes an input text file containing (x,y) coordinate pairs of points and produces an output image showing nearest neighbor connections among points.

Interactions – This program follows point extraction (**xah**). The clusters found by k-NN analysis may be used subsequently by, for instance, processing only on a region containing a particular cluster of interest.

xph

Image Polygon Hull (in Convex Hull of Polygons, Section 4.6)

Synopsis – The program determines the polygon hull of an input binary image containing a single region and produces an output image showing that polygonal convex outline.

Interactions – This program is preceded by binarization (**binarize**, **thresh***) if the image is not already binary. This is often the final analysis step, where the convex hull is the objective of interest, or that hull is tested (perhaps by image combination) to check for overlap among convex hulls of objects placed in the same space.

Notes:

- This program is applied to an image with a single region of interest.

xpm

Image Region and Polygon Moments (in Shape Analysis: Geometrical Features and Moments, Section 4.4)

Synopsis – The program determines the moments of a single region in a binary input image, producing a list of the first four region moments, principal axes, radius of gyration, shape eccentricity, and two invariant moments (see text for explanation of each).

Interactions – This program is preceded by binarization (**binarize**, **thresh***) if the image is not already binary. This is often the final analysis step, where one or more of these feature values describe the region of interest.

Notes:

- This program is applied to an image with a single region of interest.

xptstats

Point Statistics (in Spatial Statistics of Point Patterns: Distribution Functions, Section 6.2)

Synopsis – The program takes an input file from **spp** or **xah** and produces a text listing of distribution function values for the image points.

Interactions – This program follows **spp** and **xah**, and is often the last analysis step showing summary data.

xrg

Image Region Growing (in Region Detection, Section 4.3)

Synopsis – The program determines the area (number of pixels) of regions in a binary input image, returning a listing of areas. The regions of interest are indicated manually by specifying a pixel location anywhere within each region of interest.

Interactions – This program is preceded by binarization (**binarize, thresh***) if the image is not already binary. The list of areas can be used to calculate average blob size, maximum size, range of sizes, etc.

Notes:

- This program is usually used to determine the area of regions of interest in the image, rather than all regions in the image. This is because any bias of the final results by noise regions is avoided by this manual location step.

xscale

Image Scale (in Geometric Image Transformations, Section 2.4)

Synopsis – The program scales an input gray-scale image by a multiplicative factor on the *x* or *y* axis or both to produce an image of a different size.

Interactions – This program may be useful for changing the size of an image before subsequent processing to make two image sizes compatible, e.g., for programs **imgarith**, **imgbool**, or **combine**. This program may also be useful for producing larger size objects for viewing. Or it can scale *x*- and *y*-axes independently to change image aspect ratio. Since information content is not increased via this program it is not advantageous to scale before other programs other than for size or shape compatibility.

Notes:

- The pixels in the output image are interpolated during scaling to prevent a "boxed" look that would result if the pixel values were just replicated.
- The information content is not increased during scaling, however larger images may appear to have lower quality than the original just because the lower resolution features of the image are now more visible.

xsgll

Segment Group and Adjacency List (in Morphology and Topology of Line Patterns, Section 5.7)

Synopsis – The program takes a file containing line segment endpoints from **fitpolyg** and produces an output image containing line segment clusters and a file containing a segment group list.

Interactions – This program is used after PCC coding (**pcc**) and after **fitpolyg**. The resulting clusters can be processed by **eh_sgl** to produce histogram data of the clusters.

Notes:

- Line noise reduction (**linerid, structrid**) is usually a preprocessing step before this program so that the resulting clusters contain only lines of interest.

xsgt

Extract Site Geometry and Topology (in Topology and Geometry of Cellular Patterns, Section 6.3)

Synopsis – The program merges geometry and topology data from previous analyses by programs **xah** and **xvora**.

Interactions – This program follows **xah** and **xvora**.

Notes:

- All the statistics from **xah** and **xvora** are displayed in text by this program.

xvor

Image Voronoi Diagram Analysis (in The Voronoi Diagram of Point Patterns, Section 6.1)

Synopsis – The program takes an input file produced by **spp** or **xah** and produces a Voronoi diagram, optionally with Delaunay triangulation.

Interactions – This program uses data supplied by the program **spp** or **xah**. The resulting Voronoi diagram is shown for viewing and the resulting data can be written to a file for statistical analysis.

xvora

Image Voronoi Analysis Statistics (in Spatial Statistics of Point Patterns: Distribution Functions, Section 6.2)

Synopsis – The program takes an input file from **xvor** of Voronoi or Delaunay triangulation data, and produces statistics and histograms of various quantities.

Interactions – This program follows **spp** and **xvor**, and is often the last analysis step showing final analysis summary data.

Projects

Introduction

This chapter offers project suggestions that can be performed by referring to the book and using its software. The primary purpose is to provide ideas for students to undertake independent work as part of an image processing or image analysis course. We have attempted to include a range of objectives and applications that will help students learn and understand common approaches. For each project step, we list some of the programs that could be used however, the student is encouraged to try other programs from the book that he or she thinks might help accomplish the task as well. We also encourage trying different parameter values for the programs that have user-selectable settings. One of the advantages of image analysis is that the user can visually evaluate how well objectives are met. A common procedure for image researchers is to quickly test different programs and parameter values, observe results, make adjustments, and repeat until there is a good understanding of image characteristics and different programs' abilities to process them.

For more ambitious projects, we make programming suggestions. However, the programs do not need to be written from scratch; they can be started from provided programs whose names are given, and modifications made for the intended purpose. Although we make suggestions on the types of images to use for the projects, we do not provide the actual image files here. Images can be found on the web and can usually be used freely for educational purposes. All the programs work only on *tiff* image format files (*file.tif* or *file.tiff*). Many web images will be available in *tiff* format, but *jpeg*-formatted images will probably be more common (*file.jpg* or *file.jpeg*). Before using the *jpeg* image, it must be converted. To do this, open the image in a display program (such as Microsoft Paint), then save it as type *tiff*.

1. **Image Enhancement**

    The objective of this project is to process images to enhance them visually. Note that this visual enhancement may not help for subsequent machine analysis steps, in fact in some cases it may hurt because information (entropy) will be reduced. Obtain some images with poor contrast, especially pictures that appear too dark.

    (a) First look at the original histogram. If the image is truly lacking in contrast you might notice a fairly narrow width between background and foreground peaks. Or the peaks might be low at the left end of the histogram intensity axis for dark images. Save this for comparison to processed images. (**histstats**)

(b) Apply the histogram enhancement programs to the image. Try different parameter values. Can you obtain higher contrast? When you do, are some regions improved and some reduced in visual quality? Examine the histograms and understand the relationship between the processed images and their histograms as compared to the original image and histogram. (**histex, histramp, histexx**)

(c) Try the noise reduction filters. Does this help, or just blur the image? Is the blurring worse for some image types and not as noticeable for others? (**lpfltr, medfltr**)

(d) Try different filters in the frequency domain and see if these help visibility. (**fltrfreq**)

(e) *Programming Suggestion* – Contrast is a measure of visibility. This can be expressed as ($(Imax – Imin) / Irange \times 100\%$), where *Imax* and *Imin* are maximum and minimum actual values in the image or a region thereof, and *Irange* is the maximum available range, 256 for our images. Write a program that determines contrast in a sliding window across the image (**lpfltr** processes with a sliding window as do other programs). Make the window size user-selectable. Adjust the constrast throughout the image in this way. Adjust for clipping (when the result is greater than 255) and for negative values.

2. **Color Image Processing**

The objective of this project is to understand how color images are made up of three basis images, to see how these bases contribute to the overall image, and to determine to what degree feature extraction results can be obtained from fewer than three of the basis images. Obtain some color images. Some should be natural images and some should be graphically generated or "cartoon" images with regions of uniform colors.

(a) Extract single images of the red, green, and blue color planes. First, observe each basis image and see what is lost or different in these as compared to the color original. Examine their histograms and note how relative differences in peaks correspond to differences in the degree of colors in the color image. Since the colors will not be pure for the natural images, note how combinations of colors in the various separate images combine to form the colors in the original. (**rgb2gray, histstats**)

(b) Extract single images of the intensity, hue, and saturation planes. Examine each by eye and by their histograms. In which image is the greatest contrast? Do the hue and saturation images convey much information? (In some cases they will, in some not.) Review the definitions of intensity, hue, and saturation, and in comparing the original color image to these separated bases, endeavor to understand the information each conveys. If you were to choose one image that contains most of the information from the color image, which would it be? (**rgb2gray, histstats**)

(c) Transform the color image to a gray-scale image (this is the intensity plane of IHS space), then perform edge detection. Are there any edges missed in the gray-scale image that are visible in the color image? Is edge information lost in transformation from color to gray-scale? Compare the amount of information lost for natural and graphical images. (**rgb2gray, xedgefilter, bcd**)

(d) *Programming Suggestion* – Create simple stripe images of chosen red-green-blue values for each stripe, such that edges are visible in the color image, and are edge-detectable in the hue image, but not intensity. Do the same for the saturation image and each of the red, green, and blue images. This will require some trial and error and visual inspection of results, but after doing this experimentally, can you suggest general, quantitative ways that the image results can be found that meet the criteria?

3. **Facial Feature Detection**

The objective of this project is to obtain landmark features and their relative measures in a face image to be used for recognition or matching. Face recognition is often performed by first obtaining measures of facial landmarks such as mouth width, distance between pupils of eyes, vertical distance from eyebrows to mouth, etc. These measures are combined to comprise a feature vector of a person and can be compared with a database of these vectors for different people. The shortest vector distance between query and database vectors indicates the closest match – hopefully the true match if the person matching the query is in the database. Obtain a number of pairs of face images for each person.

(a) If some heads are not upright, these should be rotated so that all have the same vertical pose. (**imgrotate**)

(b) The faces will likely have different sizes that are due to the camera capture characteristics rather than the actual head sizes. So, the head sizes will have to be normalized. Choose some measure of the head with which you wish to normalize the image, such as width from ear to ear, height from top to chin, width between pupils, etc.. Scale the image so that all images have the same length of this chosen measure. (**xscale, subsample, imggrid**)

(c) Find the edges of features and measure distances between them. Because faces are captured with variable lighting conditions, edge features of the same faces might also have some variance. Try binarizing the image to see if there are different shadows in different images of the same face. (**xedgefilter, bcd, binarize, imggrid**)

(d) Use one of the optimal binarization methods and then obtain the profile. Can horizontal, vertical, or both profiles be used for distinguishing different faces? Peaks will be very dependent upon lighting. Will the optimal binarization method be robust enough to reduce this dependency? Obtain global features. Will any of these features be useful in distinguishing faces? (**threshm, threshe, threshk, threshm, thresho, profile, globalfeats**)

(e) *Programming Suggestion* – Automate some of the tasks just described. Can you use the **profile** program with **imgrotate** to automatically and iteratively adjust an angled head to vertical? Take the features extracted from the preceding steps and apply them to matching, where minimum differences between corresponding feature measures of different images determines closeness of match.

4. **Biomedical Image Segmentation**

The objective of this project is to examine how edges and regions are obtained from gray-scale images and how these features might be stronger or weaker depending upon the type of image. Obtain some biomedical images containing both well-defined structures (such as an x-ray of bones) and less well-defined, "blob-like" components (such as liver tissue or blood cells). Apply some of the segmentation and analysis techniques to find which perform better for these two different types of image components.

(a) First examine the different images with a histogram and identify the correspondence between histogram peaks and image regions. (**histstats**).

(b) Try to reduce noise with different parameter values to understand the tradeoff between noise reduction and the ultimate objective of sharply defined segmentation. (**lpfltr, medfltr**).

(c) Perform edge detection and notice which image types give strong and unbroken edges and which do not. (**xedgefilter, peak, bcd**).

(d) If you think you can isolate the image characteristic of interest by binarization, perform thresholding and then binary noise reduction. Which image types segment well using this approach? (**binarize, threshm, threshc, thresho, threshe, threshk, cellog, morph, kfill**).

(e) On images containing regions, find the contour and look at region features. (**contour, xcp, xah, xcc, xrg, xpm, xfm, xbdy, xph**).

(f) For images containing strong lines, perform line fitting to identify the major straight lines and curves. (**pcc, pccde, fitpolyg, fitline, fitspline**).

(g) *Programming Suggestion* – After segmentation, extract feature values that give some measures of the image components. For instance, for images with cellular or "blob-like" components, find the number, size, eccentricity, etc., of their contours. Use, modify, and augment the programs **xpm, xfm, xbdy**, or write your own, taking averages of the features found.

5. **Texture and Pattern Detection**

The objective of this project is to examine image textures and patterns via appropriate programs. Both textures and patterns are repeated features, but a texture is usually less ordered and of natural origin (like a picture of a field of wheat), whereas a pattern is more ordered and often man-made (like a brick wall pattern). First, obtain some images containing textures and patterns. You can search the web for "Brodatz texture" to obtain images containing different textures. In addition, obtain images that contain textures and patterns as only a part of the picture. Use the programs to identify the textured or patterned regions.

(a) Use an edge filter on images with lines, peaks, or edges to isolate these features. (**xedgefilter, peak, bcd**).

(b) Use wavelet filtering to identify the predominant scale and orientation of textures and patterns. (**gabor**)

(c) Transform the image to the spatial frequency domain to locate predominant frequencies and orientations. Once these are found, filter them out, or filter out everything but these, by performing frequency domain filtering and then transforming back to the spatial domain. (**powspec, fltrfreq**).

(d) Perform a Hough transform on a highly structed image containing lines (such as a brick wall). Compare the results with applying this program to a texture image of lines. (**hough**)

(e) Look at global features of different full-texture or full-pattern images, and compare these feature values. (**globalfeats, eh_seg, xsgll, eh_sgl**)

(f) *Programming Suggestion* – Write a segmentation program for images containing regions of known texture. Modify one or more of the programs that yield global features (**globalfeats, profile, histstats**) to operate not globally but locally in a sliding window across the image. (Several programs such as **lpfltr** use a sliding window.) Make the window size a user-selectable variable and understand the tradeoffs as window size is made larger or smaller.

6. **Line Fitting**

The objective of this project is to examine how the different line-fitting methods perform. Obtain, or hand-draw, a very simple line drawing image containing some open or closed straight lines objects, and curves or a circle.

(a) If the image is not already binary, binarize it, then thin it. (**binarize, thin**)

(b) Obtain the thin line code (PCC). (**pcc**)

(c) Use the different line fitting methods and compare their results. (**fitpolyg, fitline, fitspline**)

(d) Identify critical points on the lines (corners, beginnings, and endings of curves). Are these in the locations you think they should be? Program parameters can be changed to adjust location, however if image features occur over a range of scales, it is difficult to make a single chosen program parameter value yield good results for all features. (**fitcrit**)

(e) *Programming Suggestion* – For a well-structured drawing obtain horizontal and vertical lines, as best the programs can from the preceding steps. Then "beautify" the results by adjusting lines to be exactly horizontal and vertical, and corners to meet without gaps or intersection.

7. **Frequency Domain Features and Filtering**

The objective of this project is to identify spatial frequency and to understand how it appears in the spatial frequency domain. Obtain a few images of different global image composition. One should have fine texture, such as a closeup image of grass. One should have a pattern, such as a picket fence or a brick wall. One should be a graphically generated or a cartoon image with large regions of uniform colors. One should contain "blob-like" components, such as a biomedical image of cells. One should be a mix of components, such as a scene containing a building, trees, and sky. Obtain other images that you think have varied frequency components as well.

(a) Find power spectra for each of the images and visually locate peaks that appear as clusters. Relate spectral characteristics with characteristics of their original images. For instance, some spectra will contain strong, well-defined clusters, and some will have more dispersed results. For the well-defined clusters, note how angle in the frequency domain relates to orientation of features in the spatial domain. For the more dispersed spectral results, note the frequency bounds – some will have only lower frequency power, others higher. Relate this to the originals. (**powspec**)

(b) Try low-pass, band-pass, and high-pass filtering on the images and observe the results. Adjust the filter bands to see how these affect the resulting filtered images. (**fltrfreq**)

(c) Wavelet processing can extract frequency and orientation information as does the global frequency domain operations just described, but wavelets can also isolate localized frequency regions. So, if the entire image does not have the frequency characteristics of interest, wavelet analysis should yield better results. Try wavelet processing on the images. (**gabor**)

(d) *Programming Suggestion* – For this project, first write a program that creates vertical stripe images whose stripes are of intensities about 75 and 175, and widths are equal but user-adjustable. You will be able to see the effects of frequency and orientation filtering easily on the stripe images. First, apply low-, band-, and high-pass filtering with non-windowed filters which are filters with abrupt edges between passband and pass-stop frequency regions. What happens to the spatial domain results using these non-tapered filters? Make a filter whose pass region is adjustable in the pass-band frequency and orientation values. Note that there should be two pass-band regions in the frequency domain that are mirrored in (0,0). Apply this at different angles and frequencies and observe the transformed results in the spatial domain.

8. **Document Image Processing**

The objective of this project is to learn how image processing and analysis methods can be applied to the many, small connected components that constitute text in a document image. Obtain some gray-scale images of pages containing text. Obtain some images via a digital scanner (where lighting and perspective is controlled) and others by a handheld camera, where capture conditions are less controlled.

(a) First examine histograms of the images. Compare the sharpness of the peaks between scanned images and camera images. Look at the relative size of background to foreground (text) peaks. If the foreground peak is much smaller than the background (meaning that the text is sparse on the page), this may present difficulties to subsequent processing steps. (**histstats**)

(b) Binarize the images. Try the different binarization methods and choose which performs best. Does one method work better on scanned images and another on camera images? (**binarize, threshm, threshc, threshe, threshk, threshm, thresho**)

(c) Determine and examine the horizontal and vertical profiles of the images. Does one give sharper peaks than the other? Rotate a text image by 5, 10, and 20 degrees and look at the profiles of these. Can you suggest a method to determine when a text image is oriented correctly? (**profile, imgrotate**)

(d) Compare the region detection approaches, thinning and contour detection. Examine features from the binary, thinned, and contour images. (**contour, thin, thinw, globalfeats**)

(e) *Programming Suggestion* – Write a program that combines isolated characters into words, and words into lines of text. Start with the contour or thinned image, or with the PCC structure list from **structfeat**. Find centroids of each structure. Find closeby structures by measuring distances between centroids. Join character centroids whose distances are short and angles are horizontal into words and text lines.

9. **Fingerprint Processing**

The objective of this project is to take fingerprint images from the captured gray-scale stage to minutia features. Minutiae are features of fingerprint ridges, the black lines making up the fingerprint pattern. Minutia features are ridge endings and ridge bifurcations (three-line junctions). Noise consists of isolated lines, gaps in ridges, and bridges between two parallel ridges. Start with at least two pairs of fingerprint images, that is two images from one finger and two images from another.

(a) First examine the histograms of a few images. Do they have two distinct peaks? If so, binary image processing will be useful. (**histstats**)

(b) See if gray-scale noise reduction filters help rid noise. (**lpfltr, medfltr, xconv**).

(c) If some pairs contain fingerprints that are oriented differently from one another, rotate one to have a similar orientation as the other. (**imgrotate**)

(d) Binarize the image. Try the different programs and choose which works best for fingerprints. (**binarize, threshm, threshc, threshe, threshk, threshm, thresho**)

(e) Reduce binary image noise. (**cellog, morph, kfill**)

(f) Try subtracting the within-pair images and comparing these with between-pair image differences. You'll likely have to first rotate and translate one image with respect to the other to obtain the best subtraction. However, it is unlikely that even well-captured, matched fingerprint pairs will subtract to zero. Why is this? (**imgarith, imgrotate**)

(g) Thin the ridge lines. (**thin**)

(h) Obtain the thin line code (PCC). (**pcc**, and use **pccde** to transform back to an image)

(i) Reduce line noise. (**linerid, structrid**)

(j) Look at the minutia features by type and location and identify matching features in the image pairs. (**linefeat, structfeat**)

(k) *Programming Suggestion* – Write a simple fingerprint minutia matcher. Take the list of line features from preceding steps and match them. Corresponding minutiae will likely not match exactly in location, so try relative angles and distances between pairs and triplets of minutiae.

10. **Watermarking Creation and Detection**

The objective of this project is to insert a watermark image into a base image that you'd like to mark for some purpose, such as security or identification. There are two types of watermarks. A visible watermark can be seen in the image (albeit, it should be subtle like a watermark on a sheet of paper). An invisible watermark should not be seen. For a watermark image, use a simple binary image and for a base image try both a busy image with much detail as well as one with more uniform background.

(a) Try different ways to add the watermark image to the base image: addition, multiplication, and Boolean operations. Also change (increase, decrease, and invert) the range of intensities in the watermark image before adding it, then observe the visibility of the results. (**imgarith, imgbool, combine, inv**)

(b) On the image results that show the least visible watermarks, use some processing tools to see if these enable better detection. Do this both in the spatial and frequency domains. (**histstats, gabor, globalfeats, profile, powspec, fltrfreq**)

(c) *Programming Suggestion* – Watermarks can also be applied in the spatial frequency domain. Modify the **fltrfreq** program to multiply the transform of the base image with the transform of the watermark image. Vary a weighting factor on the watermark image to produce stronger or subtler watermarks.

Synopsis of Important Concepts

In this Appendix, we provide a synopsis of concepts and terminology that is to serve as a quick reference to key results of pertinent mathematical theory and to essential properties of various analytical tools referenced throughout the book. To be concise, we adopt a presentation that is more formal than that adopted in the main text. For those readers who wish to refresh or broaden their understanding of the formal underpinnings of the techniques used here, this synopsis will also serve as an entry point to the literature.

Section Overview

Section A.1 contains a review of the Fourier transform and its principal properties that are particularly relevant to the discussion of Chaps. 3, 4, and 7.

Section A.2 examines convolution, an operation of central importance to image processing by application of linear filters. Convolution is placed in the context of linear systems analysis in which it relates the concepts of impulse response and transfer function. These are fundamental to the design and implementation of linear filters, as discussed in Chaps. 3 and 7. A versatile implementation of convolution and autocorrelation functions by means of evaluation of serial products is also described.

Section A.3 describes the properties of important special-purpose (nonlinear) filters that are encountered in connection with advanced edge detection (Section 3.5) and template matching (3.8).

Section A.4 provides a derivation of the sampling theorem that relates a discrete sample to an original continuous (band-limited) function. This theorem expresses a concept of central importance to image analysis that, after all, relies on a discrete representation of an intensity function in the form of an array of pixels. Geometric transformations (Section 2.4) and multiresolution analysis (Section 3.7) and discrete Fourier transformation (Section 7.1) all require resampling and interpolation. The relationship of a discrete sample to the original continuous function is provided by the sampling theorem.

Section A.5 gives a concise overview of important data structures invoked in the implementation of many of our algorithms. This is supported by a library of functions for the manipulation of linked lists.

A.1 The Fourier Transform: Spatial Domain Versus Frequency Domain

Related Sections – shape features (Section 4.4), 2D discrete Fourier transform (Section 7.1).

Transformations of the general form

$$G(u, v) \equiv T\{g(x, y)\} = \iint_{-\infty}^{\infty} \{g(x, y)k(x, y; u, v)\} \, dx \, dy$$

are referred to as linear integral transforms if they satisfy linear superposition, so that the transform of the sum of two functions equals the sum of the transforms of the individual functions. Applied to functions of two spatial variables such as images, $I \equiv I(x, y)$, these transforms mediate the switching between two different representations of a given function, namely that in the spatial domain $\{x, y\}$ and that in the transform domain $\{u, v\}$. A special and prominent example of such an integral transform is the Fourier transform (FT) to whose discussion we return presently: the transform domain of the FT is the (spatial) frequency domain. A given forward transform and its inverse constitute a transform pair.

The action of a transform is specified by the form of its kernel, $k(x, y; u, v)$. Of particular practical importance are transforms that are separable and symmetric, properties respectively reflecting the following two characteristics of the transform kernel,

$$k(x, y; u, v) = k_x(x, u)k_y(y, v),$$
$$k_x(x, u) = k_y(y, v).$$

Separable 2D transforms may be evaluated by successive application of two 1D transforms. In particular, if the kernel is also symmetric, i.e., $k_x = k_y = k$, the discrete version of transform may be concisely written in the form of a matrix product, in a manner analogous to that discussed in detail in connection with the serial product formulation of the convolution in Section A.2.

The FT is a separable, symmetric integral transform defined in terms of a complex symmetric, separable kernel $kh(x, y) = \exp(-iux)\exp(-ivy)$:

$$F\{g(x, y)\} \equiv \iint_{-\infty}^{\infty} g(x, y)\exp[-i(ux + vy)] \, dx \, dy,$$

$$F^{-1}\{G(u, v)\} \equiv \iint_{-\infty}^{\infty} G(u, v)\exp[i(ux + vy)] \, du \, dv.$$

The functions $G(u, v) \equiv F\{g(x, y)\}$ and $g(x, y) \equiv F^{-1}\{G(u, v)\}$ form a transform pair.

The appearance of a complex kernel in the FT is related to the classical application of the Fourier series to the analysis of periodic functions. Fourier showed that arbitrary functions in fact also possess a representation in the form of a trigonometric series.

Table A.1. Properties of the Fourier Transform

Notation: $F\{g(x, y)\} \equiv G(u, v)$, $F\{h(x, y)\} \equiv H(u, v)$; the Operator $\Leftrightarrow$ Indicates a FT Pair: Left-Hand and Right-Hand Sides are Mutually Interconverted by Means of Fourier Transformation (See Also [Goodman, 1968], Section 2.1; [Press et al., 1988], Section 12.0).

Linearity Theorem

$$F\{ag(x, y) + bh(x, y)\} = aF\{g(x, y)\} + bF\{h(x, y)\}$$

Similarity Theorem

$$F\{g(ax, by)\} = G(u/a, v/b)/|ab|$$

Shift Theorem

$$F\{g(x - a, y - b)\} = G(u, v)\exp[-i2\pi(ua + vb)]$$

Parseval's (Total Power) Theorem

$$\iint_{-\infty}^{\infty} |g(x, y)|^2 \, dx \, dy = \iint_{-\infty}^{\infty} |G(u, v)|^2 \, du \, dv$$

Convolution Theorem

$$g(x, y) * h(x, y) = \iint_{-\infty}^{\infty} g(\xi, \eta)h(x - \xi, y - \eta) \, d\xi \, d\eta \Leftrightarrow G(u, v)H(u, v)$$

(Cross-)Correlation Theorem

$$C_{g,h}(x, y) = \iint_{-\infty}^{\infty} g(\xi, \eta)h(\xi - x, \eta - y) \, d\xi \, d\eta \Leftrightarrow G(u, v)H(u, v)^*$$

(Wiener–Khinchin) Autocorrelation Theorem

$$C_{g,g}(x, y) = \iint_{-\infty}^{\infty} g(\xi, \eta)g(\xi - x, \eta - y) \, d\xi \, d\eta \Leftrightarrow |G(u, v)|^2$$

Fourier Integral Theorem

At each point (x, y) where the function $g(x, y)$ is continuous, the following identity holds:

$$F^{-1}F\{g(x, y)\} = g(x, y)$$

The existence of fast algorithms for the numerical evaluation of the Fourier transform (Section 7.1) that have brought such prominence to the FT in a wide range of applications in fact follows from the factorization of the transform matrix into a product of very sparse matrices. This may be evaluated in time $N \log_2 N$ (see Section 7.1).

Table A.1 summarizes the essential properties of the FT. Note in particular the intimate connection to the evaluation of convolutions and correlations. The convolution theorem is central to the analysis of linear systems, described in Section A.2. It plays a fundamental role in experimental science as well as in image analysis. Another relation of great importance is the equivalence of power spectrum and autocorrelation.

Given the complex nature of the FT, two spectra may be derived from its real and imaginary components. The first, and by far the more common in experimental science because of its direct accessibility by means of diffraction measurements, is the power spectrum:

$$P(u, v) \equiv |G_i(u, v)|^2 = G(u, v)G^*(u, v).$$

The autocorrelation theorem shows that the power spectrum $P(u, v)$ and the auto-correlation function $C(x, y)$ of a function $g(x, y)$ form a transform pair. An important experimental realization of this identity connects the (auto)correlation function of the electron density of a solid, known as the Patterson function, with the set of intensities registered in an x-ray diffraction spectrum. Of lesser experimental importance is the phase function:

$$\phi(u, v) \equiv \arctan[\mathrm{Re}\{G(u, v)\}/\mathrm{Im}\{G(u, v)\}].$$

Of interest in the context of image analysis, although not widely exploited, is the fact that the phase function may provide information on singularities marking the location of topological defects in regular arrays (see Chap. 5). A sample of pictorial representations of transform pairs is given in Fig. A.1.

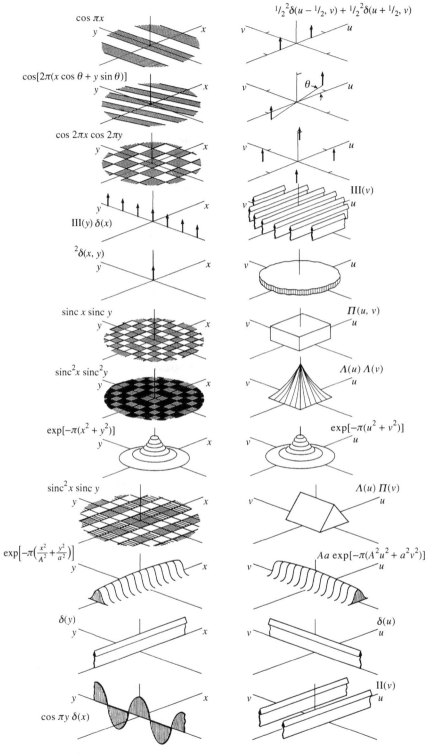

Figure A.1. Pictorial listing of 2D Fourier transform pairs (Material reproduced with Permission of The McGraw-Hill Companies from [Bracewell 86], Copyright © 1986 McGraw-Hill, Inc.).

A.2 Linear Systems: Impulse Response, Convolution, and Transfer Function

Related Sections – local operations (3.1); noise reduction (3.2); edge enhancement (3.3); edge detection (3.4); Fourier filtering (7.2).

A mapping S of a set of input functions $g_i(\mathbf{r}_1)$ onto a set of output functions $g_o(\mathbf{r}_2) \equiv S\{g_i(\mathbf{r}_1)\}$ is referred to as a linear system if it satisfies linear superposition ([Goodman, 1968], Section 2.2):

$$S\{af(\mathbf{r}_1) + bg(\mathbf{r}_1)\} = aS\{f(\mathbf{r}_1)\} + bS\{g(\mathbf{r}_1)\}.$$

Linearity facilitates the description of the action of a system on an arbitrary input by evaluating its response to input in the form of a delta function. In the example of an imaging system, a point source would represent the requisite delta function input so that, specializing to the appropriate two spatial dimensions, the system response to any input g_i would be fully specified by providing the impulse response $h(x_2, y_2; \xi, \eta)$, defined by the expression

$$h(x_2, y_2; \xi, \eta) \equiv S\{\delta(x_1 - \xi, y_1 - \eta)\}.$$

The response to an input of the form $g_i(x_1, y_1)$ is constructed as a superposition integral:

$$g_o(x_2, y_2) = \iint_{-\infty}^{\infty} g_i(\xi, \eta) h(x_2, y_2; \xi, \eta)\, d\xi\, d\eta.$$

Of particular importance are temporally and spatially invariant linear systems, so called because their impulse response function is invariant under translation (shift) in time and space. For the latter category the impulse response function assumes the simple form

$$h(x_2, y_2; \xi, \eta) = h(x_2 - \xi, y_2 - \eta).$$

and thus g_o has the form of a two-dimensional convolution of g_i with the system's impulse response. This establishes the general significance of the convolution operation in the analysis of invariant linear systems.

A.2.1 CONVOLUTION

The 1D convolution $h = f * g$ involves a mapping of $f(x)$ onto another function $h(x)$, defined by the expression $h(x) \equiv \int_{\infty}^{\infty} f(u)g(x - u)\, du$. This expression lends itself to a useful pictorial illustration, given in Fig. A.2, which emphasizes the fact that one of the two component factors is reversed before multiplication. It is this reflection operation that distinguishes convolution $f * g$ from cross correlation and self-convolution $f * f$ from autocorrelation. As defined, the convolution operation is commutative $r * s = s * r$, associative, $r * (s * t) = (r * s) * t$, and distributive over addition, $r * (s + t) = r * s + r * t$.

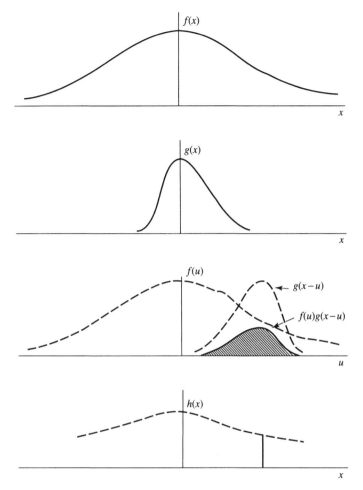

Figure A.2. Graphical illustration of the 1D convolution operation of two functions, $f * g$ (Material reproduced with Permission of The McGraw-Hill Companies from [Bracewell 86], Copyright © 1986 McGraw-Hill, Inc.).

Discrete Convolution in One Dimension: Serial Product

Pertinent to the application of the convolution operator in image processing is its discrete formulation. In the 1D case, this is obtained by consideration of two discrete samples of the functions $f = f(x)$ and $g = g(x)$ in the form $\{f_0, f_1, f_2, \ldots, f_m\}$ and $\{g_0, g_1, g_2, \ldots, g_n\}$, with $f_j \equiv f(x_j)$ and the analogous identification for $g_k \equiv g(x_k)$, the x_j representing equispaced values of x, i.e., $x_{j+1} - x_j \equiv \Delta x = \text{const}$. The convolution $\{h\} \equiv \{f\} * \{g\}$ of the two sequences $\{f\}$ and $\{g\}$ is defined as their serial product, and this is given as the sequence of coefficients obtained when $\{f\}$ and $\{g\}$ are multiplied term by term in the manner familiar from the multiplication of polynomials ([Bracewell, 1978], Chap. 3). In accordance with the definition of the convolution operation, the order of the second sequence is reversed. The result,

$$\{f_0, f_1, f_2, \ldots, f_m\} * \{g_0, g_1, g_2, \ldots, g_n\} = \{f_0 g_0, f_0 g_1 + f_1 g_0, \ldots, f_m g_n\},$$

is a sequence whose $(i + 1)$th term has the form $\sum_j f_j g_{i-j}$. For the simple example involving $\{f\} \equiv \{2, 2, 3, 3, 4\}$ of length $m = 5$ and $\{g\} \equiv \{1, 1, 2\}$ of length $n = 3$ this general expression evaluates to $\{2, 4, 9, 10, 13, 10, 8\}$ of length $7 = 5 + 3 - 1$.

Two immediate observations may be made. First, the length of the product sequence exceeds that of either input sequence and, in the general case, is in fact given by $n + m - 1$. Second, the form of the terms in the product sequence suggests a close connection of discrete convolution to matrix multiplication. This is made explicit by realizing that the desired product sequence may be generated by multiplying a matrix $\mathbf{F}_C$, composed of columns containing unit-shifted copies of the input sequence $\{f\}$, with a column vector $\mathbf{g}$, containing the second input sequence $\{g\}$ or, equivalently, by multiplying a matrix $\mathbf{G}_C$, composed of columns containing unit-shifted copies of the input sequence $\{g\}$, with a column vector $\mathbf{f}$, containing the second input sequence $\{f\}$.

The rules of matrix multiplication ensure the proper form of the serial product, which is obtained as a column vector,

$$\mathbf{h} \equiv \mathbf{F}_C \mathbf{g} = \mathbf{G}_C \mathbf{f},$$

as illustrated by the example in Table A.2. This representation of discrete convolution as a serial product is particularly useful in applications in which computational advantage can be taken of fast matrix operations that are often implemented in hardware. A further advantage lies in the generalization of the procedure to general linear transformations in which the input sequence $\{f\}$ may change between steps in the evaluation of the product: such changes are readily accommodated by modifying columns in $\mathbf{F}_C$.

A cyclic version of convolution ensures that the length of the product sequence does not exceed the greater of the lengths of the input sequences. The pertinent example in Table A.2 illustrates the concept. This formulation is most appropriate in cases exhibiting inherent periodicity in the input sequences, but is also invoked to circumvent potential problems otherwise incurred, for example, in situations limiting the memory that can be allocated for storage of the serial product. It should be noted, however, that application of a filter of length m in this case leads to corruption of elements of the output array located within a border strip of width $m - 1$ because of wraparound.

The serial product representation of the convolution operation may be readily generalized to two dimensions, provided that the convolution kernel is separable, as discussed in Section 3.1.

Convolution in Experimental Science

Convolution is ubiquitous in the experimental sciences in which it serves to describe the effect of finite instrumental response. That is, convolution models the action of any instrument, assumed to function as an invariant linear system. Such an instrument delivers not the actual value of a measured quantity $g(x, y)$ for a given set of values x, y of the pertinent input variables, but rather a weighted mean of $g(x, y)$ over a narrow range of x and y. If the instrumental response is described by a function $h(u, v; x, y)$, the measured quantity has the form of the (2D) convolution:

$$\bar{g}(x, y) \equiv \iint_{-\infty}^{\infty} g(u, v) h(x - u, y - v) \, du \, dv,$$

Table A.2. Convolution as Serial Product: An Example

Convolution of the Arrays $\{f_0, f_1, f_2, f_3, f_4\}$ and $\{g_0, g_1, g_2\}$ May be Equivalently Represented in the Form $\mathbf{h} = \mathbf{F}_C\mathbf{g}$, Yielding Explicitly ([Bracewell, 1978], Chap. 3).

$$
\begin{bmatrix} h_0 \\ h_1 \\ h_2 \\ h_3 \\ h_4 \\ h_5 \\ h_6 \end{bmatrix} = \begin{bmatrix} f_0 & 0 & 0 \\ f_1 & f_0 & 0 \\ f_2 & f_1 & f_0 \\ f_3 & f_2 & f_1 \\ f_4 & f_3 & f_2 \\ 0 & f_4 & f_3 \\ 0 & 0 & f_4 \end{bmatrix} \begin{bmatrix} g_0 \\ g_1 \\ g_2 \end{bmatrix} = \begin{bmatrix} f_0g_0 \\ f_1g_0 + f_0g_1 \\ f_2g_0 + f_1g_1 + f_0g_2 \\ f_3g_0 + f_2g_1 + f_1g_2 \\ f_4g_0 + f_3g_1 + f_2g_2 \\ f_4g_1 + f_3g_2 \\ f_4g_2 \end{bmatrix}.
$$

The commutative property of the convolution $f * g = g * f$ implies the existence of the alternative form:

$$
\begin{bmatrix} h_0 \\ h_1 \\ h_2 \\ h_3 \\ h_4 \\ h_5 \\ h_6 \end{bmatrix} = \begin{bmatrix} g_0 & 0 & 0 & 0 & 0 \\ g_1 & g_0 & 0 & 0 & 0 \\ g_2 & g_1 & g_0 & 0 & 0 \\ 0 & g_2 & g_1 & g_0 & 0 \\ 0 & 0 & g_2 & g_1 & g_0 \\ 0 & 0 & 0 & g_2 & g_1 \\ 0 & 0 & 0 & 0 & g_2 \end{bmatrix} \begin{bmatrix} f_0 \\ f_1 \\ f_2 \\ f_3 \\ f_4 \end{bmatrix}.
$$

Cyclic convolution of $\{f_0, f_1, f_2, f_3, f_4\}$ and $\{g_0, g_1, g_2\}$ may be equivalently represented in the form of a matrix multiplication, yielding explicitly

$$
\begin{bmatrix} h_0 \\ h_1 \\ h_2 \\ h_3 \\ h_4 \end{bmatrix} = \begin{bmatrix} f_0 & f_4 & f_3 & f_2 & f_1 \\ f_1 & f_0 & f_4 & f_3 & f_2 \\ f_2 & f_1 & f_0 & f_4 & f_3 \\ f_3 & f_2 & f_1 & f_0 & f_4 \\ f_4 & f_3 & f_2 & f_1 & f_0 \end{bmatrix} \begin{bmatrix} g_0 \\ g_1 \\ g_2 \\ 0 \\ 0 \end{bmatrix} = \begin{bmatrix} f_0g_0 + f_4g_1 + f_3g_2 \\ f_1g_0 + f_0g_1 + f_4g_2 \\ f_2g_0 + f_1g_1 + f_0g_2 \\ f_3g_0 + f_2g_1 + f_1g_2 \\ f_4g_0 + f_3g_1 + f_2g_2 \end{bmatrix}.
$$

or, in a useful shorthand symbolic notation [Bracewell, 1986],

$$
\bar{g}(x, y) \equiv g(x, y) * h(x, y) = h(x, y) * g(x, y).
$$

Evaluation of the functional $\bar{g}(x, y)$ requires knowledge of $g(x, y)$ for an entire range of x and y.

A.2.2 TRANSFER FUNCTION

The response of an invariant linear system, given in the form

$$
g_o(x_2, y_2) = \iint_{-\infty}^{\infty} g_i(\xi, \eta) h(x_2 - \xi, y_2 - \eta)\, d\xi\, d\eta
$$

may now be recognized as the convolution of the input function with the system's impulse response; that is,

$$g_o = g_i * h.$$

If FTs of g_i, g_o, and h are denoted by their corresponding capital letters, application of the convolution theorem (see Table A.1) yields

$$G_o = G_i H,$$

where the transfer function H represents the FT of the impulse response:

$$H(s, t) \equiv F\{h(\xi, \eta)\}.$$

The transfer function serves to specify the action of a system in the (temporal or spatial) frequency domain. For example, as elaborated in Sections 3.2, 3.3 and 7.2, the actions of low-pass and high-pass filters are particularly conveniently specified by defining appropriate transfer functions to delimit the band of frequencies to be passed.

A.3 Special-Purpose Filters

Related Sections – (advanced) edge detection (3.5),
template matching (3.8).

A.3.1 MATCHED FILTER

The typical circumstance calling for the use of a matched detection filter ([Andrews, 1970], Chap. 4) arises as follows. A signal $s_o(x, y)$ is observed that may or may not contain a contribution from a signal $s(x, y)$ of known form, in addition to otherwise present (stationary) noise $n(x, y)$. To establish the presence of $s(x, y)$, one may attempt to match the observed signal to a suitable template of the expected signal. This idea may be implemented by operating on $s_o(x, y)$ with a linear filter designed to maximize the ratio of the response in the presence of $s(x, y)$, relative to the response to (stationary) noise $n(x, y)$.

In quantitative form, the desired linear filter may be specified by requiring maximization of the ratio $\rho \equiv |s_O(x, y)|^2/|n_O(x, y)|^2$ of filter output power in response to uncorrupted signal input, $|s_O(x, y)|^2 = |\mathbf{F}^{-1}\{S_o\}|^2$, versus that in response to noise, $|n_O(x, y)|^2 = |\mathbf{F}^{-1}\{N_o(u, v)\}|^2$, where $S_o(u, v) = S_i(u, v)H(u, v)$ and $N_o(u, v) = N_i(u, v)H(u, v)$ respectively represent signal spectrum and noise spectrum of the filter with transfer function $H(u, v)$.

ρ is maximized, and optimal signal detection is ensured, by application of a linear, space-invariant filter that is matched to the expected signal $s(x, y)$ by specifying

$$H(u, v) = S_i^*(u, v) \exp[-i(ux + vy)]/|N_i(x, y)|^2.$$

Template matching is thus realized by correlation filtering: If the expected signal is present at the input, the filter transmits a field distribution proportional to the real quantity $S_i S_i^*$. That is, for the expected signal the filter performs as an autocorrelator.

A.3.2 WIENER FILTER

In the context of signal or image restoration the best estimate of an uncorrupted signal $s(x, y)$ underlying an observed signal $s_o(x, y)$ that is contaminated by random noise $n(x, y)$ of known statistical properties is provided by the Wiener filter ([Lim, 1990], Sections 6.1 and 9.2; [Pratt, 1991], Section 12.1). Its impulse response $h(x, y) \equiv \mathbf{F}^{-1}\{H_W(u, v)\}$ is specified by the requirement that its application to the measured signal, yielding the estimate $s^*(x, y) = s(x, y) * h(x, y)$, guarantee minimization of the mean square restoration error $\mathbf{E}_2 \equiv E\{e(x, y)^2\}$, where $e(x, y) \equiv s(x, y) - s^*(x, y)$ and $E\{\cdots\}$ denotes integration over all x and y.

We solve this problem by imposing the condition that the error $\mathbf{E}_2$ and any random contribution contained in the observed signal be uncorrelated and by further assuming that $s(x, y)$ and $n(x, y)$ are both random variables of zero mean. Under these assumptions, $e(x, y)$ and the observed signal $s_o(x, y)$ satisfy an orthogonality relation of the form $E\{e(x_0, y_0)s_o(x, y)\} = 0$ for all pairs of points (x_0, y_0) and (x, y). This leads to the transfer function

$$H_W(u, v) = P_{ss_o}(u, v)/P_{ss}(u, v),$$

where $P_{fg}(u, v) \equiv |\mathbf{F}\{C_{fg}(x, y)\}|^2$ denotes the (cross) power spectrum of the fields $f(x, y)$ and $g(x, y)$.

In the commonly encountered case of additive random noise $n(x, y)$, true signal and noise are uncorrelated so that $E\{s(x_0, y_0)n(x, y)\} = E\{s(x_0, y_0)\}E\{n(x, y)\}$, leading to the transfer function

$$H_W(u, v) = P_{ss}/(P_{ss} + P_{nn}),$$

which is of course identical to

$$H_W(u, v) = |S(u, v)|^2/(|S(u, v)|^2 + |N(u, v)|^2).$$

If $s(x, y)$ and $n(x, y)$ are both samples of Gaussian random processes, $H_W(u, v)$ represents the optimal estimator of the mean square restoration error.

A.4 The Whittaker–Shannon Sampling Theorem

Related Sections – geometric interpolation (2.4); subsampling (3.6); multiresolution analysis (3.7).

The sampling theorem, originally formulated for the 1D case, makes an important statement about band-limited functions, i.e., functions whose spectrum vanishes above a certain cutoff frequency f_c. The theorem holds that the entire original function may be reconstructed, essentially by means of interpolation, from a discrete set of the values it assumes, provided that samples are taken at intervals T chosen such that $1/2T$ equals the upper cutoff (Nyquist) frequency in the spectrum of the original function. The argument given here addresses the 2D case pertinent to image analysis and follows the lucid exposition presented by Goodman (1968, Chap. 2.3).

Consider a rectangular lattice of samples of the (continuous) function $g = g(x, y)$, as illustrated in Fig. A.3(a). With the comb function ([Bracewell, 1986], Chap. 4), this is compactly represented in the form

$$g_s(x, y) = \text{comb}(x/X)\text{comb}(y/Y)g(x, y).$$

That is, $g_s(x, y)$ consists of a rectangular array of δ functions, spaced with respective periods X and Y in x and y directions; the area under each δ function is proportional to the value assumed by the original function $g(x, y)$ at that lattice point. Application of the convolution theorem and appeal to the connection between sampling and replicating properties of the comb function, as expressed by its invariance under Fourier transformation,

$$F\{\text{comb}(x/X)\} = X\text{comb}(Xu),$$

facilitate the evaluation of the Fourier transform of $g_s(x, y)$. With $G(u, v)$ denoting the FT of $g(x, y)$, and $G_s(u, v)$ denoting the FT of $g_s(x, y)$:

$$\begin{aligned} G_s(u, v) &= F\{\text{comb}(x/X)\text{comb}(y/Y)\} * G(u, v) \\ &= XY\,\text{comb}(Xu)\text{comb}(Yv) * G(u, v), \end{aligned}$$

and hence, given the replicating property of the comb function,

$$G_s(u, v) = \sum_{n=-\infty}^{n=\infty} \sum_{m=-\infty}^{m=\infty} G(u - n/X, v - m/Y);$$

$G_s(u, v)$ represents the spectrum of the original function $g(x, y)$, replicated at each node $(n/X, m/Y)$ in the (u, v) plane, as illustrated in Fig. A.3(b).

Since the function g is by assumption band limited, its spectrum is nonzero over only a finite region in the (u, v)-plane; for the sampled function, this is the cone-shaped region sketched in Fig. A.3(b) surrounding each lattice node $(n/X, m/Y)$. Inspection of Fig. A.3(b) makes clear that cones will not touch if the intervals X and Y are chosen to be sufficiently small: that is, overlap between (cone-shaped) regions of finite spectral weight associated with adjacent lattice nodes may always be avoided by a proper choice

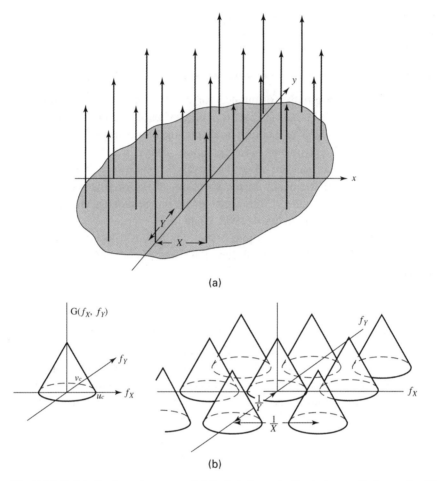

(a)

(b)

Figure A.3. (a): Rectangular array of delta functions, to illustrate proof of sampling theorem; (b, left): cone illustrating the shape of the spectrum G(u, v) of the original function g(x, y); (b, right): spectrum $G_s(u, v)$ of the sampled function $g_s(x, y)$, generated by replication of G(u, v) at all nodes of the reciprocal lattice (Material reproduced with Permission of The McGraw-Hill Companies from [Goodman 68], Copyright © 1968 McGraw-Hill, Inc.).

of X and Y. In that case, however, one recovers the original spectrum $G(u, v)$ from $G_s(u, v)$ by applying a linear filter that transmits the term $(n = 0, m = 0)$ while excluding all others: one so obtains the spectrum shown on the left of Fig. A.3(b), a single cone for the example sketched there.

The maximum permissible separation between samples, say $\hat{X}$ and $\hat{Y}$, to ensure avoidance of overlap between replicated spectral representations, is readily determined by inspection of Fig. A.3(b): Overlap is avoided if the following conditions are satisfied:

$$1/\hat{X} \geq 2u_c,$$
$$1/\hat{Y} \geq 2v_c,$$

where u_c and v_c represent upper limits for the spatial frequencies in the spectrum of the function g; in Fig. A.3, the dimensions $2u_c$ and $2v_c$ would be those of the smallest rectangle completely enclosing the nonzero spectral region of each replica of $G(u, v)$. That is, the maximum spacings of the sampling lattice to ensure exact recovery of the original function from its discrete representation are $2u_c^{-1}$ and $2v_c^{-1}$.

Having ensured that, under appropriate conditions, the original spectrum may be recovered from that of the discrete sample, we now proceed to specify the transfer function that describes the filter through which sampled data should be passed for recovery. In practice, any filter will do that meets the essential requirement to pass the sample of the spectrum located at the origin ($n = 0$, $m = 0$) of reciprocal space while rejecting all others. One function that certainly satisfies this condition is simply

$$H(u, v) = \text{box}(u/2u_c)\text{box}(v/2v_c).$$

Then, by construction,

$$G(u, v) = G_s(u, v)H(u, v),$$

and, by inverse Fourier transformation to direct space,

$$g(x, y) = [\text{comb}(x/X)\text{comb}(y/Y)g(x, y)] * h(x, y),$$

where h denotes the impulse response of the filter:

$$h(x, y) \equiv F^{-1}\{H(u, v)\} = 4u_c v_c \text{ sinc}(2u_c x)\text{sinc}(2v_c y).$$

Applying the sampling property of $\text{comb}(x)$ and $\text{comb}(y)$ to $g(x, y)$, one derives the explicit form of the sampling theorem:

$$
\begin{aligned}
g(x, y) &= [\text{comb}(x/X)\text{comb}(y/Y)g(x, y)] * h(x, y) \\
&= XY \sum_{n=-\infty}^{n=\infty} \sum_{m=-\infty}^{m=\infty} [g(nX, mY)\delta(x - nX, y - mY)] * h(x, y) \\
&= 4XYu_c v_c \sum_{n=-\infty}^{n=\infty} \sum_{m=-\infty}^{m=\infty} g(nX, mY)\text{sinc}[2u_c(x - nX)]\text{sinc}[2v_c(y - mY)].
\end{aligned}
$$

Finally, letting the sampling intervals assume their maximal values, $X = \frac{1}{2u_c}$, $Y = \frac{1}{2v_c}$ one obtains the classic form of the sampling theorem:

$$g(x, y) = \sum_{n=-\infty}^{n=\infty} \sum_{m=-\infty}^{m=\infty} g(n/2u_c, m/2v_c)\text{sinc}[2u_c(x - n/2u_c)]\text{sinc}[2v_c(y - m/2v_c)].$$

A.5 Commonly Used Data Structures

Related Sections – circular region segmentation (4.3),
shape features (4.4, 4.5), convex hull (4.6) PCC and TLC(5.2);
polygonalization (5.3) segment clusters (5.7); Voronoi
diagram (6.1), geometry and topology of cellular patterns
(6.3), k-NN problem (6.4).

As an entry point into the extensive specialized literature, this appendix provides a brief introductory overview of data structures. Our intent here is not to be comprehensive, but to review those data structures that are encountered in the implementation of procedures discussed in the foregoing chapters, notably linked lists in a variety of manifestations. Structured programming languages such as C provide the tools, including pointers, arrays, and structures, to introduce new, user-defined, structured data types [Kernighan and Ritchie, 1988]. Their competent use can greatly simplify the implementation of complex applications; specifically, they are indispensable for the efficient and transparent coding of many of the algorithms described in this book. To further encourage and facilitate their use, we provide here a set of simple routines, implementing a set of primitives to handle common list operations.

A.5.1   LINEAR DATA STRUCTURES: LISTS

An array provides storage for an ordered set of data. That is, a block of memory of specified length is reserved to provide for contiguous storage of array elements in preassigned order. Along with the dynamic memory allocation capability of the C language, an array provides a convenient device to handle a variety of tasks and is therefore ubiquitous in C code. However, as illustrated in the example below, more general situations frequently arise in which there is no *a priori* knowledge of the number of data items to be collected nor of the order in which they will be encountered.

To motivate the list concept, we refer back to Section 4.3, which introduced a simple algorithm for the fast determination of centroids of circular disks of known (or estimated) average size. The strategy discussed in that section relies on a coarse raster scan with an appropriately chosen vertical step size to ensure that each disk in the set will be intersected by at least n line scans, $n \geq 2$. Each intersection generates a pair of edge points for the corresponding disk. For each disk, the set of edge point coordinates, collected in successive line scans and properly ordered, serves as a basis for an estimate of the centroid location. In general, the number of disks that will be intersected in a given line scan and the total number of disks in the set are not known *a priori*, and memory must therefore be allocated dynamically.

While the task just sketched could conceivably be handled by arrays, this would be very inefficient from the point of view of memory usage. Furthermore, one may wish to maintain the list of disks in a particular order different from that in which disks are encountered in the raster scan; this will require insertion of new entries in arbitrary positions. Insertion of an element in an arbitrary position in an existing array would require global rearrangements of all subsequent elements and would thus cause serious inefficiencies.

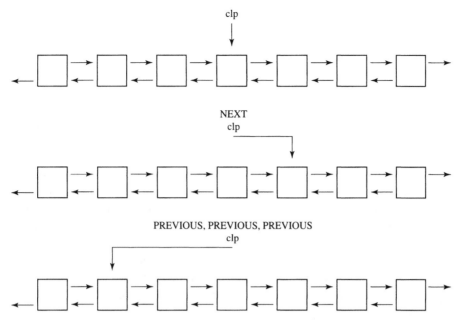

Figure A.4. Graphical illustration of a doubly linked list. Also indicated is the action of the operations *next* and *previous* on the *current list pointer*, clp (Reproduced with Permission from [Sessions 89], Copyright © 1989 by Prentice-Hall, Inc.).

The generalization of the array concept that is suitable for situations such as the one above is that of a self-referential structure: Each entry, also referred to as a node or a link, contains, in addition to (the memory location of) an actual data item, pointers to the preceding and/or the succeeding node(s). This information is required for maintaining a record of memory locations as nodes are dynamically added or deleted: This is so because, in contrast to the representation of array elements, memory locations of list entries are no longer contiguous. As illustrated in Fig. A.4, lists are linear data structures: Each node has exactly one predecessor and one successor [with the exception of first (head) and last (tail) entries]. Lists are designed to accommodate the information as it is being collected while invoking dynamic memory allocation to manage memory usage in an efficient way.

Our accommodation of edge pairs in the disk-scanning algorithm thus invokes two types of lists: One is a list of disk chords (pairs of matched edge points), which is initiated anew at the beginning of each line scan; a second is a list of disks that is being maintained for the entirety of the raster scan. It is updated following the completion of each line scan and serves to assign each newly collected set of chords from the edge point list to the appropriate disk structure in the disk list. The latter grows as the raster scan proceeds and newly encountered disks are entered; concomitantly, disks already present in the list are monitored and maintained in Active or converted to Inactive status, according to an appropriate criterion.

Special Cases: Queue, Stack, Cache
Special cases of linked lists are encountered with sufficient frequency so as to warrant special names. The primary distinction among the three cases mentioned here

is the order in which items are entered and retrieved from the structure [Sessions, 1989].

A *queue* is a linear structure with the property that items are retrieved in the same order in which they are entered, in other words, F(irst)I(n)F(irst)O(ut). Thus, in a queue of print jobs, submitted to an output device on a multiuser computer system, requests are handled in the order in which they are received. This strict rule may be modified if priorities are assigned to the entries in the queue, which is then referred to as a priority queue.

A *stack* is a linear data structure with the property that items are retrieved (popped) in reverse order compared with the order in which they were entered (pushed), in other words, L(ast)I(n)F(irst)O(ut). An example, familiar to all programmers, is the stack in which reentry points are saved as nested subroutines are invoked in a running program: As the sequence of nested calls is made, corresponding pointers are pushed on the stack and these are popped in reverse order during the subsequent ascent back to the calling routine.

A *cache* is a linear data structure with the property that items are entered and retrieved according to a predefined (and then unmodified) rule or set of priorities.

The implementation of these special cases on the basis of the collection of linked list operations is discussed by Sessions (1989).

Table A.3. Context-Independent Implementation of (Doubly) Linked Lists

```
struct lt                          /* link (node) type */
{
         struct lt *next;
         struct lt *prev;
         char *item;               /* pointer to data element */
};


struct ll                          /* linked list */
{
         struct lt *head;          /* ptr to first element in list */
         struct lt *tail;          /* ptr to last element in list */
         struct lt *clp;           /* ptr to current element in list */
         int itemsize;
         int listlength;
         int (*match) ();          /* func ptr to "comparison" func */
};
```

A.5.2   LIST PRIMITIVES

To provide a context-independent implementation of operations to manipulate lists, it is desirable to define the requisite structures without explicit reference to the particular requirements of the actual task at hand. This is accomplished by the definitions given in Table A.3. They isolate the layout of the structures from the content of the data entries: The latter are referred to only by means of a pointer to a standard type, here *char, whose size is independent of the arbitrarily complex nature of the data item, itself commonly a structure, such as the disk in the example discussed above. The function pointer to the match function permits the coexistence of several list constructs that might require different comparisons among their respective entries: The appropriate function may be set as needed. As stressed by Sessions, the implementation of the primitives provided here to manipulate linked lists is iterative rather than recursive. Fundamental operations performed on lists include: insert(); delete(); set_current_list_pointer(), etc. (see also, list primitives in code compilation).

Table A.4. Context-Independent Implementation of Binary Tree

```
struct btnt                         /* binary tree node type */
{
            struct btnt *left;
            struct btnt *right;
            char *item;             /* pointer to data element */
};


struct bt
{
            struct btnt *root;      /* pointer to first element in tree */
            struct btnt *cnp;       /* pointer to current tree node */
            int itemsize;
            int travdirn;           /* direct of traversal (left, right)*/
            int (*less) ();         /* func pointers to */
            int (*eq) ();           /* "comparison" funcs */
};
```

A.5.3 BRANCHED DATA STRUCTURES: TREES

A further generalization of data structures beyond the list concept of the previous section is often required: This involves the introduction of branched data structures, or trees. The distinction between linear and branched data structures is based on their topology or, more precisely, on the topology of the corresponding graph. The generalization of linear to branched data structures is based on relaxing the constraint that a node must not have more than one successor. Thus, if an array is the simplest linear data structure, with one successor for each element, the binary tree represents the simplest branched data structure: Here, each node retains a single predecessor but is allowed two successors (or descendants) (see, e.g., [Kernighan and Ritchie, 1988]). The corresponding node structure is shown in Table A.4. Many types of tree structures are commonly encountered, including A(del'son)V(el'skii)L(andis) Trees and Tries and B Trees [Kruse et al., 1991].

An implementation of primitives to operate on branched data structures is generally based on recursion, although an iterative implementation invoking a stack is also possible ([Kruse et al., 1991], Subsection 8.5.5). A small set of basic primitives for binary trees, proceeding from the most fundamental operation of tree traversal, was developed by Sessions ([Sessions, 1989], Section 11.3). A more extensive exposition is available, ([Kruse et al., 1991], Chaps. 9 and 10). A classic optimization problem arising in the construction of branched structures is that of maintaining a balanced distribution of data. The implementation of a solution that relies on a temporary transcription of a given tree into a linked list and the subsequent recursive construction of a reconfigured tree has been discussed.

References

[Andrews, 1970] H. C. Andrews, *Computer Techniques in Image Processing* (Academic, New York, 1970); Chap. 4.

[Bracewell, 1986] R. N. Bracewell, *The Fourier Transform and Its Application* – 2nd Edition (McGraw-Hill, New York, 1986), Chap. 3.

[Gonzalez and Woods, 1992] R. C. Gonzalez and R. E. Woods, *Digitial Image Processing* (Addison-Wesley, Reading, MA, 1992).

[Goodman, 1968] J. W. Goodman, *Introduction to Fourier Optics* (McGraw-Hill, San Francisco, 1968).

[Kernighan and Ritchie, 1988] B. W. Kernighan and D. M. Ritchie, *The C Programming Language*, 2nd ed. (Prentice-Hall, Englewood Cliffs, NS, 1988).

[Kruse et al., 1991] R. L. Kruse, B. P. Leung, and C. L. Tondo, *Data Structures and Program Design in C* (Prentice-Hall, Englewood Cliffs, NJ, 1991).

[Lim, 1990] J. S. Lim, *Two-Dimensional Signal and Image Processing* (Prentice-Hall, Englewood Cliffs, NJ, 1990).

[Press et al., 1988] W. H. Press, B. P. Flannery, S. A. Teukolsky, and W. T. Vetterling, *Numerical Recipes in C* (Cambridge U. Press, New York, 1988).

[Pratt, 1991] W. K. Pratt, *Digital Image Processing* (Wiley-Interscience, New York, 1991).

[Sessions, 1989] R. Sessions, *Resusable Data Structures for C* (Prentice-Hall, Englewood Cliffs, NJ, 1989).

Index

Please note:
This CD-ROM contains files conforming to the Joliet file naming convention and is optimized for Windows-95, 98, and NT computers and LINUX systems capable of reading Joliet CD-ROM filesystems.
MacIntosh and other non-Windows computers may not read these filenames correctly, or be able to find links to these files.

For technical help and answers to questions regarding the software in this book and CDROM, please go to the website
www.mlmsoftwaregroup.com